Sustainable Water

Sustainable Water

Challenges and Solutions from California

Allison Lassiter

Afterword by Peter Gleick

UNIVERSITY OF CALIFORNIA PRESS

University of California Press, one of the most
distinguished university presses in the United States,
enriches lives around the world by advancing scholarship
in the humanities, social sciences, and natural sciences. Its
activities are supported by the UC Press Foundation and
by philanthropic contributions from individuals and
institutions. For more information, visit www.ucpress.edu.

University of California Press
Oakland, California

Library of Congress Cataloging-in-Publication Data
    Library of Congress Cataloging-in-Publication Data
Sustainable water : challenges and solutions from
California / [edited by] Allison Lassiter ; afterword by
Peter Gleick. -- First edition.
        pages cm
    Includes bibliographical references and index.
    ISBN 978-0-520-28536-1 (cloth : alk. paper) — ISBN
    0-520-28536-0 (cloth : alk. paper) — ISBN 978-0-520-
    28535-4 (pbk. : alk. paper) — ISBN 0-520-28535-2
    (pbk. : alk. paper) — ISBN 978-0-520-96087-9 (ebook)
    — ISBN 0-520-96087-4 (ebook)
    1. Water-supply—California.   2. Water-supply—
California—History.   3. Water security—California.
I. Lassiter, Allison, 1980– editor.
    TD224.C3S865 2015
    333.91009794—dc23
                                            2014040941

Manufactured in the United States of America

24   23   22   21   20   19   18   17   16   15
10   9   8   7   6   5   4   3   2   1

Contents

Contributors

JOHN T. ANDREW is assistant deputy director of the California Department of Water Resources, where since 2006 he has overseen the department's climate change activities. His previous organizational affiliations include the Stege Sanitary District, the CALFED Bay-Delta Program, the California Department of Health Services, the Lawrence Berkeley National Laboratory, and the U.S. Environmental Protection Agency. Andrew has over twenty-five years of experience in water resources and environmental engineering, and holds degrees in civil engineering and public policy from the University of California, Berkeley.

CAROLINA BALAZS is a postdoctoral fellow in the Department of Environmental Science and Policy at the University of California, Davis, and has worked as a research scientist with the Community Water Center for many years. Her research focuses on environmental health, environmental justice, and regional planning issues related to access to clean drinking water in rural California. Dr. Balazs received her PhD and MS from the University of California, Berkeley, Energy and Resources Group, and her BS from Brown University.

CELESTE CANTÚ is general manager of the Santa Ana Watershed Project Authority and former executive director of the California State Water Resources Control Board. She comes to water resources management as an urban planner with a BA from Yale University and an MPA from Harvard's Kennedy School of Government.

JULIET CHRISTIAN-SMITH is a climate scientist with the Climate and Energy Program at the Union of Concerned Scientists, based in the California office. Dr. Christian-Smith is the lead author of *A Twenty-First Century U.S. Water Policy* (Oxford Press, 2012) and an editor of the journal *Sustainability Science*. The focus of her work is providing California and U.S. policymakers and the public with robust, timely, accessible, and policy-relevant information on climate science and climate impacts.

MATTHEW DEITCH is a senior environmental scientist at the Center for Ecosystem Management and Restoration. He has worked for the past twelve years on water management issues in coastal California. His work today focuses on how water can be managed in ways that meet human and environmental needs in coastal California. He received his PhD in environmental planning from the University of California, Berkeley.

CAITLIN S. DYCKMAN is an associate professor in the Planning Program within Clemson University's Department of Planning, Development, and Preservation. A native Californian, she received her JD from the UC Davis King Hall School of Law, and her PhD from the Department of City and Regional Planning at UC Berkeley, specializing in environmental planning and water law. Her current research areas include environmental policy, law, and planning, with an emphasis on the land–water nexus.

HOWARD FOSTER received a master's degree in landscape architecture in 1974 and practiced as a professional landscape architect for ten years, then received a PhD in environmental planning from the University of California, Berkeley, in 1993. Since then he has served primarily as a researcher at UC Berkeley, where he developed a number of public-participation, Web-based geographic information systems. He is currently developing architectures for federating geographic information systems in support of emergency management.

JULIAN FULTON is a PhD candidate in the Energy and Resources Group at the University of California, Berkeley, and a research affiliate with the Pacific Institute Water Program. He aspires to continue working on challenging water issues in California and globally through a career in teaching, research, writing, and civic engagement.

PETER GLEICK is president of the Pacific Institute in Oakland, California, a member of the U.S. National Academy of Sciences, a MacArthur fellow, and a leading expert on global climate and water issues. He is also the author or editor of ten books, including *A 21st-Century US Water Policy* (Oxford University Press). He is a graduate of Yale University and the University of California, Berkeley, where he received an MS and a PhD.

BRIAN E. GRAY is a professor of law at the University of California's Hastings College of the Law, where he has received the William Rutter Award for Excellence in Teaching and the Hastings Outstanding Professor Award. Professor Gray is the author or coauthor of numerous articles and books on water resources law and policy, including *Managing California's Water: From Conflict to Reconciliation* (Public Policy Institute of California, 2011).

MAURICE HALL is the science and engineering lead for the Nature Conservancy's California Water Program, where his work focuses on developing and implementing integrated water management projects to better meet the needs of water-dependent ecosystems while improving water supply reliability for people, and on helping improve policies and institutions to meet our current and future water needs. Before joining the Nature Conservancy in 2008,

Dr. Hall's experience in water resources included more than fifteen years in consulting, university teaching, and research, and state service in California, Oregon, Colorado, and Virginia. He received his BS in chemical engineering from the University of Tennessee at Chattanooga, and his PhD in watershed sciences from Colorado State University, and he is a registered professional civil engineer in California.

ELLEN HANAK is an economist and senior fellow at the Public Policy Institute of California, where she manages an interdisciplinary research program on water policy involving scholars from across the state.

MICHAEL HANEMANN holds the Wrigley Chair in Sustainability in the Department of Economics at Arizona State University, and is Chancellor's Professor Emeritus in the Department of Agricultural and Resource Economics and the Goldman School of Public Policy at the University of California, Berkeley. A member of the National Academy of Sciences and a leading expert in environmental economics and the economics of water, he has long been involved in California water issues.

SASHA HARRIS-LOVETT is a doctoral student in the Energy and Resources Group at the University of California, Berkeley, and a National Science Foundation Graduate Research Program fellow. Her research focuses on the legitimacy of potable water reuse and decision-making for innovative water management.

MATTHEW HEBERGER is a researcher at the nonprofit Pacific Institute in Oakland, California. Mr. Heberger holds a BS in agricultural and biological engineering from Cornell University and an MS in water resources engineering from Tufts University. He's spent the last fifteen years working on water issues as a hygiene and sanitation educator in West Africa, and as a consulting engineer and policy analyst in Washington, DC, and California.

G. MATHIAS "MATT" KONDOLF is a fluvial geomorphologist and environmental planner, specializing in environmental river management and restoration. As professor of environmental planning at the University of California, Berkeley, he teaches courses in hydrology, river restoration, and environmental science. His research concerns human–river interactions broadly, with emphasis on management of flood-prone lands, sediment management in reservoirs and regulated river channels, and river restoration.

JAY LUND holds the Ray B. Krone Chair in Environmental Engineering and is director of the Center for Watershed Sciences at the University of California, Davis. He specializes in the management of water and environmental systems.

DAMIAN PARK received his PhD in agricultural and resource economics from the University of California, Berkeley, in 2011. His research focuses on California water markets and water rights. He currently teaches economics at Santa Clara University and lives in Oakland, California, with his wife and daughter.

KRISTEN PODOLAK is a conservation planner working in the Sierra Nevada for the Nature Conservancy. Her work focuses on increasing the pace and scale of forest fuels reduction in National Forests and protecting and restoring meadows. She received her PhD in environmental planning with a focus on river restoration from the University of California, Berkeley.

JOHN RADKE is a professor in the Department of Landscape Architecture and Environmental Planning and the Department of City and Regional Planning at the University of California, Berkeley. His research involves developing analytical methods and metrics that measure and recognize spatial structure and change in complex landscapes. His metrics detect and recognize patterns in data-rich environments, often generated by sophisticated sensors that record and map spatial distributions of phenomena. He models a variety of real-world problems, from fire risk in the wild-land/urban interface, to predicting landscape hazards in highly erodible terrains, to measuring the consequences of infrastructure failure due to climate change and sea level rise. In addition, Professor Radke teaches several courses in geographic information science.

ISHA RAY is an associate professor in the Energy and Resources Group at the University of California, Berkeley, where her research focuses on access to water and equity in national and international settings. Prior to joining the group, she was an analyst on economics and institutions at the Turkey office of the International Water Management Institute. She has a BA in philosophy, politics, and economics from Somerville College, Oxford University, and a PhD in applied economics from the Food Research Institute at Stanford University.

DAVID SEDLAK is the Malozemoff Professor in the Department of Civil and Environmental Engineering at the University of California, Berkeley. He is also co-director of the Berkeley Water Center and deputy director of the National Science Foundation's Engineering Research Center for Reinventing the Nation's Urban Water Infrastructure (ReNUWIt). He is the author of *Water 4.0: The Past, Present and Future of the World's Most Vital Resource* and a contributor to the U.S. National Research Council's 2012 report on water reuse.

FRASER SHILLING is a research scientist at the University of California, Davis, specializing in water, transportation, and policy. He advises states and others on ways to measure and improve sustainability and co-directs the Road Ecology Center at UC Davis. Fraser received a PhD in ecology from the University of Southern California in 1991.

DANIEL WENDELL is associate director of groundwater for the Nature Conservancy in California. Mr. Wendell has thirty years of professional experience dealing with groundwater resource issues, including well design, conjunctive use assessments, and development of conceptual and numeric flow, transport, and subsidence models. He has conducted groundwater investigations throughout California, including the Central Valley and major groundwater basins in the San Diego, Los Angeles, Ventura, Santa Barbara, San Francisco, and Monterey areas. Mr. Wendell has MS degrees in geology and hydrogeology from the University of Nevada, Reno, and is a registered geologist and certified hydrogeologist in California.

ROBERT WILKINSON is an adjunct professor at the Bren School of Environmental Science and Management, and a senior lecturer in the Environmental Studies Program, at the University of California, Santa Barbara. Dr. Wilkinson's teaching, research, and consulting focus is on water and energy policy, climate change, and environmental policy issues. Dr. Wilkinson is also a senior fellow with the Rocky Mountain Institute. He co-chairs the U.S. Sustainable Water Resources

Roundtable and serves on a number of advisory boards, including the Water and Energy Team for the California Climate Action Team and the advisory committee for California's State Water Plan; and he is an advisor to agencies including the California Energy Commission, the California State Water Resources Control Board, the California Department of Water Resources, and others on water, energy, and climate issues.

CLEO WOELFLE-ERSKINE studies water's entanglements with humans, other species, and landscapes. His work includes grass-roots graywater and potable water systems in the United States and Mexico, environmental justice policy work in California, and academic research into urban water transitions in India and efforts to balance rural water use and salmonid recovery on California's North Coast. His books include *Creating Rain Gardens* (Timber Press, 2012) and *Dam Nation: Dispatches from the Water Underground* (Soft Skull, 2007).

SARAH YARNELL is a faculty researcher with the Center for Watershed Sciences at the University of California, Davis. Her studies focus on integrating the traditional fields of hydrology, ecology, and geomorphology in the river environment. She is currently conducting research that applies understanding of river ecosystem processes to managed systems in California, with a focus on the development and maintenance of riverine habitat under current and future climate conditions. She received a PhD in hydrologic sciences and an MS in geology from UC Davis.

Acknowledgments

This book is a product of chance encounters, leaps of faith, generous help, and many people's hard work. First and foremost, I owe an incredible debt of gratitude to my intellectual home, the Department of Landscape Architecture and Environmental Planning at the University of California, Berkeley. They helped birth the book concept in 2012, when they sponsored my lecture series, Planning for Water Sustainability in the 21st Century. They helped send the book into the world two years later, with support for graphic production through a grant from the endowed Beatrix Jones Farrand Faculty Research Fund.

Several key people contributed to shaping this project from a collection of ideas into a published volume. Thank you to Andrew Fahlund for suggesting that the lecture series seed this book. Thank you to Edward Wade for setting the ball in motion that led me to University of California Press. Thank you to Louise Mozingo, Ed Dobb, and Pam Lassiter for teaching me to navigate the world of book proposals. Thank you to David Feldman, Jerry Mitchell, and the anonymous reviewers of both the proposal and the draft manuscript for your invaluable suggestions on the book's content and tone. Thank you to Kyle O'Konis for managing the graphic content of the book—I really have no idea what I would have done without you. Thank you to Merrik Bush-Pirkle for your patient and cheerful guidance. And thank you to our editor, Blake Edgar, for believing in both the book and me.

Of course, the book never would have happened without the authors. Thank you for enduring years of emails, multiple drafts, and word-count harping. Even more importantly, thank you for tirelessly dedicating yourselves to the thorny and unending problem of California water, ensuring that we have a future that protects our communities, economy, and ecosystems. I am grateful that I was able to work with you.

FIGURE 0.1. California and selected cities.

Shasta lake
Sacramento river

Tehama-Colusa canal

Feather river
Lake Tahoe
American river
Pardee reservoir
Mono lake
Hetch Hetchy reservoir
Delta-Mendota canal
San Joaquin river
Friant-Kern canal
Los Angeles
aqueduct
Kern river

California
aqueduct
Colorado river
aqueduct
Colorado river
Salton sea
All American canal
San Diego aqueducts

▨▨▨ Federal aqueduct
▬▬▬ State aqueduct
▬▬▬ Local aqueduct

FIGURE 0.2. Select water resources and infrastructure in California. Adapted from *Major Water Conveyance Facilities in California*, courtesy of Metropolitan Water District of Southern California.

Introduction

ALLISON LASSITER

California is in a drought, one of the worst in its recorded history. In 2014, wells ran dry in the city of East Porterville, forcing residents to depend on bottled water rations. Department of Fish and Wildlife officials moved salmon stranded by drying rivers downstream in trucks. Water reservoirs hit record lows, and the Sierra Nevada snowpack measured just 18 percent of average. For the first time in its 54-year history, the state's largest water-delivery system had no water to give its urban and agricultural customers. Turning to less desirable, alternative water sources, many farmers protected their crops by pumping already overtaxed groundwater reserves at unprecedented and unsustainable rates.

In response to pervasive drought conditions, state and local agencies scrambled to produce new policies and programs. Governor Jerry Brown declared a state of emergency twice, issuing proclamations in both January and April of 2014. To reduce urban water use, the state passed legislation allowing local water agencies to levy $500-per-day fines on urban water-wasters, while cities issued their own water restrictions. To protect rapidly depleting aquifers, Governor Brown signed the historic and contentious Sustainable Groundwater Management Act, requiring localities to draft groundwater management plans by 2025. As conditions worsened over the following year, in April 2015 Governor Brown issued a first-ever executive order mandating a 25 percent reduction in urban water use across the state. Yet, many of the new policies and programs are not only late, they are also not enough to assure water supplies if drought endures.

Protecting water resources today is difficult: California is drier now than it was in 1976–77, previously the most significant modern drought.[1] It is drier than in the Dust Bowl era, and drier than any year in the state's early days, when drought forced settlers to pack up and move to wetter regions. California may, in fact, be facing its most significant drought in 1,200 years (Griffin and Anchukaitis 2014).

Research from paleoclimatologists indicates that extreme droughts will become more frequent. California's precipitation has always been cyclical—there are annual cycles (rainy winters and dry summers), decadal cycles (La Niña and El Niño), and century-scale climate cycles. Recent research indicates that California developed during a wet era, and may be turning toward a new, dry era (Ingram and Malamud-Roam 2013). Precipitation variation and extremes will likely be compounded by climate change (Ault et al. 2014).

Yet, precipitation—its timing, volume, and patterns—is only a piece of California's many water crises. Drought is a magnifying glass, revealing that California's water supply system is inflexible and brittle. It is also stressed by urban population growth, increased agricultural production, outdated infrastructure, legacy institutions, and siloed management. Though there are ongoing revisions and improvements in California water, the stressors are not going away and adaptation is necessary.

Sustainable Water uncovers opportunities to build resilience. The chapters are written by leading experts from water agencies, think tanks, nonprofits, and academia, all actively contributing to California water management. With perspectives from policy, law, economics, hydrology, ecology, engineering, and planning, *Sustainable Water* illustrates California's many, concurrent water crises. Layering these crises together, while looking across and between disciplines, exposes connections and intersections.

The stories in this book will sound familiar in many places around the world. From Toledo to Atlanta, Australia to India, countless water supply systems are vulnerable and uncertain. California water offers both cautions and solutions for any region pursuing resilient and secure water amid a dynamic society and environment.

UNTANGLING CRISES, FINDING SOLUTIONS

Because of its deeply interdisciplinary nature, managing water in California is knotty and complicated. To tease apart problems, *Sustainable*

Water develops six categories of critical challenges facing California water: adapting to climate change, defining water access and ownership, managing risks, integrating institutions, representing vulnerable communities, and reducing consumption. Each category is illustrated in two or three chapters but encompasses a far wider set of discussions than can be included a single book.

The six challenges are ordered by geographic scale. The chapters discuss statewide policies, then regional systems, then communities, and finally personal water use. Crises are nested and interacting, while solutions cut across disciplines and scales.

Adapting to Climate Change

In the first chapter, John Andrew gives an overview of the many impacts of climate change on water resources in California. He provides evidence that California's climate is becoming increasingly variable, with average precipitation and water stores diminishing over time. Simultaneously, he highlights that increased flood risk may pose a more immediate danger than water scarcity for many of California's residents and for agriculture. To simultaneously confront scarcity and flood, Mr. Andrew asserts that it will be necessary to adapt behaviors. Through adaptation, managing climate change and water is possible.

Building on the relationship between water and climate, in chapter 2 Robert Wilkinson discusses the water–energy nexus. He details the points at which California's water system consumes energy and where the energy system consumes water. He suggests that each water and energy source has a different efficiency profile and climate impact. Dr. Wilkinson examines methods of quantifying water and energy intensity to help reveal when drawing from a different source of energy or water may lessen climate impacts. Finally, he evaluates policy approaches and plans that may produce multiple benefits across water, energy, and climate management objectives.

Defining Water Access and Ownership

Adapting to change and dynamically shifting among water sources is more complex than it may initially seem. Water allocation is hotly contested, and rerouting and substituting water sources may have legal implications. The next two essays reveal the thorny issue of water rights, addressing who has access to water and why.

In chapter 3, Michael Hanemann, Caitlin Dyckman, and Damian Park peer into California history and describe the evolution of the state's two types of water rights: riparian and appropriative. They give particular attention to appropriative rights, which assign water ownership. Of note is that ownership status was never recorded by volume. Instead, a system of "first in time, first in right" evolved. This manifests today in an incoherent water system, where junior rights holders may not receive allocations in a dry year. The authors assert that the costs of this mismanagement are an inability to attain environmental quality objectives and a growing vulnerability to climate change impacts.

Brian Gray digs further into water rights in chapter 4, specifically discussing the mandate of "reasonable use," one of the major legal points of contention in the state. Mr. Gray reveals that the definition of reasonable use is relative to environmental and social conditions. As context changes, reasonable use is disputed again. The impacts of this slippery, dynamic definition spill over into other water debates. For example, a contested reasonable use conveys uncertainty to water rights, which raises problems when establishing water markets. Mr. Gray demonstrates that, without major rights reform, reasonable use will lead to ongoing revisions to water rights in the courts.

Managing Risks

Because of a fluctuating climate and legal landscape, availability and access to water throughout the state are always changing. The next three chapters further explore change, evaluating shifting demands, broken infrastructure connections, and unreliable supplies. They each examine different methods of mitigating risk if—or when—expected water sources are not available. To evaluate possible future scenarios and interventions, all three chapters rely on modeling and simulation.

In chapter 5, Juliet Christian-Smith and Matthew Heberger discuss population growth, climate change, and the challenge of ongoing, increasing demand for municipal water. They forecast future demand, and then explore patterns of high and low urban water users throughout the state. They review policies and programs intended to reduce residential consumption, noting that a significant barrier to implementing comprehensive water conservation programs is that reduced use often leaves utilities with a revenue shortage. Ultimately, Dr. Christian-

Smith and Mr. Heberger conclude with a suite of suggested pricing strategies for closing the revenue gap.

Connecting water supplies and demands is a massive infrastructure of dams, levees, pipes, and pumps. In chapter 6, Howard Foster and John Radke discuss the critical infrastructure in California's Sacramento–San Joaquin Delta and the impacts of its failure. To prepare for an anticipated levee breach, Dr. Foster and Dr. Radke discuss methods of coordinating emergency responses. They focus on data sharing, an essential component of communication where many agencies are working at different levels of government on multiple time horizons. Preparing for failure is fundamental to mitigating risk and addressing crisis.

In chapter 7, Ellen Hanak and Jay Lund further address risk management. They suggest confronting drought, legal rulings, and infrastructure failure with a diversified portfolio of water supply and demand management strategies. Drawing on a diversity of water strategies can ensure a reliable water source, despite myriad uncertainties. Yet, there are both technical and institutional barriers to effective portfolio management. Technically, evaluating when to draw on each water source requires complex modeling. Institutionally, there are legal impediments to effectively managing some water sources, and there are physical obstacles to conveying water. Drawing from modeling results, Dr. Hanak and Dr. Lund suggest urban conservation, groundwater banking, and water marketing as priorities for portfolio development.

Integrating Institutions

The next three chapters explore the details of integrated management solutions and diversified water portfolios. They reveal that integrated management brings together not only multiple physical systems and watershed basins but also social systems. Collaboration and stakeholder-based processes are essential.

Daniel Wendell and Maurice Hall dive into groundwater banking in chapter 8. They review different methods of using groundwater while simultaneously managing municipal water supply and ecological populations. The chapter details the hydrological connection between surface water and groundwater, cautioning that it is necessary to acknowledge the inherent tradeoffs between groundwater pumping and surface water flows. To balance the two, Mr. Wendell and Dr. Hall suggest defining basin management objectives with local stakeholders.

In chapter 9, Celeste Cantú discusses integrated regional water management in the Santa Ana watershed. Declining groundwater quality was one of the major drivers for establishing the Santa Ana Watershed Project Authority (SAWPA), a joint powers authority with five member water agencies. Ms. Cantú reveals the watershed's vulnerability to many different stressors, including high rates of urbanization, dependence on imported water, and salt contamination in the aquifer. SAWPA works to create a resilient water system in the face of vulnerabilities by integrating aspects of water management that are often unconnected. It includes initiatives focusing on energy recovery and production facilities, recycled water, biosolids and fertilizer treatment, and managing forest and wetlands for water capture and quality. Though the mutual benefits of cooperation and increased system resilience are clear, Ms. Cantú also discusses some of its limitations, including the difficulty of solving more complex issues.

In the following chapter, Sasha Harris-Lovett and David Sedlak chronicle the history of water reuse in California. Prominent in this story is the Orange County Water District, one of the member agencies of SAWPA. In Orange County, imported water is not sufficient to meet demands. To augment supplies, the water district introduced highly treated wastewater into their drinking water aquifer. Other urban water districts have followed their example by pursuing water reuse for both potable and nonpotable uses, with varying degrees of success. The authors suggest that to further expand water reuse, water utilities and regulators will need to overcome technical, social, and institutional challenges.

Representing Vulnerable Communities

The next three essays focus on specific cases of stakeholder conflict and cooperation in water management decisions. Case studies of a town and two rivers reveal trade-offs among management choices. These chapters demonstrate that power is a critical component of water negotiation. Marginalized populations often include the poor and nonhuman species, who are less able to advocate for themselves.

In chapter 11, Carolina Balazs and Isha Ray discuss the small community of Tooleville, California. Located in the Central Valley, Tooleville is a low-income, primarily Hispanic, agricultural community with one drinking water well. The well is contaminated. Tooleville is interested in sharing drinking water supplies with a nearby, wealthier community, but is not seen as an attractive annexation opportunity. Dr. Balazs and Dr. Ray evaluate the relationships between race, class, and drinking water access exemplified in Tooleville through a drinking water disparities framework. They assess feedbacks between the natural environment, the built environment, and the sociopolitical environment that impact water quality. Ultimately, they seek to define a just water policy for the twenty-first century.

In chapter 12, Matthew Deitch and G. Mathias Kondolf discuss a different vulnerable community, California's native salmon. Listed under the Endangered Species Act, several salmon populations are currently in danger of extinction. Salmon's critical habitat requirements—cold water flows and the ability to migrate upstream to spawning habitat—often conflict with water resource development and land-use change. Alteration of river channels and urbanizing watersheds reduce the quality of habitat, while dams and other barriers reduce access to habitat. Dr. Deitch and Dr. Kondolf examine the case study of the Russian River. In the Russian River watershed, salmon directly compete with human users for water. When frost is imminent, grape growers protect their crops by misting vines with water. When growers throughout the watershed simultaneously draw on the region's water, river levels drop, leaving salmon stranded.

Kristen Podolak and Sarah Yarnell describe conflict and cooperation between diverse water users in chapter 13. They discuss the stakeholders concerned with hydropower dams and the opportunity presented by dam relicensing. Several major hydropower dam licenses are coming up for renewal by the Federal Energy Regulatory Commission, which has jurisdiction over many of the nation's hydropower projects. Because existing hydropower licenses were established before major environmental legislation, relicensing will include major renegotiations in water flows. Stakeholder groups have the opportunity to influence aspects of the negotiated flow regime. Dr. Podolak and Dr. Yarnell discuss two cases: the Mokelumne River Project and the Rock Creek–Cresta Project. Both involved gridlock among the various stakeholders as recreational and ecological interests competed with water storage and energy generation. Dr. Podolak and Dr. Yarnell suggest an approach to resolving stakeholder conflict and creating ecologically protective plans through adaptive licensing, a flexible system that uses monitoring and scientific understanding to inform license conditions.

Reducing Consumption

The final two chapters turn toward personal water use. As urban regions grow, and the number of residents in California continues to rise, reducing per capita water use remains an important component of managing stress on California's water system.

In chapter 14, Cleo Woelfle-Erskine critically evaluates using home graywater for irrigation. Mr. Woelfle-Erksine finds that graywater systems do not always reduce total household water consumption, but discusses how graywater may contribute to changing the culture of water use. His experience working among graywater advocates suggests that graywater helps citizens become more cognizant of consumption and encourages participation in water politics. Furthermore, residential graywater and rainwater catchment diversify supply, building a more resilient urban water portfolio.

In the last chapter, Julian Fulton and Fraser Shilling connect consumer goods to state and global water resources. They introduce the concept of a *water footprint*, the total amount of water required to support each person's lifestyle. Beyond a water utility bill, the water footprint encompasses the water required to grow food, make goods, and produce the energy for transit and housing. In California, the average water footprint is approximately ten times as large as direct consumption of municipal water. Mr. Fulton and Dr. Shilling reveal that California relies on importing a substantial amount of water through commodities, which exposes both vulnerability and opportunity.

STRATEGIES FOR A FLEXIBLE FUTURE

In the afterword, Peter Gleick shares his positive vision for the future of California water. He suggests that while California has reached the real limits of water resources, living within available supplies is possible with increased efficiency. He discusses modernizing urban and agricultural practices, integrating ecological concerns and water policy, and reforming institutions. California's sustainable water future, which he calls the "soft path," is economically productive, while protecting communities and ecosystems.

All of the authors in *Sustainable Water* put forward solutions and provocations that increase flexibility. Many chapters propose integration: combining management objectives and bringing stakeholders together. While the components of California's water delivery and supply system were once conceived separately, jointly managing critical systems is necessary to withstand change. A resilient supply portfolio that provides for all of California's residents will cross traditional boundaries: land and water, surface and subsurface water, drinking water and wastewater, and the needs of diverse water users.

In all likelihood, managing California's finite water supplies will forever be contentious. Inevitably, some water users will be at the bottom of the allocation list, the last to get water in dry years. Even so, California water does not have to be defined by its crises, reacting with emergency during drought. The chapters in *Sustainable Water* reveal paths toward a flexible and resilient water future.

NOTES

1. There are many different methods of measuring drought. By precipitation alone, 2014 was California's third-driest year on record. But 2014 was also the warmest year on record. When accounting for both ongoing precipitation deficit and temperature, 2014's can be considered the worst drought. This is reflected in the Palmer Drought Severity Index, a measure of latent soil moisture published by the National Oceanic and Atmospheric Administration that is commonly used in assessing relative drought. January, February, May, June, July, August, September, October, and November 2014 were each the driest month on record, respectively, in 119 years of record keeping. In 2015, latent soil moisture may have improved slightly, though it is too early to tell at the time of this writing.

REFERENCES

Ault, Toby R., Julia E. Cole, Jonathan T. Overpeck, Gregory T. Pederson, and David M. Meko. 2014. "Assessing the Risk of Persistent Drought Using Climate Model Simulations and Paleoclimate Data." *Journal of Climate* 27:7529–49.

Griffin, Daniel, and Kevin J. Anchukaitis. 2014. "How Unusual is the 2012–2014 California Drought?" *Geophysical Research Letters*, published online December 30, 2014, doi:10.1002/2014GL062433.

Ingram, B. Lynn, and Frances Malamud-Roam. 2013. *The West without Water.* Berkeley: University of California Press.

CHAPTER 1

Adapting California's Water Sector to a Changing Climate

JOHN T. ANDREW

Californians have indeed always coped with wide swings in climate, habitually forgetting one extreme as the climate veered to the other. But human-induced climate change—in part from the dumping of greenhouse gases (GHGs) and other pollutants into our atmosphere—will likely expand that variability beyond what we have experienced, bringing with it more extremes and ultimately more uncertainty. In fact, climate change has already altered our hydrology and sea levels, presenting a significant challenge to California water management.

Even in a changing climate, however, California will still be blessed with lots of water—sometimes too much—so the parable of climate change and water only descends into damnation if we fail to choose management, efficiency, and stewardship. Moreover, as climate change conveys more uncertainty, our responses could actually result in enhanced resilience. Though there is often a drought of optimism in California water management, alarmism is absolutely the wrong answer and message. Californians need not flee, nor perish, in response to climate change; with collaboration, varied options, and smart decisions, they can adapt.

The views expressed here do not necessarily reflect those of the California Department of Water Resources, the California Natural Resources Agency, or the State of California.

THE CLIMATE CONTEXT FOR CALIFORNIA'S WATER RESOURCES AND USE

Strong on climate, California is notoriously weak on weather.

—Carey McWilliams (1949)

California genuinely enjoys a Mediterranean climate, a foundation for our agricultural, tourism, and recreational economies, our relatively low residential energy demand, and our overall quality of life. Most of our precipitation—whether it is rain or snow—falls in the winter and spring. Geographically, precipitation is concentrated in the north and in the east, along the Sierra Nevada. Large parts of the rest of the state actually meet the definition of desert. For most of the year, no rain falls for months at a time, constituting California's annual drought.

As Carey McWilliams (1949) observed, the state's water demand patterns are "upside-down," both in time and space. Water falls freely from the heavens in the winter and spring, but it is primarily desired in the summer and early fall, especially for agriculture and all those green lawns in the suburbs. Most of the urban demand is located along the parched portions of the Pacific coastline, and some of the most productive and diverse farms in the world are in the arid inland areas of Central and Southern California.

To match precipitation with population (and its food and fiber), local, state, and federal governments have invested in nation-scale, inter-regional infrastructure to pipe precipitation from where it falls to where it is demanded. These interbasin projects generally transport water a very long way, from north to south and from east to west—and in the case of Southern California, in both directions. Groundwater is the state's other major water supply, though this source is often limited by contamination and overdraft.

The specter of California running out of water has spawned many a tome, but too much water is actually a far greater danger. Accordingly, governments at all levels have also constructed extensive infrastructure to defend against high water; yet there remain seven million people and over half a trillion dollars of property at risk from flooding (California Department of Water Resources 2013).

The inter-regional water transfer schemes are undoubtedly integral to California's public health, economy, and way of life. To paraphrase former House Speaker Tip O'Neill, though, all California water management is local. That is, various local and regional entities—an untidy, decentralized democracy of, really, thousands of water and wastewater utilities, irrigation districts, flood control agencies, reclamation and levee districts, cities and counties, and various other stakeholders—manage most of the water in California. That is the reality that we must recognize and leverage in responding to climate change.

THE SCIENCE OF A CHANGING WORLD

Climate data and science overwhelmingly show that Mother Earth is warming—resulting in the reduction of snow and ice, other changes in the water cycle, and the rising of the seas—and that it is "extremely likely" that human activity is the main cause of the warming since the 1950s (Intergovernmental Panel on Climate Change 2013). The fundamental findings of climate science are widely endorsed by both professional societies and academies of science around the world, including the national academy of science of every major industrialized nation. In fact, "since 2007, no scientific body of national or international standing has maintained a dissenting opinion" (California Department of Water Resources 2012).

According to the National Oceanic and Atmospheric Administration (NOAA), 2012 roasted as the warmest year on record for the continental United States, a full degree Fahrenheit warmer than the previous high—which was just in 1998. Though there are many ways to contextualize the climate science and the records broken, Seth Borenstein of the Associated Press vividly framed NOAA's finding: "The last time the world had a cooler than average year was 1976. . . . That means that more than half of the people on Earth haven't lived during a cooler than normal year for the globe."[1]

In 2012, the state of California released its Third Assessment of Climate Change, noting that average temperatures statewide had increased by 1.7 °F since 1895, and projecting a range of warming of 4.1–8.6 °F by 2100 (Moser, Ekstrom, and Franco 2012). Specific to water resources, the Third Assessment found likely increases in extreme precipitation, in a place where the climate already tends toward extremes.

CLIMATE CHANGE IMPACTS ON CALIFORNIA'S WATER

We are likely already witnessing the impacts of a changing climate on California's water—in precipitation, snowpack, rivers, and estuaries—and should not be surprised if these impacts amplify, accelerate, or abruptly adjust in the coming decades. Beyond direct hydrologic impacts, climate change may also affect water quality, water demand, hydroelectric generation, and water institutions.

Precipitation

During the twentieth century, the average amount of California's precipitation remained steady—neither more nor less (California Environmental Protection Agency 2013). But when describing water in California, "average" often means little. For instance, precipitation is about more than the absolute volume of water; details matter, such as precipitation's form, location, timing, intensity, duration, and variability. So though average volume has not changed thus far, the form of precipitation already has, with more precipitation falling as rain rather than snow (California Department of Water Resources 2014).

Snowpack

In a state where water storage—above or below ground—is a reliable source of conflict, the Sierra snowpack may be the only form of storage upon which we can all agree. The California Department of Water Resources (DWR) uses April–July runoff as a proxy for snowpack; that surrogate indicates that the snowpack provides an average of 15 million acre-feet (maf)[2] of storage per year. When it melts, this water meets 35 percent of net total agricultural and urban demand (Roos and Anderson 2006). For context, consider that all the surface storage ever constructed by humanity in the Sacramento Valley sums to 14 maf, with another 11 maf in the San Joaquin Valley (see Figure 1). So for all the storage that was ever built, in either basin, Providence provides that much for free, as snow.

Unfortunately, this form of storage—the only one with a consensus—is disappearing. The DWR (2008) estimates that approximately 10 percent of the April 1 Sierra snowpack storage has already vanished, reducing annual spring runoff on the Sacramento River by 1 maf. As a rule of thumb, for every 1 °C (1.8 °F) increase in temperature, the snowline rises by 500 feet (Roos and Anderson 2006). So, for a 3 °C (5.4 °F) temperature increase—well within projections for California by 2100—the Sierra snowline would rise by 1,500 feet, resulting in a loss of about 4–5 maf of snowpack storage. This warming also results in more rain falling on the watershed, directly running off, and leading to more intense flooding, rather than falling as snow, stored as snowpack, and running off slowly or seeping into the ground.

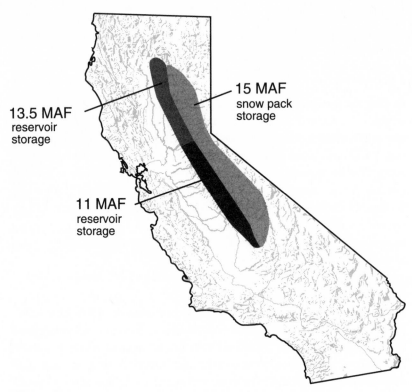

13.5 MAF
reservoir
storage

15 MAF
snow pack
storage

11 MAF
reservoir
storage

FIGURE 1.1. Climate change may significantly reduce Sierra snowpack storage, which is comparable in size to constructed storage: a moderate 3°C increase in temperature is projected to result in an increase in snow elevation of 1,500 feet. (Roos and Anderson 2006).

Hydrology

Because of its snowpack-dominated hydrology, much of California's water supply stands vulnerable to climate change. In fact, across the state, rivers are already exhibiting changes in runoff timing, reflecting changes in the snowpack (California Environmental Protection Agency 2013). As just one example, spring runoff on the Sacramento River declined from constituting approximately 45 percent of annual runoff at the beginning of the twentieth century, to about 33 percent at the century's close (California Environmental Protection Agency 2013). This trend is based upon good old-fashioned data—no computer models, no future-GHG-emission scenarios, and no other assumptions leading to layers of uncertainty about the future.

This shift in hydrology is troublesome for California's communal surface water storage system, which must meet multiple objectives: water supply, flood management, hydroelectric generation, environmental protection, and recreation. During winter, reservoirs in the Central Valley are deliberately drained to create reserve capacity, in order to intercept flood peaks and thereby protect people and places downstream. By early spring, these flood reservations are lifted, allowing water managers to harvest the spring runoff, which is necessary to carry California through the dry months of summer and early fall. However, snow now melts earlier in the year, likewise shifting runoff to earlier in the year, when often there is already too much water to manage.

The state's other major source of water, groundwater, will not be immune from the effects of climate change, either direct or indirect. For example, highly variable flow in rivers and streams may reduce groundwater recharge.[3] Further, when surface water dries up, Californians are likely to mine their aquifers, as they have historically during droughts. In this regard, agriculture and rural communities may be the most affected, through their particular dependence on groundwater.

Although there is anxiety about aridity, too much water remains the clear and present danger. During the twentieth century—again, without relying upon computers, scenarios, or assumptions—it is clear that California's rivers experienced larger flood peaks (Roos and Anderson 2006). For instance, on the American River, the four largest flood peaks on record have struck since 1950, including those in 1986 and 1997 (see Figure 2). For the operators of Folsom Dam on the American River, charged in part with the protection of the state's capital and its metropolitan region, this trend means that the hydrologic record used to design the dam (from the first half of the twentieth century) is distinctly different from the hydrology experienced during operation in the second half.

California's flood infrastructure is aging, and much of it was intended to protect farms instead of cities. Flooding immediately impacts human safety, drinking water quality, and crops, and the long recovery from flooding prolongs the misery, adding ordeals such as disease transmission, mold, housing displacement, business resumption, and mental health distress (California Natural Resources Agency 2014). The increased flood risk from climate change may fall hardest on the poor, because of where they live and their lack of resources and mobility.

In all, these hydrological impacts undermine the foundation of the state's hydraulic empires. As Maurice Roos (2005, 2), California's state hydrologist, observed: "By and large, reservoirs and water delivery

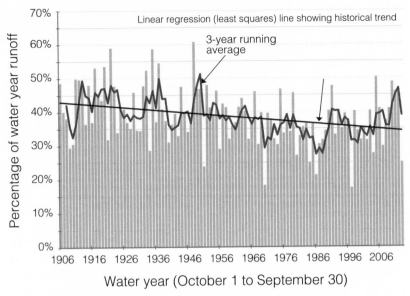

FIGURE 1.2. Spring runoff on the Sacramento River fell during the twentieth century (Roos and Anderson 2006).

systems and operating rules have been developed from historical hydrology on the assumption that the past is a good guide to the future. With global warming, that assumption may not be valid."

Water Quality

Considerable computing power has been invested in calculating the impact of climate change on water supply, while impacts to water quality—a growing concern—remain little examined. There is already an established link between heavy rainfall and waterborne disease outbreaks, and heavier runoff due to a changing climate will wash more non-point source pollution (i.e. contamination running off farms and cities) into water supplies. Conversely, climate change will probably also lead to more periods of low runoff, especially in the summer and fall, which may result in higher concentrations of some contaminants that may pass through water treatment processes, in turn affecting potability and reuse. Warmer weather will also lead to warmer water, and combined with lower summer flows, may increase the growth of algae, diminish dissolved oxygen, and generally affect aquatic habitat.

Water quality also faces the phenomenon of flood following fire, wherein summer and fall wildfires—a risk expected to increase due to climate change—scorch the earth, just in time for the early season rains, resulting in erosion and mudflows. Such runoff may deposit sediment in reservoirs, obscure water clarity with turbidity (cloudiness) that can interfere with disinfection, and cause greater wear and tear on water system equipment.

Beyond potential direct climate change impacts, water quality also serves as a good example of the potential indirect impacts of climate change, including what policy wonks call "unintended consequences," which can be as important as direct impacts. For example, the sound operation of water systems, including water treatment plants, may confront trade-offs with other climate-sensitive concerns, such as energy generation and use, fisheries, environmental water quality, and recreation.

Water Demand

Most climate change studies spotlight the supply side of water resources. However, climate change is also expected to sharpen the competition for already over-allocated water sources among the urban, agricultural, and environmental sectors. In a warmer and perhaps drier climate, water demand may rise, especially for outdoor water use, both in cities (for land-scaping) and on farms (due to evaporation and plant water use).[4] Across three future scenarios of water use for the state, the *California Water Plan Update 2009* projected that in 2050, climate change may increase statewide water demand by up to 2-3 maf per year (with significant regional variability), exacerbating the potential water supply shortages from climate change (California Department of Water Resources 2010).

Water and Energy

Water demand is also the water sector's main source of GHG emissions. Water and wastewater systems use energy as an input to convey, treat, and deliver high-quality water to customers and to collect, treat, and safely reuse and dispose of wastewater, to protect both public health and the environment.[5] Though at times there is much mania about the movement of water in California and its related GHG emissions, the movement of goods and people is actually the largest source (40 percent) of the state's carbon footprint (California Air Resources Board 2014). An inconvenient truth, for many, is that the movement of water actually generates zero-GHG-emissions hydroelectricity, which is the largest source of renewable energy in California, accounting for an average of 15 percent of the state's electricity (California Air Resources Board 2014). In fact, the Assembly Bill 32 Scoping Plan—the state's strategic plan to reduce GHG emissions—depends fundamentally upon hydroelectricity to meet the state's 2020 mandate of rolling back GHG emissions to 1990 levels, by reducing emissions and also easing the integration of other, more intermittent renewable energy sources (e.g. wind and solar) into the grid.

Regrettably, climate change may limit some of the current benefits of hydroelectric generation. For example, high-elevation hydroelectric facilities may lack the storage necessary to cope with a shifting hydrology (Moser, Ekstrom, and Franco 2012). Reduced hydroelectric generation may also coincide with expected increases in summer energy demand, which may instead be met by fossil fuel–based energy generation that emits GHGs.

Real and immediate opportunities to reduce GHG emissions in the water sector primarily involve urban water users. According to a foundational California Energy Commission study (2005), the most energy-intensive part of the water sector is the end use of water (e.g. water heating), accounting for three-quarters of the electricity demand and nearly all of the natural gas demand related to water. Collectively, water customers account for 10 percent of the total energy use in the state, and urban water conservation presents the greatest opportunity to reduce the carbon footprint of water in California (California Air Resources Board 2008; California Department of Water Resources 2014).

Sea Level Rise

During the twentieth century, the tidal gauge at the Golden Gate—source of the longest continuous record of sea level in the United States—measured an incremental rise of about seven inches in sea level (California Environmental Protection Agency 2013). Sea level rise has two main components: the thermal expansion of the oceans (water expands when it warms), and the addition of water to the seas from the melting of land-based ice masses (e.g. glaciers and ice caps). In 2012, the National Research Council projected that sea levels along California's coast may rise another 5–24 inches by 2050—and the seas will hardly stop rising on that date (see Figure 3). In the near future, though, the slowly rising level of the seas is less a concern than higher storm surges and coastal flooding, due to the combination of extreme tides, winds, and rain.

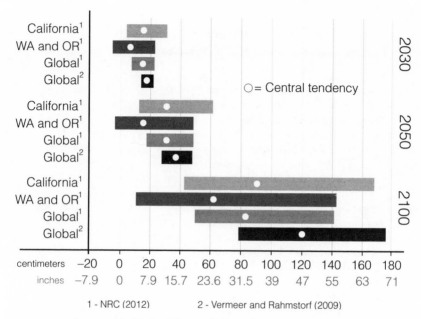

FIGURE 1.3. By 2050, the National Research Council projects approximately one foot of sea-level rise for the California coast south of Cape Mendocino (National Research Council 2012).

THE TRINITY OF WATER, CLIMATE, AND FAITH

Climate change may thus bring biblical blights like droughts, floods, and rising seas to California's water resources. Who knows? Other sectors could suffer famine, plague, and possibly some locusts. Alas, these scourges may be an appropriate judgment day for California water management, an often faith-based effort where decision-makers sometimes act more like evangelicals, defaulting to a sacrosanct set of beliefs (as in "I *believe* in dams" or "I *don't believe* in desalination") rather than rationally applying good information, science, or policy. This may be preordained for a state that named its two major rivers for the Holy Sacrament and a saint. For the general public, its daily experience with water is also largely an article of faith—it comes from the tap, always. How we generally talk about climate change completely complements this approach to water management, again, as something that you "believe in"—or not.

Despite ample observations and copious research, in all honesty we do not know exactly where climate change is going. Precipitation, in particular, remains deeply uncertain for California. Nonetheless, taking trends

An Exceptionally Vulnerable Region: The Sacramento–San Joaquin Delta

The Sacramento–San Joaquin River Delta may be the region of California most vulnerable to climate change. Part of the West Coast's largest estuary, the Delta lies at the confluence of the Sacramento and San Joaquin Rivers, which drain nearly half of California's watersheds. It is where the swamp meets the sea, a maze of rivers and sloughs—eerily similar to Southern Louisiana—with roads aptly named Netherlands and Holland, Noah and Ark. The Delta is genuinely a unique region, home to half a million people, farms and other businesses, and a distinctive culture. Along with the Suisun Marsh, the Delta covers a 1,300-square-mile natural and artificial landscape, with an urbanizing fringe. Although its ecosystem conditions have been declining for decades, it is also still home to birds, fish, and other wildlife, and especially important to the international Pacific Flyway. For the greater state, the region is also a crossroads of critical infrastructure, as Foster and Radke expound in chapter 6. Well known as a "switching yard" for California's state and federal water systems—upon which 25 million people (two-thirds of all Californians) and more than three million acres of farmland depend for water—the Delta is also vital to the state's transportation (shipping, highways, railroads) and energy systems, and thus its economy.

From flooding rivers to rising seas, climate change is affecting the Delta from literally all directions. Many of the 57 islands and tracts that compose most of the Delta have subsided below sea level—some deeply so—as a result of historic reclamation and a century of farming. Today, these islands are continuously protected by 1,100 miles of levees of varying and sometimes questionable materials, construction, and foundations. These levees are now challenged not only by subsidence on the land side but also by greater floods and rising seas on the water side. Potential impacts of climate change in the Delta include salinity intrusion, alteration of aquatic habitat, and increased risk of levee failure, with the latter threatening the inundation of communities and possibly the interruption of water exports statewide. Just as we once assumed a static climate, we have also wrongly assumed—or wished for—a static Delta.

together with studies, California seems to be in for more variability and extremes. Indeed, the preponderance of evidence indicates that, as Kathy Jacobs, director of the National Climate Assessment, once said: "Water will be the delivery mechanism for the impacts of climate change."

Though we do not know exactly the destination of this pilgrimage, we already know more than enough to act and to change the path we are on, given the long lead times required for infrastructure planning, design, and construction. Continuing the religious analogies, there will be no miracles nor forgiveness to save California's water sector in a changing climate, only Puritanical hard work. Adaptation may be costly; it may require sacrifice, and temporary and planned retreat; indeed, people, organizations, and regions may have to get along with one another. Mercifully, California is blessed with options and choices to better manage its water, even with climate change.

POLICY RESPONSES TO CLIMATE CHANGE

California is best known for—and has invested by far the most in—its mitigation response to climate change (i.e. reduction of GHG emissions and sequestration of carbon). In 2006, Governor Arnold Schwarzenegger and the California Legislature enacted AB 32, the state's Global Warming Solutions Act, which grants sweeping authority to the California Air Resources Board to reduce GHG emissions to 1990 levels. Yet, even with immediate, substantial, and vigilant global action to shrink GHG emissions, there will remain a need to adapt to climate change— that is, changes to our infrastructure, behavior, and society to cope with a changing climate or take advantage of opportunities it may present.

A water analogy may be useful here. In the world of water quality, there are *legacy pollutants*, those contaminants from previous eras of mining, farming, and manufacturing. Though the activities that caused the contamination are long gone, their consequences persist. Similarly, for climate change, there are *legacy emissions*, those GHG pollutants released since the dawn of the Industrial Era, which are already changing our climate and will continue to do so for some time. Mitigation is obviously not an option to address past emissions, and even if the world drastically reduced its carbon footprint in the first half of this century, the atmosphere would continue to warm for centuries, and the seas to rise for millennia (see figure 1.4).

ADAPTING WATER MANAGEMENT TO CLIMATE CHANGE

Adapt or perish, now as ever, is nature's inexorable imperative.
—H. G. Wells (1945, 19)

With this sobering reality in mind, in 2008, the DWR proposed a comprehensive strategy to adapt California's water sector to a changing climate, titled *Managing an Uncertain Future*. A month later, the Public Policy

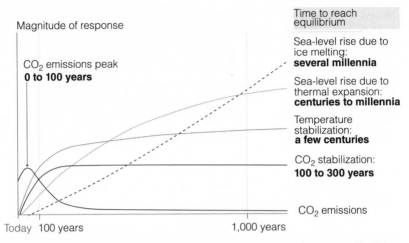

FIGURE 1.4. Even with global action to reduce GHG emissions, adaptation will still be necessary (Intergovernmental Panel on Climate Change 2001).

Institute of California issued its own report on the very same subject (written by Drs. Ellen Hanak and Jay Lund, who also provided chapter 7 of this book), covering much the same ground as the DWR's report and mirroring many of its findings. One conclusion from these analyses is that we should spend less time trying to precisely predict the future, and more on getting ready for it. As Podolak and Yarnell present (chapter 13), an adaptive management approach—iteratively studying, acting, monitoring, succeeding (and failing), learning, and sharing—generally works well for addressing complex water management predicaments, including adapting to climate change. Though there is no such thing as a no-regrets strategy—there are always opportunity and transaction costs, trade-offs, and compromise—we can take many actions now and in the near future that work across a range of possible water futures, giving us the flexibility and capacity to respond, no matter how climate change unfolds.

Regional Self-Reliance

Recognizing that local and regional entities are the principal actors in California water management, the water sector's primary tool for adapting to climate change is something called integrated regional water management (IRWM).[6] As a long-term planning process, IRWM is a strategic, statewide initiative that embraces the concept that people at the local level are best situated to know their region's priorities, capacity, and ultimately values with regard to water management. Partnerships among such stakeholders can leverage and integrate resources, take advantage of economies of scale, and produce solutions to water problems that yield multiple benefits and are sustainable.

Yes, California is one state, but in reality there are many Californias, including the North Coast, the Sierra Nevada, the Great Central Valley, and the South Coastal Plain. Different regions will be impacted differently by climate change. As noted earlier, some parts of the state are relatively wet, others dry. Some depend upon the Sierra snowpack, while others are located along the rising seas of California's 1,100-mile coastline. The impacts of sea level rise are actually regional as well, contingent upon local factors such as land movement.

Thus, a "one size fits all" approach will not work. To create more self-reliance, each region of California should develop and implement its own water management plan, reflecting a formal assessment of its climate change vulnerabilities. These regional plans should include a diverse portfolio of strategies (e.g. aggressive water conservation, storage, transfers, recycling, stormwater, desalination) appropriate for the region. Developing local supplies to meet local demands is good insurance not only for drought but also for other emergencies, such as earthquakes. Though IRWM admittedly predates the state's response to global warming, it is even more relevant in a changing climate. Fortunately, IRWM is well underway in California, with most of its land area and virtually all of its population covered by 48 IRWM regions. Moreover, many—though far from all—regional water management groups are already substantively incorporating climate change into their planning (Conrad 2012).

That said, we must concede that IRWM as strategy for adapting to climate change may have its limits. Though water problems can be less complex at smaller scales, climate projections are exceptionally uncertain at local scales and regarding extreme events. As Cantú notes (chapter 9), IRWM requires patience, trust, and ownership—qualities that are often as scarce as water in a drought. Frankly, many regions of California currently depend upon other regions for their water, a situation that is unlikely to change soon. One prominent commentator, perhaps inadvertently, once summarily dismissed a portfolio approach as "nothing heroic," a "nickel and dime" strategy (Reisner 1986).

An overemphasis on regional self-reliance also perilously perpetuates the mythology of the West, which arguably is the U.S. region most dependent upon "the state" for its development, necessitating its very own federal agency, the U.S. Bureau of Reclamation (Worster 1985).

Moreover, California itself is part of a larger region that shares the Colorado River system, which confronts its own climate change challenges. No region is a republic apart—not even California.

Thus, with IRWM, there still remains a need for continued and expanded inter-regional cooperation. Further, a role for the state and federal governments persists, specifically, to align their analysis, planning, guidelines, policies, and regulations, to support integration at the regional level. In spite of many admonitions, adapting to climate change will transpire watershed by watershed, which well fits an IRWM approach.

Statewide Integrated Water Management

California is infamous for dividing the waters. In contrast to the rest of the United States, groundwater is mostly legally separate from surface water in California, despite the clear hydrologic connection.[7] Whether it is surface or groundwater, water supply management is generally distinct from flood management—though it is often the very same water. And in both water supply and flooding, planning is often disconnected from our ecosystems, something that is "mitigated for" at the end of a water project or part of permitting therefor—although ecosystems are a critical part of the water cycle. The largest disparity of all may be that water supply, flooding, and the environment, all are often planned for and managed separately from land use, the historic and prized purview of cities and counties. On the one hand, poor land-use planning drives development into areas of little water, hardening the landscape, exacerbating water supply availability; on the other hand, it drives development into areas of too much water, putting people in harm's way.

In a changing climate, we will not have the luxury of such artificial divisions in our water planning and management. In taking an integrated approach to water management, healthy headwaters, forests, and wetlands would be protected, managed, and valued for their ability to buffer, cleanse, and store water. Rivers would be reacquainted with their floodplains, to reduce flood peaks, conserve open space, and connect habitats. An environmental stewardship ethic in water resources engineering would incorporate and support substantive ecosystem benefits in every project—equal to and as integral as any other project element. Overall, existing infrastructure can be reoperated more dynamically—and with additional storage, decision support tools, and some cooperation, surface, ground, and flood waters can be managed conjunctively. Sensible land-use planning—just integrating existing water and land-use planning and zoning—could enable more effective flood management and groundwater and habitat protection.

Inter-regional and statewide systems are also required to supply a framework of stability for regional self-reliance, facilitate water transfers and markets, and provide redundancy in the face of uncertainty. Particularly pressing statewide needs include robust flood and drought preparedness, response, and recovery planning. The dilemma in the Delta— important as a region itself and to many other regions—deserves a solution that meets both local and statewide needs.

Climate change pushes us to recognize the interrelatedness of our actions, and the need to incorporate uncertainty into our planning. Notwithstanding regional integration, we are admittedly a very long way from integrated water management at a statewide level. However, we will soon need to make real efforts in this regard if California's water sector is going to survive, much less prosper in, a changing climate.

Information and Science for Decision-Making

Making good water management decisions requires knowledge of our climate, water sources, and water uses. If the past can no longer be our sole guide to the future, then one response is to increase our situational awareness of water supply and demand in the present. For instance, the collection of real-time data on water, weather, and the environment could be better incorporated into the real-time decision-making of water operations, especially flood management (e.g. forecast-based operations). Preservation of long-term data records is also vital; without this we would have little indication of whether the climate is changing (and how), and we would lack the ability to formally evaluate the effectiveness of our responses. One initial step toward integrated water management would be the integration or coordination of existing water data collection and management systems. Whether real-time or long-term, monitoring programs represent relatively low-cost, high-value investments in adapting to climate change.

Good decision-making is also dependent upon further climate change research, even though there is unrelenting uncertainty. But that research cannot be for research's sake; climate science must focus on user needs, produce actionable findings, and be undertaken in full partnership with practitioners. And more disappointing is the state of policy analysis regarding climate change, which sadly can still be summed up as: ". . . and climate change will make things worse."

As we learn from science, we can also learn from each other, and specifically from a mistake the early East Coast settlers of the Sacramento Valley made. As Robert Kelley (1989) writes, the native peoples of the Central Valley well knew that the region regularly flooded into an inland sea—but few of the newcomers "knew or talked with the Indians." Today, this experience is valued as tribal ecological knowledge, which draws upon native peoples' cultural, spiritual, and practical connections to the land and the water.

Institutions

Climate change may precipitate trials of our water institutions as well as our infrastructure. Just as California's water management infrastructure assumes a static climate, so do many water-related regulatory processes, including water quality control plans, water rights, wastewater discharge permits, endangered species protection, reservoir operating rules, dam licensing, and flood insurance. Undeniably, most water laws and regulations are still playing catch-up with climate change science, planning, and policy. That said, some long-standing institutions have already recognized and substantively incorporated climate change, for example the state's flagship water planning process, the California Water Plan Update. Actually, there is considerable flexibility to accommodate climate change in other institutions as well—that is, if we choose to use it.

Institutions must also fill the critical need for climate literacy among the general public. The uncertainty inherent in climate change, combined with politics across the spectrum, has bred a denial of science, not unlike the reaction to the teaching of evolution nearly a century ago. In the absence of climate literacy efforts, many have turned to the gospel of talk radio for their science. Climate change presents an opportunity—again, if we choose to seize it—for institutions of all kinds to communicate fundamental scientific principles and processes, using weather and climate events as teachable moments. It is unlikely that climate change will be the only scientifically complex issue facing Californians this century, so both the bureaucracy and the academy do no service to the public by "dumbing down" the science; instead, we need to concentrate on "smarting up" the public.

The most significant institutional issue may be people. Adaptation to climate change requires professionals with an interdisciplinary skill set and a particular aptitude for translating science into action, talents not well recognized, compensated, or supported in the bureaucracy. Moreover, we should fully expect that climate change will bring more emergencies for which we will need to mobilize personnel of many talents and types, placing them in additional danger.[8]

Investment

Budgeting for California's future may be even scarier than climate change. We will have to pay to adapt, and we cannot just expect that funding will fall from the heavens like our water does (or used to). Climate change mitigation promises a "green economy" of new technologies, businesses, and jobs. In contrast, adapting to climate change may require very large, upfront, public investments—to avert catastrophe and thus even larger costs later—which will compete with equally considerable investments in other worthy societal needs like education, healthcare, and public safety. Nonetheless, the status quo is hardly free of costs, and continuing to invest in it could eventually lead to adverse financial consequences for insurance, real estate, mortgages, and local tax bases.

California water management has greatly benefitted from voter generosity with the approval of a series of water bonds over the past 40 years, and from federal largesse specifically for irrigation, flood management, and wastewater treatment. However, given the fiscal condition of our state and federal governments, compounded by the fiscal mood of the electorate, the days of such state-level generosity may be coming to a close, as for sure is the era of federal bounty. Thus, with a local and regional approach to water and climate change, so, too, may appropriately follow the responsibility for financing water management, which should result in smarter and more efficient funding decisions.

Regardless of where the money comes from, the wise use of taxpayer dollars demands that we explicitly account for climate risk in all our investments. Economics should also be used to better manage water—but in so doing, we must ensure that disadvantaged communities are not priced out of the basic human right to water and sanitation services.[9] In all, climate change is a long-term challenge to California's water resources; it deserves a long-term, stable source of funding to adapt to it.

ADAPT—FLEE IF YOU MUST—BUT PERISH IS NOT AN OPTION

Climate is our state—it is key to agriculture, energy management, recreation and tourism, our natural heritage, and our overall quality of life. For those of us in water management, climate is also our business. So, along with population growth and consumption, climate change presents one of the most significant challenges to what we do. Indeed, we ignore climate change at the risk of undermining the water sector's business model.

But if the grimmest projections of climate change come true, California will still have a lot of water—and at times, too much. Some of that water will become wastewater, which is a resource for myriad nonpotable uses. The state's aquifers hold an order of magnitude more water than the diminishing Sierra snowpack, so groundwater—if strategically and conjunctively managed with surface water, notwithstanding the inevitable holy war over how to do so—can be a fundamental adaptation option as well. With 1,100 miles of coastline—and the vast majority of California's urban demand along those shores—cities will have the option of ocean desalination, which could give inland ecosystems a break, and given its high costs, also serve as an implicit definer of reasonable water use. Most of all, regardless of where the water comes from, we must be extraordinarily efficient in its use.

When it comes to climate change in California, the state benefits from the perfect storm of innovative people, universities, businesses, and even bureaucrats. In fact, California's approach to climate change and water is already serving as a model for other states (Augustyn and Chou 2013). Political support for climate change action is strong, too—at least at the grass-roots, certainly not at the federal level. Recent polling indicates that most Californians agree that global warming is underway, it is a threat to California's future, and steps need to be taken to counter it (Public Policy Institute of California 2013). In addition, adaptation to climate change can often be incorporated into existing actions, because it also addresses other water management objectives; in fact, such an approach is especially important when other factors interact with or intensify climate change impacts. Also, the water sector can leverage a long history of formal water planning, which has resulted in an incredible system of interconnected infrastructure. Further, the water sector is well funded—water confabs involve golf courses, not bake sales—relative to other sectors like public health and biodiversity, which may continue to hurt for resources in a changing climate.

California's water sector thus has a solid foundation, and enormous potential for adaptation. With this inheritance, though, comes the danger that it will be squandered on boutique issues, like renaming, rebranding, or reorganizing resource agencies, producing "carbon-free" water, and restoring Hetch Hetchy Valley, when at the very same time many disadvantaged communities lack safe drinking water and sanitation,[10] and species are on the brink of extinction. We also do not have the luxury of limiting our options—as we have in the past—to only those we believe in; that way lays maladaptation. While the water sector needs to do its part to reduce the state's carbon footprint, as that old Berkeley bumper sticker used to say, it needs to "think globally" (mitigation) but "act locally" (adaptation). In the short term, emergency planning and response and vigorous water conservation are critical; in the long term, more actionable information, regional groundwater management, and much better land-use planning are necessary. Just as there is no average water year in California, notions of a "new normal" are flawed as well; we should be planning for change, not normal—old or new. Though exodus may be needed from some areas or at some times—and climate refugees from elsewhere may flee to California, as they did from the Dust Bowl—adaptation is clearly possible.

Responding to climate change and ensuring the sustainability of our water resources requires us to keep the faith across generations. We should be held accountable by our children for how we leave the state's water resources and world's climate for them. Thus, sustainability means that we need to stop stealing from our kids—in terms of water, climate, and money—so that they may have the basis and capacity to make their own decisions about the water and climate challenges they will inherit. As a more youthful governor Jerry Brown once said, "the world still looks to California": we must lead by example, and serve as a model for other states, future generations, and the whole world to follow.

NOTES

1. Borenstien, Seth. 2013. "World Warm Last Year, but Not Like Record US Heat." *The Big Story*, January 15, http://bigstory.ap.org/article/world-not-hot-us-2012-squeaks-top-10.

2. An acre-foot is literally what it sounds like: one acre flooded to a depth of one foot. A legacy of the agricultural origins of California's water systems, one acre-foot is almost 326,000 gallons—plenty of water for two urban families for a year.

3. Wendell and Hall (chapter 8 in this volume) provide an in-depth treatment of groundwater and its management.

4. See Christian-Smith and Heberger (chapter 5 in this volume) for a detailed description of urban water demand.

5. Wilkinson (chapter 2 in this volume) offers another, expanded view of this subject.

6. Comprehensive discussions of IRWM and integration are available nearby from Wilkinson (chapter 2), Hanak and Lund (chapter 7), and Cantú (chapter 9).

7. See Gray (chapter 4 in this volume) for more on the "antiquated separation" of groundwater and surface water law.

8. Foster and Radke (chapter 6), Cantú (chapter 9), and Podolak and Yarnell (chapter 13) also highlight the importance of people and interpersonal relationships in water resources management.

9. Hanak and Lund (chapter 7 in this volume) discuss at length the role of economics in water management.

10. Balazs and Ray (chapter 11 in this volume) say more on this frequently neglected issue.

REFERENCES

Augustyn, Fay, and Ben Chou. 2013. *Getting Climate Smart: A Water Preparedness Guide for State Action.* Washington, DC: American Rivers and Natural Resources Defense Council.
Brown, Edmund G. ["Jerry"], Jr. 1982. State of the State Address, January 7. http://governors.library.ca.gov/addresses/s_34-JBrown7.html
California Air Resources Board. 2008. *AB 32 Scoping Plan.* Sacramento: California Air Resources Board.
———. 2014. *AB 32 Scoping Plan Update.* Sacramento: California Air Resources Board.
———. 2008. *Managing an Uncertain Future: Climate Change Adaptation Strategies for California's Water.* Sacramento: California Department of Water Resources.
———. 2010. *California Water Plan Update 2009.* Sacramento: California Department of Water Resources.
———. 2012. *Climate Action Plan, Phase I: Greenhouse Gas Emissions Reduction Plan.* Sacramento: California Department of Water Resources.
———. 2013. *California's Flood Future: Recommendations for Managing the State's Flood Risk.* Sacramento: California Department of Water Resources.
———. 2014. *California Water Plan Update 2013.* Sacramento: California Department of Water Resources.
California Energy Commission. 2005. *Integrated Energy Policy Report.* Sacramento: California Energy Commission.
California Environmental Protection Agency. 2013. *Indicators of Climate Change in California.* Sacramento: California Environmental Protection Agency.
California Natural Resources Agency. 2014. *Safeguarding California Plan.* Sacramento: California Natural Resources Agency.
Conrad, Esther. 2012. *Climate Change and Integrated Regional Water Management in California: A Preliminary Assessment of Regional Approaches.* Berkeley: Department of Environmental Science, Policy and Management, University of California.
Hanak, Ellen, and Jay Lund. 2008. *Adapting California's Water Management to Climate Change.* San Francisco: Public Policy Institute of California.

Intergovernmental Panel on Climate Change. 2001. *Synthesis Report, Third Assessment Report*. Switzerland: Intergovernmental Panel on Climate Change.

———. 2013. *Summary for Policymakers, Working Group 1, Fifth Assessment Report*. Switzerland: Intergovernmental Panel on Climate Change.

Kelley, Robert. 1989. *Battling the Inland Sea: Floods, Public Policy, and the Sacramento Valley*. Berkeley: University of California Press.

McWilliams, Carey. 1949. *California: The Great Exception*. Berkeley: University of California Press.

Moser, Susanne, Julia Ekstrom, and Guido Franco. 2012. *Our Changing Climate, 2012*. Sacramento: California Climate Change Center.

National Research Council, Committee on Sea Level Rise in California, Oregon, and Washington. 2012. *Sea-Level Rise for the Coasts of California, Oregon, and Washington: Past, Present, and Future*. Washington, DC: National Academies Press.

Public Policy Institute of California. 2013. *Californians and the Environment: Statewide Survey*. San Francisco: Public Policy Institute of California.

Reisner, Marc. 1986. *Cadillac Desert: The American West and Its Disappearing Water*. New York: Viking Penguin.

Roos, Maurice. 2005. *Accounting for Climate Change: California Water Plan Update 2005*. Vol. 4. www.waterplan.water.ca.gov/docs/cwpu2005/vol4/vol4-globalclimate-accountingforclimatechange.pdf.

Roos, Maurice, and Michael L. Anderson. 2006. "Monitoring Monthly Hydrologic Data to Detect Climate Change in California." Poster presented at the Third Annual Research Conference on Climate Change, Sacramento, CA, September 13–15.

Wells, H. G. 1945. *Mind at the End of its Tether*. London: W. Heinemann.

Worster, Donald. 1985. *Rivers of Empire: Water, Aridity, and the Growth of the American West*. New York: Pantheon.

CHAPTER 2

The Water–Energy–Climate Nexus in California

ROBERT WILKINSON

The water–energy nexus is the relationship between the use of energy to extract, treat, deliver, and use water, and to collect and treat wastewater, and the use of water to extract, convert, and use energy. The climate link involves both impacts of water and energy systems on climate, and the impact of climate change on water and energy systems. The water–energy–climate nexus is an opportunity as well as a challenge. This chapter explores the relationships among them and provides some conceptual and methodological approaches to managing both water and climate for sustainability.

New approaches and innovations in water and energy present interesting and important synergies. The water–energy–climate nexus is an opportunity for integrated management for multiple benefits. In many cases, improving water-use efficiency provides significant energy savings and related greenhouse gas emissions reductions. Innovations in technology and technique for water management, including efficiency improvements and source shifting (e.g. using rainwater or recycled water where appropriate), have the potential to yield multiple economic, environmental, and social benefits. On the other side of the equation, advances in energy systems such as solar photovoltaic and wind energy can reduce or even eliminate water inputs.

In California, government agencies are integrating water and energy policies to respond to climate change as well as to environmental challenges and economic imperatives. The integration of water, energy, and climate policy and planning, including policy processes at the state's Energy Commission, Public Utilities Commission, Department of Water Resources, Water Resources Control Board, and Air Resources Board, is moving forward. Methodologies to account for embedded energy in water systems—from initial extraction through treatment, distribution, end use, wastewater treatment and discharge—and water use by energy systems, have been developed and are outlined below. Institutional collaboration between energy, water, and other authorities is also evolving; and, encouragingly, the governor's 2014 *Water Action Plan* includes reference to concepts like the water–energy nexus, multiple benefits, and integrated approaches (California Natural Resources Agency, California Environmental Protection Agency, and California Department of Food and Agriculture 2014).

Just in the past decade California has recognized that water is one of the largest electricity uses in the state. Water systems account for approximately 19 percent of total electricity use and about 33 percent of non–power plant natural gas use (California Energy Commission 2005; GEI Consultants and Navigant Consulting 2010). The largest single electricity user in California is the Edmonston Pumping Plant, with the largest single water lift in the world. It pumps water almost 2,000 feet in elevation over the Tehachapi Mountains between the Central Valley and urban Southern California. The Edmonston plant is part of the State Water Project, which in total (with a series of pumping plants and other facilities that lift and move water over hundreds of miles) is the largest electricity user in the state.

The California Energy Commission and the California Public Utilities Commission (CPUC) have both concluded that energy embedded in water presents large untapped opportunities for cost-effectively improving energy efficiency and reducing greenhouse gas emissions. The Energy Commission commented in its 2005 *Integrated Energy Policy Report* that "the Energy Commission, the Department of Water Resources, the CPUC, local water agencies, and other stakeholders should explore and pursue cost-effective water efficiency opportunities that would save energy and decrease the energy intensity in the water sector." This corresponds well with the formal water plans of the Department of Water Resources (2013).

This chapter is structured to address three issue areas, each of which can be put as a question. (1) The water–energy–climate nexus: What is the relationship between water and energy, and how will climate change influence that relationship? (2) System dynamics and methodologies for quantification of key elements of the water–energy–climate nexus: How do we quantify key aspects of water, energy, and climate in a consistent way in order to compare options, understand implications, and identify leverage points to intervene in the system? (3) Integrated policy strategies to achieve multiple benefits: How can we integrate policy approaches and strategies to maximize benefits?

The water, energy, climate nexus: What is the relationship between water and energy, and how will climate change influence, and be influenced by, that relationship?

Water supplies for human use are provided through diversions and extraction of surface and groundwater. We often refer to this as *developed* water. Energy supplies and use are facilitated by conversion of primary energy (e.g. fuels, solar radiation, wind) to usable energy services at a point of use. Historically, water and energy were locally provided. In modern times, they are often transported over long distances from their sources to the place where they are ultimately used. As technological capacity developed over the past century, surface water diversions, groundwater extraction, and conveyance systems increased in volume and geographic extent. Interbasin transfers supplement water available within natural hydrological basins or watersheds. Similarly, energy systems have evolved from largely local sources a century ago to continent-wide electricity grids and pipeline networks, and to global supply lines for fuels.

The focus of technology development and policy for much of the past century has been on the supply side of both the energy and water equations. That is, the emphasis was on extracting, storing, converting, and conveying water and energy from sources to users. Water and energy policy throughout the world has generally been designed to facilitate the development and use of these supply-side technologies. In the last few decades, however, scientific developments and technological innovation have increasingly been applied to improve of the *efficiency of use* of energy and water resources. (*Efficiency* as used here describes the useful work or service provided by a given amount of water or energy.) Significant economic as well as environmental benefits have been cost-effectively achieved through efficiency improvements in water and energy systems. Various technologies, from electric motors and lighting systems to sensors, pumps, and plumbing fixtures, have vastly improved end-use efficiencies.

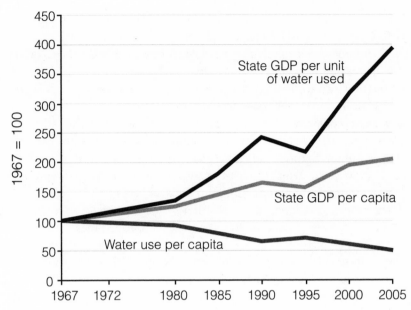

FIGURE 2.1. California gross domestic product and population per unit of water used.
Source: Hanak et al. (2012).

Dramatic improvements have been made in energy and water efficiency in California. Electricity use per capita, for example, has remained level at around 7,000 kWh per person in California since the mid-1970s, while the U.S. figure is around 13,000 kWh per person (U.S. Energy Information Administration 2013). Per capita water use has declined over the same time frame. Technical innovations, price signals, and policy measures have enabled California to *quadruple* its gross domestic product per unit of water since the 1960s (figure 2.1).

Today, the main constraints on water extraction are not technological. Indeed, there is significant spare capacity for extraction and conveyance in many areas. The limits are increasingly imposed by competing claims on scarce water resources, legal constraints, and environmental impacts. Costs of building and maintaining infrastructure have also risen dramatically. The maintenance costs for existing water and wastewater systems are staggering. The American Society of Civil Engineers (n. d.) estimates that over $39 billion will be needed for safe drinking water and $29.9 billion for wastewater treatment systems in California over the next 20 years, along with the need to deal with 807 "high hazard" dams in the state. This does not count the tens of billions of dollars proposed for various new projects.

Technology development and policy to meet water needs are therefore increasingly focused on more efficient use and on water treatment. Innovation and technology development in the areas of end-use water applications and water treatment have progressed rapidly. Techniques and technologies ranging from laser leveling of fields and drip and micro-spray irrigation systems to the improved design of membrane filters, plumbing fixtures, industrial processes, and treatment technology have changed the demand side of the water equation. Water supply systems (e.g. treatment and distribution) are also becoming more efficient. For example, geographical information systems and field technologies enable improved capabilities to locate leaks in buried pipes.

Climate change poses important water and energy management challenges. Research indicates that the rate and magnitude of temperature increases, and related impacts, are increasing. These trends and findings are validating science that has been conducted for decades. Climate models consistently indicate a hotter future for the U.S. West. Winter temperatures in the Sierra Nevada in the instrumental record indicate a rise by almost 2 °C during the second half of the twentieth century, and trends toward earlier snowmelt and runoff to the Sacramento–San Joaquin Delta over the same period were identified by scientists at Scripps 20 years ago (Dettinger and Cavan 1994). California's Department of Water Resources and the California Energy Commission have been supporting research and tracking climate change science since the 1980s.

Climate change impacts are driving policy responses, and energy and water are central to the issue. The opening line of AB 32 is: "Global warming poses a serious threat to the economic well-being, public health, natural resources, and the environment of California" (California Global Warming Solutions Act 2006). Water is the leading impact area. Energy use is the leading cause. Integrated policy, planning, and management of water resources and energy systems can provide important opportunities to respond effectively to challenges posed by climate change. Both mitigation strategies (e.g. to reduce greenhouse gas emissions) and adaptation strategies (to deal with impacts) are being developed. One key opportunity is to reduce energy use and greenhouse gas emissions by improving water-use efficiency and switching from energy-intensive water sources to ones that require less. While both energy and water managers have used integrated planning approaches for decades, the broader integration of water and energy management in the context of climate change is a relatively new and exciting policy area (California Department of Water Resources 2014).

California has been a leader in energy planning, from setting the first building and appliance efficiency standards in 1976 to consistently outperforming the rest of the nation in per capita electricity consumption improvement for the past 40 years (Foster et al. 2012). California's continued focus on energy efficiency has saved the state an estimated $65 billion and helped make California more energy-independent (Brown 2013).

As the California Council on Science and Technology (2014) concludes in a recent report on California's water future, "California has a long history of success in leveraging innovations in science, technology, management and implementation strategies to improve its resource management, including its continued leadership in energy efficiency. The State's best strategy for dealing with its water challenges, both current and future, lies in taking a system management approach to water similar to the approach used for energy system management. As with energy, innovative water technologies represent a business opportunity for California."

System dynamics and methodologies for quantification of key elements of the water–energy–climate nexus: How do we quantify key aspects of water, energy, and climate in a consistent way in order to compare options, understand implications, and identify leverage points to intervene in the system?

Water and energy systems are interconnected in important ways. Water systems provide energy (e.g. through hydropower), and they often consume energy through pumping, thermal, and other processes. Energy systems use water for extraction, processing, cooling, and other parts of the energy cycle (Wilkinson 2011). While water and energy are interrelated, they are not necessarily linked inextricably. Indeed, multiple benefits may be achieved by decoupling some of the links.

Energy Inputs to Water Systems

California's water systems are often energy-intensive. Moving large quantities of water over long distances and significant elevation lifts, treating and distributing it within communities, using the water, and collecting and treating wastewater, all together account for a major use of energy. The *energy intensity* of water is the total amount of energy, calculated on a whole-system basis, required for the use of a given amount of water in a specific location.

The total energy embedded in a unit of water used in a particular place varies with location, source, and use. Pumping water at each stage is often energy-intensive. Other important energy inputs include thermal energy (heating and cooling) at the point of use, and aeration in wastewater treatment processes.

There are three broad categories of energy elements of water systems, which correspond directly to the water-use cycle:

1. *Primary water extraction, conveyance, storage, treatment, and distribution.* Extracting and lifting water is highly energy-intensive. The pumping of surface and groundwater requires significant amounts of energy, depending on the depth of the source. Where water is stored in intermediate facilities, energy is often required to store and then recover the water. Within local service areas, water is treated, pumped, and pressurized for distribution. Local conditions and sources determine both the treatment requirements and the energy required for pumping and pressurization. Some distribution systems are gravity-driven, while others require pumping.

2. *Water use (on-site water pumping, treatment, and thermal inputs).* Individual water users require energy to further treat water supplies (softeners, filters, etc.), circulate and pressurize water supplies (e.g. building circulation pumps), and heat and cool water for various purposes.

3. *Wastewater collection, treatment, and discharge.* Finally, wastewater is collected and treated by a wastewater system (unless a septic system or other alternative is being used) and discharged. Wastewater is sometimes pumped to treatment facilities where gravity flow is not possible, and the standard treatment processes require energy for pumping, aeration, and other purposes.

The schematic flow diagram in figure 2.2 is based on work originally supported by the California Institute for Energy Efficiency through the Lawrence Berkeley Lab, and it is the basis for the California Energy Commission's 2005 analysis. The methodology is applicable to water sources ranging from surface and groundwater supplies to ocean desalination and recycling. It has now been used as the basic approach to calculating the energy intensity of water supplies by a number of entities including the California Energy Commission, the Canadian govern-

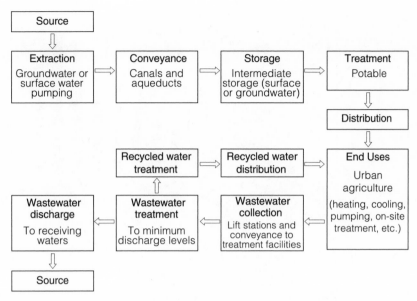

FIGURE 2.2. Flow diagram of energy inputs to water systems. Based on Wilkinson (2000).

ment, and the WateReuse Research Foundation, and it is available for free in a user-friendly computer model (Cooley and Wilkinson 2012). An open-access model, WESim (www.pacinst.org/publication/wesim/), was developed by the Pacific Institute and the Bren School at UC Santa Barbara based on this methodology.

The energy intensity of water varies considerably with the geographic location of both end-users and sources. Water use in certain places is highly energy-intensive due to the combined requirements of conveyance, treatment and distribution, and wastewater collection and treatment processes. Large energy savings are possible through water efficiency improvements and through source switching (e.g. using recycled water in place of other sources for appropriate purposes), in part because embedded energy is saved at multiple steps in the process. For example, replacement of old water-wasting devices with new high-efficiency options not only reduces the energy required to extract, convey, treat, and deliver the water to the user; it also reduces the energy required to collect and treat the wastewater.

Figure 2.3 shows actual data for the energy intensity of major water supply options for inland and coastal locations in Southern California. Each bar represents the energy intensity, including conveyance, of a

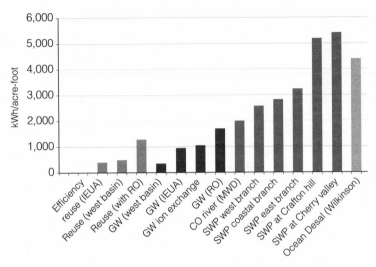

FIGURE 2.3. Energy intensity of selected water supplies in Southern California: Efficiency, reuse, groundwater, imports, and ocean desalination. For imported water from the Colorado River and State Water Project (SWP) treatment energy is not included. The figures are for untreated water delivered to urban Southern California. IEUA: Inland Empire Utilities Agency. RO: reverse osmosis. GW: groundwater. MWD: Metropolitan Water District. Desal: desalination. Source: Wilkinson (2008).

specific water supply source used at a selected location in Southern California. Water conservation—for example, not using water in the first place—avoids additional energy inputs along all segments of the water use cycle. Consequently, cost-effective water use efficiency is often the preferred water resource option from an energy perspective. For all other water resources, there are ranges of actual energy inputs that depend on many factors, including the quality of source water, the energy intensity of the technologies used to treat the source water to standards needed by end-users, the distance water needs to be transported to reach end-users, and the efficiency of the conveyance, distribution, and treatment facilities and systems (Wilkinson 2000). In many cases, as indicated by the examples in figure 2.3, the treatment and use of local water supplies such as groundwater, seawater, brackish water, and wastewater requires much less energy than imported supplies. Innovations in treatment processes, including membranes, pressure recovery, and other aspects, are further reducing the energy requirements of treatment. This trend is expected to continue.

Water Intensity of Energy Systems

The other side of the water–energy nexus is the water used in the extraction, refining, production, conversion, and use of energy. Water inputs to energy systems can be quantified to understand where water is used and how much is required for different energy sources and conversion technologies. The water intensity of energy is essentially the inverse of the energy intensity of water: it is the total amount of water, calculated on a whole-system basis, required to produce and use a given amount of energy in specific locations.

Water inputs to energy systems are highly variable. They depend on the primary energy source and on the conversion technologies employed at each step in the process. For example, primary fuels such as coal and biomass often require water for production. Biofuels often require water for irrigation of crops as well as for production processes. It is important to note that both renewable and nonrenewable energy sources can be either water-thrifty or water-intensive depending on a number of factors including the technologies deployed. Every water input at each step needs to be accounted for to develop a comprehensive water-intensity metric.

Water is increasingly viewed as a limiting factor in thermal power plant siting and operation. The U.S. Geological Survey estimates in its most recent analysis that 48 percent of all U.S. freshwater and saline-water withdrawals are used for thermoelectric power (Hutson et al. 2005). Although cooling systems account for most of the water used in power generation, water is also used in other parts of the process such as mining and "fracking" (for both oil and gas), processing, transporting fuels (e.g. coal slurry lines), and other steps. These processes may also have important local impacts on water quality.

The U.S. national laboratories have been working for years on the energy–water nexus, as reflected in *Energy Demands on Water Resources: Report to Congress on the Interdependency of Energy and Water* (U.S. Department of Energy 2006). As with other analyses of the issue, the report finds that some energy systems are highly dependent on large volumes of water (and thereby vulnerable to disruption), while others are relatively independent of water. Water use for renewable and nonrenewable forms of energy varies substantially. Solar photovoltaics, wind turbines, and some geothermal, cogeneration, and landfill gas-to-energy projects use little water. In contrast, irrigated bioenergy crops can consume exponentially more water per unit of energy provided. Finally, although reservoirs often have multiple purposes (e.g. flood control, water storage, recreation), evaporative losses from hydroelectric facilities per unit of electricity are higher than in many other forms of generation.

Thermoelectric freshwater withdrawal per unit of energy generated, and the impact of this withdrawal, depend largely on the cooling technology used. Currently there are two main types of cooling technologies used in power plants: once-through and recirculating (Hutson et al. 2005). Once-through cooling systems withdraw water from a natural water body, use it for heat exchange, and return it to the water body at a higher temperature after one cycle of use. Recirculating (closed-loop) technologies include wet cooling towers and cooling ponds. Wet recirculating systems use water over multiple cooling cycles and have much lower withdrawals than once-through cooling systems. Most new plants, especially those built after 1970, use some form of recirculating cooling, which require less water to be extracted from surface or groundwater sources once the recirculating system is filled (Macknick et al. 2011). Although thermodynamically there are larger evaporative losses from recirculating systems than from once-through systems (due to the larger temperature increase in the cooling water), the adverse environmental impacts of the combination of thermal barriers and thermal pollution associated with the return of heated water to the natural system are the reason that once-through cooling is largely not utilized today. Thermoelectric cooling technologies that use no water, or smaller amounts of water than recirculating cooling, are available (dry cooling and hybrid wet/dry cooling systems), but their use involves an "energy penalty" due to impacts on back-pressure and auxiliary loads.

Integrated policy strategies to achieve multiple benefits: How can we integrate policy approaches and strategies to maximize benefits?

In California we have designed and built water and energy systems responding to our opportunities and constraints—physical, economic, social, and environmental—as perceived at different stages of history. Some of yesterday's solutions are today's challenges. This historical context is important to an understanding of current strategies for sustainability.

Infrastructure Choices and Supply-Side Thinking

California's early systems often provided both water and electricity. Los Angeles tapped the Owens River, east of the Sierra Nevada, and later the Mono Lake tributaries, for both water supply and power. San Francisco also looked to the High Sierra, damming the Hetch Hetchy Valley in Yosemite National Park at about the same time. These early interbasin systems were designed to use gravity not only to deliver water to California's two largest urban areas at the time but also to generate power for the cities.

In the 1930s the Colorado River Aqueduct was built to bring Colorado River water to urban Southern California. Unlike the Los Angeles Aqueduct, it required large pumps and significant energy inputs to lift water over mountains and move it across the desert from Parker Dam. The Central Valley Project, built in the 1930s and 1940s, includes not only hydropower generation but also large pumping facilities, as does the State Water Project, built in the 1960s and 1970s.

Several aspects of these major water projects are worth noting. First, few environmental constraints existed (at least in law) in the first half of the century, so planners and designers were not concerned that rivers like the San Joaquin would be entirely dewatered for long segments. Second, while energy generation was a key factor in the early systems, by mid-century there was a notion that energy was becoming less important as a limiting factor. Indeed, the idea that nuclear energy would be "too cheap to meter"[1] led to plans and designs for water systems that required huge energy inputs. Serious plans were developed to replumb North America (tapping Alaska's rivers for the U.S. Southwest and Mexico) in systems that would require massive amounts of energy and a willingness to ignore environmental impacts.

It turns out that nuclear power is expensive and plagued with serious problems (only one plant is left running in California). The full costs of energy include climate change and other environmental impacts. Those costs are not too cheap to meter either. California developed laws in the 1960s and 1970s to address the environmental impacts of water systems and other projects. These limiting factors—cost, environmental impacts, and public acceptance—have been significant for both water and energy systems.

Demand-Side Approaches and Decentralization

Technology development and policy strategies to meet water needs are increasingly focused on more efficient use and on water treatment technologies. Innovation and development of technology in the areas of end-use water applications and water treatment have progressed rapidly.

End-uses of water now require much less volume to provide equivalent or superior services. Rainwater capture for groundwater recharge, and other innovative water-capture strategies, are also enhancing water supply reliability (Garrison, Wilkinson, and Horner 2009). Water supply systems (e.g. treatment and distribution) are also becoming more efficient.

Instead of looking over the next mountain range for more water and energy supplies, we are increasing looking locally. Efficiency is the first option. For both water and energy, the challenge is to get more from less by improved design and operation. Next are decentralized options like rooftop photovoltaic panels for electricity (costs are now at grid parity in many areas); and for water, rainwater harvesting and recycling are providing increasing fractions of the portfolio of supplies. In fact, most water agencies in California are planning to increase efficiency and local supplies while reducing demands for imported water.

Integrated Policy and Management

The California Department of Water Resources (2012) defines the *integrated water management* approach as "a philosophy and practice of coordinating the management of water and related resources for the purpose of maximizing economic and societal benefits while maintaining the sustainability of vital ecosystems." California is currently integrating water, energy, and climate response strategies and policies to tap multiple benefits and to respond to climate change. Water managers are looking at water delivery system and end-use efficiency improvements, source switching, and other measures that save energy by reducing pumping and other energy inputs.

New approaches to the integration of water and energy planning, including policy processes at the California Energy Commission, Public Utilities Commission, Department of Water Resources, and State Water Resources Control Board are being developed. Methodologies have been developed for accounting for embedded energy, from initial extraction through treatment, distribution, end-use, and wastewater treatment and discharge (Wilkinson 2000). Institutional collaboration between energy and water management authorities is also evolving.

Improvements in urban water-use efficiency have been identified by the Department of Water Resources for over a decade in its State Water Plans as California's *largest new water supply* for the next quarter-century, followed by groundwater management and reuse. These are also the most reliable sources, as indicated by the darker gray bars (figure 2.4).

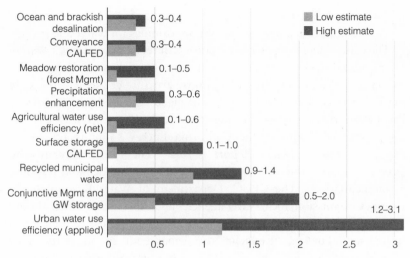

FIGURE 2.4. California Water Plan estimates for new water supplies over the next several decades. Source: California Department of Water Resources (2009).

The Department of Water Resources (2005) identified efficiency improvements as a key element of the state's future water supply. Observing the water planning process, the Energy Commission noted: "The *2005 Water Plan Update* mirrors the state's adopted loading order for electricity resources" (Klein 2005). Though a "loading order" approach to water management has not yet been formally developed, it is clear that improving efficiency, managing and using groundwater more effectively (including rainwater harvesting to recharge groundwater), and recycling and reusing water are high-priority approaches. A version of a loading order for water management would be useful and should be explored.

Integrated Planning at the State Level

An important shift in planning and governance is occurring in both the state agencies managing water and energy and at the local and regional levels. California has developed several forms of integrated planning, ranging from interagency efforts to respond to climate change (e.g. WET-CAT—the Water-Energy Team of the state's Climate Action Team[2]) to regional integrated water management planning processes supported by state grants and administered collaboratively by the Department of Water Resources and the State Water Resources Control Board (California

Department of Water Resources 2014; Spanos 2012). These efforts build on the identification a decade ago of opportunities for water and energy.

The California Energy Commission's staff report, *California's Water–Energy Relationship,* noted: "As California continues to struggle with its many critical energy supply and infrastructure challenges, the state must identify and address the points of highest stress. At the top of this list is California's water–energy relationship" (Klein 2005). One of the top recommendations in the California Energy Commission's 2005 *Integrated Energy Policy Report* is this: "The Energy Commission strongly supports the following energy efficiency and demand response recommendations: The CPUC, Department of Water Resources, the Energy Commission, local water agencies and other stakeholders should assess efficiency improvements in hot and cold water use in homes and businesses." Further, "Reducing the demand for energy is the most effective way to reduce energy costs and bolster California's economy" (California Energy Commission 2005).

Tapping Multiple Benefits

When the costs and benefits of a proposed policy or action are analyzed, we typically focus on accounting for specific costs, and then we compare those costs with a specific, well-defined benefit such as an additional increment of water supply. We often fail to account for other important benefits that accrue from well-planned investments that solve for multiple objectives. With a focus on *multiple benefits,* we account for various goals achieved through a single investment. For example, improvements in water-use efficiency—meeting the same end-use needs with less water—also typically provide related benefits such as reduced energy requirements for water pumping and treatment (with reduced pollution related to energy production as a result) and reduced water and wastewater infrastructure capacity and processing requirements. The impacts of extraction of source water from surface or groundwater systems are also reduced. Water managers often do not receive credit for providing these multiple benefits when they implement water efficiency, recharge, and reuse strategies. From both an investment perspective and the standpoint of public policy, the multiple benefits of efficiency improvements and recharge and reuse should be fully included in cost–benefit analysis.

Policies that account for the full embedded energy of water supplies have the potential to provide significant additional public- and private-sector benefits. Economic and environmental benefits are potentially available through new policy approaches that properly account for the energy intensity of water. As outlined above, energy savings may be achieved both upstream and downstream of the point of use when the energy consumption of both water supply and wastewater treatment systems are taken into account. Methods, metrics, and data are available and are being further refined to provide a solid foundation for policy approaches to account for energy savings from water management options.

The Role of Price Signals Coupled with Policy

One reason the focus of technological innovation has shifted from supply development to improving efficiency and local sources is economics. When water is cheap, there is little incentive to design and build water-efficient technologies or to treat water for reuse. As the cost of water increases, technology options for reducing waste, enabling reuse, and providing greater end-use efficiency become more cost-effective and even profitable. Technologies for measuring, timing, and controlling water use, and new innovations in the treatment and reuse of water, are growing areas of technology development and application.

Impetus for scientific inquiry and technology innovation and development has been provided by both price signals (increasing costs) and public policy (e.g. requirements for internalization of external costs). Public policy is increasingly incorporating these costs, including those of climate change, into resource prices. As water and energy prices continue to reflect costs, including environmental costs previously externalized, these prices increase.

At the same time, technology has provided a wide range of options for expanding utility value through efficiencies (less water and energy being required to perform a useful service). Broader application of these technologies and techniques can yield significant additional energy, water, economic, and environmental benefits.

Public policy can be designed to encourage improved water and energy management practices by both suppliers and users. Appliance efficiency standards (for both energy and water) and minimum-waste requirements are examples. Policy measures have also been used to frame and guide market signals by implementing mechanisms such as increasing tiered pricing structures, meter requirements (some areas do not even measure use), and other means to utilize simple market principles and price signals more effectively. It is worth noting that toilets are regulated nationwide by the Energy Policy Act, signed by President Bush in 1992.

In an economic and resource management sense, efficiency improvements are now considered *supply* options, because permanent improvements in the demand-side infrastructure provide reliable water and energy savings. Coupling technology options such as efficient plumbing and energy-using devices to economic incentives (e.g. rebates) and disincentives (e.g. increasing tiered rate structures) is a good strategy. The coupling provides both the means to improve productive water and energy use and the incentive to do it.

Conclusion: Toward More Sustainable Water and Energy Management through Integrated Strategies
Tapping Multiple Benefits

California faces formidable challenges in providing water and energy to its citizens in sustainable ways in the face of scarcity and variability, rising costs, security threats, climate change, and much else. We are fortunate to have the scientific, technological, and managerial capacity, and the institutions of governance, to take on these difficult challenges. Innovation and integrated strategies will be essential to shifting toward a sustainable approach.

Integrated water and energy management strategies, with a focus on vastly improved end-use efficiency, and careful consideration of alternative technology opportunities provided by advances in science and technology, can provide multiple significant benefits to society. Cost-effective improvements in energy and water productivity, with associated economic and environmental quality benefits, and increased reliability and resilience, are attainable.

Methodologies and metrics exist to tap the multiple benefits of integrated water–energy–climate strategies, though they can be improved. The policies required to incentivize, enable, and mandate integrated water and energy policy exist and are being refined to tap ample opportunities to improve both the economic and the environmental performance of water and energy systems.

With better information regarding energy and water use, public policy combined with investment and management strategies can dramatically improve productivity and efficiency. Potential benefits include improved allocation of capital, avoided capital and operating costs, and reduced burdens on ratepayers and taxpayers. Other benefits, including restoration and maintenance of environmental quality, can also be realized more cost-effectively through policy coordination. The full benefits derived through water–energy–climate strategies have not been adequately quantified or factored into policy.

To quantify and realize the benefits, we should utilize multiple-benefit analysis to determine the cost-effectiveness of investments in water and energy systems and in climate response strategies. This will enable stronger cofunding strategies for water and energy agencies. Policymakers need to craft supportive policy structures to enable water and energy entities to tap linked water–energy–climate improvement opportunities.

NOTES

1. This phrase "too cheap to meter" comes from the chairman of the United States Atomic Energy Commission, Lewis Strauss, in a 1954 speech to the National Association of Science Writers.

2. The Water-Energy Team of the Climate Action Team (www.climatechange.ca.gov/climate_action_team/water.html) is tasked with coordinating its efforts on both greenhouse gas emissions reduction and adaptation actions affecting the portion of the energy sector that supports the storage, transport, and delivery of water for agricultural, residential, and commercial needs.

REFERENCES

American Society of Civil Engineers. n. d. *Report Card for California's Infrastructure.* www.infrastructurereportcard.org/a/#p/state-facts/california.

Brown, Edmund G., Jr. 2013. *State of the State Address.* Remarks as prepared January 24, 2013. Sacramento, CA.

California Council on Science and Technology. 2014. *Achieving a Sustainable California Water Future through Innovations in Science and Technology.* http://www.ccst.us/news/2014/0409water.php.

California Department of Water Resources. 2005. *California Water Plan Update 2005.* Bulletin 160–05. Sacramento: California Department of Water Resources.

———. 2012. *Strategic Plan for the Future of Integrated Regional Water Management in California: Development Approach.* Sacramento: California Department of Water Resources.

———. 2013. *California Water Plan Updates.* Bulletins 160–05/09/13. Sacramento: California Department of Water Resources. www.waterplan.water.ca.gov/cwpu2013/prd/.

———. 2014. *Integrated Regional Water Management.* www.water.ca.gov/irwm/grants/.

California Energy Commission. 2005. *Integrated Energy Policy Report, November 2005.* CEC-100-2005-007-CMF. Sacramento: California Energy Commission.

California Global Warming Solutions Act of 2006, AB32, Section 38501(a).

California Natural Resources Agency, California Environmental Protection Agency, and California Department of Food and Agriculture. 2014. *California Water Action Plan: Actions for Reliability, Restoration and Resilience.* http://resources.ca.gov/docs/Final_Water_Action_Plan.pdf.

Cooley, Heather, and Robert Wilkinson. 2012. *Implications of Future Water Supply Sources for Energy Demands, and Computer Model with WESim User Manual.* Pacific Institute and Bren School, University of California, Santa Barbara. www.pacinst.org/publication/wesim/

Dettinger, Michael D., and Dan R. Cayan. 1994. "Large-Scale Atmospheric Forcing of Recent Trends toward Early Snowmelt Runoff in California." *Journal of Climate* 8:606–23.

Energy Policy Act of 1992, 102nd Congress H.R.776.ENR, abbreviated as EPACT92. http://thomas.loc.gov/cgi-bin/query/z?c102:H.R.776.ENR:.

Foster, Ben, Anna Chittum, Sara Hayes, Max Neubauer, Seth Nowak, Shruti Vaidyanathan, Kate Farley, Kaye Schultz, and Terry Sullivan. 2012. *2012 State Energy Efficiency Scorecard.* Report #E12C. Washington, DC: American Council for an Energy-Efficient Economy.

Garrison, Noah, Robert C. Wilkinson, and Richard Horner, 2009. *A Clear Blue Future: How Greening California Cities Can Address Water Resources and Climate Challenges in the 21st Century.* Natural Resources Defense Council and Water Policy Program, Bren School of Environmental Science and Management, University of California, Santa Barbara. www.nrdc.org/water/lid/.

GEI Consultants and Navigant Consulting. 2010. *Embedded Energy in Water Studies.* Study 1: Statewide and Regional Water-Energy Relationship. Study 2: Water Agency and Function Component Study and Embedded Energy-Water Load Profiles. Robert Wilkinson, Lead Technical Advisor, through the UC Office of the President. Energy Division, California Public Utilities Commission, managed by California Institute for Energy and Environment. www.cpuc.ca.gov/PUC/energy/Energy+Efficiency/EM+and+V/Embedded+Energy+in+Water+Studies1_and_2.htm

Hanak, Ellen, Jay Lund, Barton "Buzz" Thompson, W. Bowman Cutter, Brian Gray, David Houston, Richard Howitt, Katrina Jessoe, Gary Libecap, Josué Medellín-Azuara, Sheila Olmstead, Daniel Sumner, David Sunding, Brian Thomas, and Robert Wilkinson. 2012. *Water and the California Economy.* Public Policy Institute of California. www.ppic.org/main/publication.asp?i=1015.

Hutson, Susan S., Nancy L. Barber, Joan F. Kenny, Kristin S. Linsey, Deborah S. Lumia, and Molly A. Maupin. 2005. *Estimated Use of Water in the United States in 2000.* Circular 1268 (released March 2004; revised April 2004, May 2004, February 2005). U.S. Geological Survey. http://water.usgs.gov/pubs/circ/2004/circ1268/.

Klein, Gary. 2005. *California's Water–Energy Relationship: Final Staff Report, Prepared in Support of the 2005 Integrated Energy Policy Report Proceed-*

ing. 04-IEPR-01E, November 2005, CEC-700-2005-011-SF. Sacramento: California Energy Commission.

Macknick, Jordan, Robin Newmark, Garvin Heath, and KC Hallett. 2011. *A Review of Operational Water Consumption and Withdrawal Factors for Electricity Generating Technologies*. Golden, CO: National Renewable Energy Laboratory.

Spanos, Katherine 2012. *The Climate Has Changed: Now What? Integrated Regional Water Management and Climate Change: Planning a Coincidental or Inevitable Union?* Paper presented at the 30th Annual Water Law Conference of the American Bar Association, February 22–24, 2012, San Diego, California. www.water.ca.gov/climatechange/articles.cfm.

U.S. Department of Energy. 2006. *Energy Demands on Water Resources: Report to Congress on the Interdependency of Energy and Water*. www .sandia.gov/energy-water/congress_report.htm

U.S. Energy Information Administration. 2013. *State Profiles and Energy Estimates*. www.eia.doe.gov/emeu/states/_seds.html.

Wilkinson, Robert C. 2000. *Methodology for Analysis of the Energy Intensity of California's Water Systems, and an Assessment of Multiple Potential Benefits through Integrated Water-Energy Efficiency Measures*. Exploratory Research Project. Ernest Orlando Lawrence Berkeley Laboratory, California Institute for Energy Efficiency.

———, 2008. *Invited Testimony to Congress: Water Supply Challenges for the 21st Century, Committee on Science and Technology, United States House of Representatives*. http://www.bren.ucsb.edu/people/Faculty/documents /wilkinson_US_house_testimony_000.pdf.

Wilkinson, Robert. 2011. "The Water–Energy Nexus: Methodologies, Challenges, and Opportunities." In *The Water-Energy Nexus in the American West*, edited by Douglas S. Kenney and Robert Wilkinson. Cheltenham: Edward Elgar.

CHAPTER 3

California's Flawed Surface Water Rights

MICHAEL HANEMANN, CAITLIN DYCKMAN,
AND DAMIAN PARK

IN THE BEGINNING: A CHAOTIC SITUATION

California sprang into existence following the discovery of gold in 1848. Aside from domestic use, the first major use of water in California was in mining. The first mining consisted of placer mining of alluvial deposits in stream beds throughout the Sierra foothills. As those deposits were depleted, hydraulic mining arose, in which high-pressure jets of water were used to remove overlying earth from upland gold-bearing deposits. That type of mining, first employed in 1853, required substantial water diversions.

When California entered the Union in 1850, the English common law was adopted as the "rule of decision" in courts, including the doctrine of riparian rights for surface water[1] (it was also the governing doctrine in the rest of the Union). Riparian rights entitle the owner of land bordering a surface water body ("riparian" land) to use the water on his or her riparian land. This is a right to *use* water, not a right of ownership, and it inheres only in riparian lands. Riparian rights remain with the riparian land regardless of changes in ownership. Water under a riparian right cannot be used on nonriparian land.[2] The right is shared equally among all riparians: they own access to the stream as "tenants in common." They can divert water as long as this does not impair the

Dedicated to the memory of Joe Sax. A fuller version of this material is in preparation.

rights of other riparians. No specific quantity attaches to a riparian right. If a riparian originally applied X, this does not preclude him from applying $5X$ later. Nonuse does not terminate the right. There is no recording of the volume diverted. No institution administers the riparian right. Disputes are resolved through litigation among riparians.

The riparian doctrine was logical where it originated, in a humid region with plentiful streamflow. Streamflow is treated as a common pool to be shared among all riparian landowners. But in an arid region like California, where rivers can run dry by the late summer and annual streamflow can vary by an order of magnitude, there needs to be a specific mechanism for allocating limited streamflow. The riparian right lacks this.

Using water for hydraulic mining violated riparian requirements. In most cases, the deposits being mined were not located on riparian land. And the miners did not own the land where water was being diverted or used—these were public lands. Consequently, a new type of water right was developed, adapted from the rules developed by miners for the right to a mining claim. The miners "met and organized mining districts, adopting rules for the definition of their property rights. . . . These rules limited the size of claims. . . . They required miners to post notices of their claims and to record them with district recorders. . . . To retain their claims, miners had to work them with diligence. . . . When questions of right arose, they were settled by reference to priority . . . first in time, first in right." With the emergence of hydraulic mining, "the miners applied the same rules to water as they had to land—first in time, first in right. To perfect the right, ditches had to be dug with diligence and the water applied to beneficial use" (Dunbar 1983, 61).

In 1851, the California legislature endorsed the mining-camp rules as state law. Subsequently, district courts applied the principle of first possession to water cases. In 1855, the California Supreme Court endorsed what became known as the appropriative water right. The right to divert water is based on the time and quantity of the initial diversion creating that right. The link between ownership of land and ownership of water is severed. The locations of water diversion and application can be different. If there is too little streamflow, the senior appropriators divert their full quantity until the stream is exhausted, while the remaining (junior) appropriators receive nothing.[3]

The Supreme Court's rulings did not extinguish the concept of a riparian right, and California courts continued to uphold it. In 1866 and 1870, Congress gave recognition to appropriative rights. In 1872, California's legislature formally recognized appropriative rights and codified the procedures for acquiring them. The codification maintained a dual system of appropriative and riparian rights.

While the appropriative water right was modeled after the right to a mining claim, crucial differences existed between the two resources that rendered the right less well suited to water than to mining. The nature of the economic activity was not the same, and the institutions for recording and enforcing the property right functioned very differently.

According to Clay and Wright (2005, 163), gold mining was "a race to find a small number of high payoff claims." As they note (157): "Typically a miner worked a claim only long enough to determine its potential. If he decided it was a relatively low-value claim—as most were—he continued the search [elsewhere]. . . . Because miners were continually looking for new and better sites even as they worked their present holdings, mining district rules were as much concerned with procedures for the abandonment and repossession of claims as they were with protection of the rights of existing claimholders." Mining was thus different from a production-oriented activity such as irrigated agriculture. Mining rules sought to ensure the "orderly turnover" of mining sites to maximize the chance of a bonanza discovery, not to promote land settlement.

The interactions among miners were fundamentally different from those among water users, and played out over a much larger spatial scale. For a miner, the question was "Is someone else working a claim at this location? If not, I will." For an irrigator, it was not enough to know whether someone else was diverting water at this location: it was also necessary to know whether other diversions were occurring on the same stream.[4]

Mining districts provided a nongovernmental apparatus that was quite effective in recording mining claims. Posting a claim at the site and recording it with the mining district was a reasonable procedure. Furthermore, the mining districts played some role in mining claim enforcement and dispute resolution. Mining district codes typically specified procedures for settling disputes over contested claims (Clay and Wright 2005, 163–67), although those procedures were not necessarily final. While imperfect, the system based on mining districts, the posting of claims, and the right of first possession was relatively coherent and promoted its objective: orderly and rapid exploration of mining sites.

The situation with appropriative water rights was entirely different. In an arid region, land is worthless without water, and the objective was to ensure continued access to water. Because of the spatial scale of potential interactions among competing water users, posting a claim along a river bank on a two-foot stake was a much less transparent means of recording a property right for water than for mining. Property right quantification was inherently less precise. The spatial area claimed for exploration was essential to mining. With water, the volume of water diverted during some time at a particular location comprised the claim. But volume is a problematic measure of this right, because the diversion occurs intermittently; the exercise of the property right is not uniform over time, as with a land claim.

In additional to being poorly quantified and not transparently recorded, another crucial difference between an appropriative water right and a mining claim was that the former lacked any administrative apparatus for verifying or enforcing the priority date or the amount of the right. The mining districts played no role in the recording or enforcement of appropriative water rights, even for hydraulic mining, and there was no other entity, governmental or nongovernmental, that performed this role. In the event of a dispute among water users, whether about the seniority or the quantity of a right, the only recourse was litigation.

Litigation has many weaknesses as a method of dispute resolution. Litigation is time-consuming. And water use for irrigation is especially time-sensitive—crops need water during the growing season. Litigation could not resolve a dispute in time to save that year's crop. Litigation was costly. The decree bound only the parties to the litigation, not other water users omitted from the litigation.[5] Finally, there was no mechanism to enforce a judicial decree resulting from litigation, except further litigation for contempt of court (Chandler 1913, 149).

Given the differences in the nature of the resource and the way it was used, mining claims provided a poor analogy for water use. The system of appropriative water rights based on the right of first possession was considerably less coherent than the system of mining claims based on the same principle.

The 1872 code changed things marginally. Under that code, in addition to posting the claim at the river bank, a water user was required to file a copy with the county recorder and to commence construction of the diversion facility within 60 days of posting. But the county recorder played no role in verifying the claim to an appropriative water right, checking whether there was sufficient streamflow for that amount to have been diverted as claimed, checking whether construction was initiated (or completed), monitoring diversions to ensure subsequent conformity with the water right, or sharing information about appropriative rights claimed with other counties bordering the same stream.

Moreover, *nonstatutory* appropriations were still legal—made, as before 1872, by posting a notice at the site of the diversion and without recording the claim with the county recorder. A property rights scheme lacking effective recording and enforcement is a contradiction in terms. Yet that characterizes the appropriative right to water in California until 1914.

WHAT OTHER STATES DID

After the California Gold Rush, other major discoveries of gold occurred in Colorado (1858–1859), Nevada (1859), Idaho (1860), Montana (1862–1864), and Arizona (1863), leading to immigration into those states. Like California, they adopted the English common law and, with it, riparian water rights. In those gold rushes, as in California, the miners organized mining districts and adopted rules to protect claims to mining rights and to water. As in California, the mining was being conducted on public lands and required something other than a riparian right. The new mining districts copied California in adopting an appropriative right to water based on the right of first possession, which was subsequently recognized by the state courts and legislatures.

Colorado was the first western state to enact laws for the administration of surface water rights. This came about gradually. In 1861, in its first session, the territorial legislature endorsed "the records, laws and proceedings of each mining district," and also enacted a statute authorizing the appropriation of water for irrigation of both riparian and nonriparian lands. Under this law, there was no requirement to record the appropriation. The first court decision dealing with appropriative rights, in 1872, upheld this as a necessity in Colorado's climate. In 1876, the state constitution stated that "priority of appropriation shall give the better right as between those using water for the same purpose." The question of whether riparian rights still existed in Colorado was answered in 1882 when the state Supreme Court declared that the riparian doctrine was "inapplicable to Colorado." Eventually, all the mountain states followed Colorado in establishing prior appropriation as the exclusive right to surface water.[6]

At first, following the 1861 legislation, appropriative water rights in Colorado functioned in the same incoherent manner as in California, with no recording requirement and no state administrative apparatus for verification, enforcement or monitoring. As in California, there was chaos, and "it was impossible to determine the number and priorities of the appropriations of a stream except through an expensive adjudication lawsuit" (Dunbar 1983, 87). The drought of 1874 triggered interest in a new legislative solution when diverters along the Cache la Poudre River failed to agree on how to divide the depleted streamflow, and upstream diverters with junior rights "took what they wanted, depriving the downstream appropriators of their legitimate supply of irrigation water" (87).

Legislation in 1879 established ten water districts around the state, each with a water commissioner who was to enforce the distribution of water based on prior rights. The water commissioners had no powers to determine the priority of rights or resolve conflicts regarding rights. Instead, judges could initiate an inquiry into a water right and make a finding. This legislation was subsequently challenged because it gave judges the power to initiate an inquiry on their own authority without waiting for someone to file a suit in the conventional manner. The 1881 Adjudication Act resolved this: it required irrigators with *existing* appropriative rights to file their claims for priority with district courts by June 1881 to determine the priority and quantity of their right.

The Colorado system was only partly successful. The administrative system for enforcing court water rights decrees worked well and was widely admired and emulated. But the reliance on judges was problematic: they did not consult with the state engineer, lacked engineering training, and typically did not verify the accuracy of data presented in court. Consequently, irrigators made extravagant claims, which judges then accepted. The water rights decreed by the courts varied erratically, with no rationale or relation to actual use (Meade 1903, 149–55).

The situation improved slowly. In 1887, statewide recording of appropriative rights claims was initiated (Meade 1903, 144). In 1899, the water commissioners received additional powers, including the power to arrest and prosecute anyone violating orders for the opening or closing of head-gates. In 1903 and 1919, the Colorado legislature completed the adjudication procedure. The 1903 Adjudication Act provided the courts with authority to adjudicate all other appropriative water rights in the same manner as irrigation rights. The 1919 Adjudication Limitation Act was designed to settle the priorities of all water rights. It required any claimant to an appropriation to submit the claim for adjudication by January 1921; failure to do so caused a presumption of abandonment (Hobbs 1999, 9).

In 1886 and 1888, the neighboring state of Wyoming largely copied Colorado's legislation of 1879 and 1881. It soon found the same defects as in Colorado: "there was no central register of appropriation claims. . . .

Many of the streams were overappropriated, and few had been adjudicated. Of those that had been [adjudicated] the decreed rights were excessive and inconsistent" (Dunbar 1983, 106). Reacting to this situation, on attaining statehood in 1890, the Wyoming legislature created an administrative system for the control of water rights. The administrative apparatus both conferred water rights (handled by courts in Colorado) and administered them.

The Wyoming system created water divisions and a Board of Control consisting of the State Engineer and the superintendents of the water divisions. A person wishing to appropriate water applied for a permit from the State Engineer. If the State Engineer determined that unappropriated water was available and the diversion was not "detrimental to the public welfare," the permit was granted. No appropriation after 1890 was valid without a permit. The division superintendents monitored diversions and enforced priority. The Board of Control adjudicated streams, subject to appeal to the district courts.

The permit procedure did not apply retroactively to water rights acquired before 1890, but the 1890 legislation specified a procedure to address those rights. Under the stream adjudication process, the owners of all water rights, including those acquired before 1890, were required to file their claim with the State Board of Control. Failure to do this extinguished the right (Squillace 1991, 97). Under this scheme, all of Wyoming's streams were adjudicated by 1922; any users who failed to claim a pre-1890 right lost it (Squillace 1989, 324).

Over the period of 1895–1909, other western states adopted versions of the Wyoming system.[7] Only Oregon copied Colorado's court-based determination of water rights. The Wyoming system aroused controversy for vesting the determination of water rights in an administrative board rather than the courts. Engineers supported this, but lawyers and some water users opposed it. After 1902, the federal government advocated it as a precondition for receiving water projects from the new Bureau of Reclamation, which proved decisive.

In summary, all the other western states moved to systems for the conferral, recording, and administration of appropriative rights that, while not perfect, were comprehensive and orderly. Those systems generated usable records of rights holders and their seniority. Moreover, there was a local administrative apparatus for monitoring diversions, ensuring conformity with the decreed appropriative right, and enforcing seniority in the event of shortage. Of all the states, Colorado was the most successful at verifying water rights. In some other states, gaps still remain between the amounts of water claimed and actually put to beneficial use (Tarlock 2000, 882). Nevertheless, there is a relatively coherent system for recording, monitoring, and enforcing appropriative water rights.

WHAT CALIFORNIA DID

While other western states regularized the administration of appropriative rights, California did nothing, even as the use of water for irrigation grew dramatically. The acreage irrigated in California increased from roughly 60,000 acres in 1870 to 300,000 acres in 1880, and then to 1.4 million acres in 1900, 2.7 million acres in 1910, and 4.2 million acres in 1920 (Rhode 1995, table 1). The expansion was partly associated with the assemblage of large land holdings, often obtained through dubious acquisitions of Spanish land grants and fraudulent acquisitions of public lands disposed under the 1850 Swamp Act and the 1877 Desert Land Act. Public land purchases were legally restricted to 320 acres per capita under the former and 640 acres under the latter, but those restrictions were blatantly evaded. For example, Henry Miller, the largest landowner in California, acquired a 100-mile swathe of riparian land along the San Joaquin River and 50 miles of riparian land along the Kern River, much of it obtained fraudulently under the Swamp Act.

The abusive acquisition of land was accompanied by a stretching, if not abuse, of water rights. "Owners of riparian land have. . . . rented and sold water claimed under the riparian doctrine to those who irrigate non-riparian lands, and the right to do this has been sustained in repeated judicial decisions" (Meade 1903, 194). Other landowners claimed appropriative rights and used those to monopolize land that they did not own. They could do this because of the permissive system for claiming appropriative rights and the legal ambiguity then existing: Was the amount of the right the amount *actually* being used, the amount that *could* be used given the canal capacity, or the amount the appropriator *aspired* to use in future? (Pisani 2002, 38). For example, while the average flow of the Kings River varied from 5,000 to 10,000 cfs in flood season and from 500 to 1,000 cfs during the low-flow period, the claims to Kings River water amounted to 750,000 cfs, exclusive of multiple claims to the entire river flow. On the San Joaquin River, six entities each claimed the entire average flow, and the remaining claims totaled 8 times its maximum flow—152 times its average flow (Meade 1903, 190).

James Haggin and two partners owned 400,000 acres of land in Kern County by 1878, but claimed appropriative rights for water to irrigate two million acres, many times more than the Kern River ever carried. As their holdings grew, they collided with the downstream riparian rights of Henry Miller. During a severe drought in 1877, their upstream diversions dewatered Miller's lands. In May 1879, Miller sued Haggin and others. The case was tried in 1881, leading to a decision for Haggin. The California Supreme Court heard the case in 1883 and 1884 and ruled for Miller. The majority opinion held that riparian rights were still valid in California. This generated immense public controversy, and the court agreed to rehear the case. The final decision, in April 1886, again favored Miller. The court ruled that riparian rights counted as property rights under common law, and property rights, once vested, could not be taken without compensation.[8]

California's dual system of inconsistent water rights was thus permanently enshrined. Riparian rights were inherently unquantified. Appropriative rights were quantified incoherently, if at all, and unregulated. The only mechanism for resolving disputes, which abounded, was litigation. But, as Pisani (1984, 338) notes: "The legal system resolved few water rights conflicts. . . . In the absence of a state engineering office, the courts relied almost entirely on biased witnesses for hydrographic information. . . . In any case, court tests rarely included all interested parties, so the decisions were invariably incomplete. Then, too, enforcing a court decree was no easy matter; contempt proceedings were expensive and subject to the same delays as water rights suits." Some litigation was epic in its scale. The Kings River was notorious. Litigation began in the drought year of 1876 and escalated, totaling 137 suits by 1917. The piecemeal judgments in those suits produced some striking anomalies, such as places where "A had rights superior to B, who had rights superior to C, who had rights superior to A" (Governor's Commission 1978, 24).

Support for water law reform grew during the drought of 1898–99, but it was blocked by water users. Change finally came after the election of a reform governor (Hiram Johnson) and a reform legislature in 1910. In 1911, the legislature declared that "All water or the use of water within the State of California is the property of the people of the State of California" (Cal. Stats. 1911, 821). It created a State Conservation Commission to examine the need for new laws to control use of the state's natural resources. The commission's recommendations were enacted in the Water Commission Act of 1913. Challenged by water users, the legislation was put on the ballot. It was approved by the voters and took effect in December 1914.

The legislation established a State Water Commission, with the power to regulate unappropriated surface waters of the state. A person wishing to appropriate water after December 1914 applied to the commission for a *permit*. If the commission determined that surplus water was available, the permit was granted. The permit holder then had the right to take and use the water according to the permit terms. Upon compliance with the permit terms and demonstration of beneficial use, the commission issued a *license* which confirmed the appropriative right. The commission's authority was initially nondiscretionary: if the applicant followed the prescribed procedures and unappropriated water was available, the permit had to be issued. In 1917, the commission was given discretion to refuse applications deemed detrimental to the public welfare. In 1921, it was given the power to grant a right "under such terms and conditions" as it judges "in the public interest" and to reject applications not in the public interest.

Once it had issued a permit or a license, the commission had only "a limited role in resolving disputes and *enforcing* rights of water holders, a task left mainly to the courts."[9] Thus, while the 1913 act allowed for the administrative conferral of an appropriative (post-1914) right, as in Wyoming, the resolution of any subsequent disputes among water right holders was still left to the courts, as in Colorado. There were two routes by which the commission might enter such disputes. First, under the court "reference" procedure, a court was permitted to transfer a case to the commission to act as referee. Second, upon its own initiative or upon request of a water right holder, the commission could conduct a statutory adjudication of the stream, determining all appropriative rights to water, whether issued before or after 1914. If they wished, water users could obtain a judicial review prior to a final decree.

The California system had two key differences from those of Colorado and Wyoming. First, Colorado and Wyoming arranged for appropriative rights predating the reform legislation to be brought under that legislation and adjudicated, whereas California's commission lacked authority over pre-1914 rights, except in the case of a court reference or statutory adjudication. Second, while Colorado's water districts and Wyoming's water divisions provided an administrative apparatus to supervise the distribution of water, monitoring diversions to ensure conformity with water rights, no such arrangement existed in California. Supervision of water distribution was opposed by water users at the

Conservation Commission's hearings; it was dropped by the commission chair "in a spirit of conciliation."[10] At the time, commentators saw the omission of public supervision of water distribution as a serious flaw.[11]

The Water Commission had no authority over riparian rights. The Conservation Commission had wanted to abolish riparian rights but felt unable to do this (Miller 1985, 12). Instead, the 1913 act stipulated that unused riparian water would be forfeited after 10 years of nonuse. However, the Water Commission had little power to enforce this—riparians could seek relief with the courts—and it was declared unconstitutional in 1935.[12]

Once in operation, the Water Commission issued biennial reports. The first report, in 1917, identified two weaknesses in the 1913 act: lack of detail regarding procedures for a statutory adjudication, and the lack of power to supervise water distribution. The legislature responded by enacting details for an adjudication but did nothing regarding supervision of water distribution. The commission's second biennial report, in 1918, noted the consequences: "The irrigation season of 1918 has been one of unusually low run-off. For most streams in northern California, at least, the run-off this season is the lowest recorded. A number of requests have been received asking the Commission to send a representative to take charge of the distribution of the water of streams" (17)—but it lacked authority to do so. The legislature still did nothing.

The drought worsened during the winter of 1919–20. It coincided with a dramatic increase in rice acreage in the Sacramento Valley, which required more water than other crops. The rice acreage grew from 100 acres in 1910 and 15,000 acres in 1914 to 154,700 acres in 1920 (California State Water Commission 1921, 71). In February, the commission issued a warning of an impending water shortage. The situation was exacerbated by the lack of reliable data on water rights in the Sacramento Valley (153). The problem was eventually solved by voluntary action, without the commission's intervention and without enforcement of seniority. Various agencies organized an Emergency Water Conservation Conference, which included representatives of water users in the Sacramento Valley. The conference persuaded growers to reduce rice plantings by 50,000 acres and to ensure that water was used "with all due economy. . . . Irrigation water was never handled so carefully in the Sacramento Valley as it was during the summer of 1920" (154–55). Though the crisis passed, it highlighted "the necessity for an early determination of the underlying rights to divert water from the Sacramento River." "Without such a determination," the commission warned, "there is absolutely no basis for a diversion of water among the various claimants in periods of shortage" (155).

The commission's 1921 report noted a deluge of applications for new permits, "far in excess of the natural summer flow" of all the streams in California, which "has now been fully appropriated and put to use." Consequently, "the greater portion of the required additional supply must be developed by the construction of storage reservoirs and regulation of stream flow, holding the flood runoff for use during periods of low natural flow" (12). Despite the interest in new reservoir projects, such projects faced the challenge of "how storage water can be released into a natural stream with assurance of its escaping illegal diversion before reaching its destination, or how such a reservoir can be operated to the satisfaction of prior and vested rights of downstream water users" (13). To answer those questions, the commission believed that California needed "a complete water code"; still lacking were "detailed provisions for the public supervision of the distribution of water ... in accordance with defined rights, and the appointment of water masters when needed" (15). Accordingly, it proposed amendments to the 1913 act. There was partial success. The appointment of water masters to control the use of water was authorized, but only upon written request of the owners of at least 15 percent of the diversion facilities in the region. The recommendation for supervision of water distribution was ignored.

Thus, when California did finally act, it conceded to politically powerful interests. Compared to other western states, the California commission was weakened by the exclusion of riparian rights and pre-1914 appropriative rights, and had little power to enforce post-1914 appropriative rights.

CONSOLIDATION

In July 1921, the commission ceased existence. A new Department of Public Works came into existence, and the commission's functions and duties were assumed by its Division of Water Rights. In its 1924 report, looking back to the 1913 act, the division observed that, when the commission started, it had "faced a most difficult situation" because of "the maze of legal entanglement" associated with riparian and pre-1914 appropriative rights. It conceded that the legal situation had not changed, "nor has litigation over water matters been done away with. It has, however, been greatly reduced" (9).

Given the lack of authority over riparian and pre-1914 appropriative rights, and the small quantity of post-1914 appropriative rights, how was litigation reduced? Two tools existed for determining pre-1914 rights (the court reference procedure and statutory adjudication), and one tool for supervising the distribution of water under existing rights (a water master). But either a court or the water users had to request these actions.[13] All the commission or division could do on its own authority was to conduct special hydrological investigations; field investigations were also part of reviewing permit applications and determining whether unappropriated water was available. Those investigations turned out to be crucial.

The investigations had a "moral effect," because "technical and legal information on points formally so obscure can now be secured from an authoritative and impartial source. The assistance of the Division is sought in controversies over water matters not necessarily within the scope of the [1913] act" (Division of Water Rights 1924, 10). As the division noted:

> In any legal controversy over water it is most often the questions of fact which are at issue. . . . Whether or not the Division has any quasi-judicial function, it can, if it is in possession of the facts regarding conditions on a stream, make known these facts to the interested parties in the issue and the matter is then susceptible of compromise. . . . The Division acts in the nature of a bureau in answering questions regarding water right principles. In this, it has been of much service in settling difficulties, in clearing up a number of intricate water tangles, and in bringing together those who desired an equitable settlement of their difficulties, but were in doubt as how best to proceed. It is believed that much useless and expensive litigation has been avoided through this service. (Division of Water Rights 1922, 9)

Thus, the division saw "its largest function" as "a fact finding and recording body" (9).

Three factors helped the division bring some order to the tangle of water rights. First, in assessing permit applications it bypassed "paper" water rights and focused on whether unappropriated flow was available. Frequently, "those claiming vested rights admit that there is unappropriated water available to a new appropriator in the source from which they are already diverting and do not object to a new diversion, provided their prior rights are respected" (Division of Water Rights 1924, 38). Second, in 1923 the division was allowed to modify its procedure for protested permit applications: it could now hold a hearing before the decision, instead of afterwards. This had substantial procedural and psychological impacts. "Particularly in the larger and more important cases, the action of the Division is expedited, as each of the contesting parties assumes a greater share of the burden in preparing his own case and presents it in better shape, thus overcoming the tendency for such matters to drag on. . . . It may also be noted that the hearing procedure through its formality discourages trivial protests and has tended toward the adoption of a new mental attitude on the part of those whose interests are jeopardized by proposed appropriations."[14] Third, the division extended its influence through its readiness to conduct informal investigations. "While some [requests for a hydrographic investigation] come to the Division as requests for adjudication, others come as a request for an informal physical investigation. In some instances the existing rights are so complicated . . . that it is felt that the formal adjudication procedure might not be successful; however, if the physical facts can be determined by investigation this will suffice" (Division of Water Rights 1922, 14).

Thus, a degree of order came to California's water rights administration. But this occurred only if the water users invited it. Thus, water users on the Kings River decided in 1917 to end their litigation wars and request a water master, because they wanted a dam (the Pine Flat Dam), for which a determination of existing water rights was needed (California State Water Commission 1921, 14). There were other cases where the water users wanted no intervention; there, the courthouse remained the only venue for dispute resolution.

WATER MANAGEMENT THROUGH THE COURTS

One dispute involved upstream diversions and downstream water quality in the Delta. While the Delta waters are tidal, they are not saline, except in late summer and fall when low outflow from the Sacramento and San Joaquin Rivers permits saltwater to advance inland. During the 1920 drought, the Sacramento River flow dropped by about 90 percent, to a record low of 420 cfs, coupled with record salinity intrusion into the Delta. In July 1920, the Delta town of Antioch sued upstream irrigators in the Sacramento Valley to stop their diversions from causing salinity to reach the intake for the town's water supply (itself a diversion of less than 1 cfs). At least 3,500 cfs of Delta flow was needed for an acceptable salinity level. In 1922 the California Supreme Court rejected the suit because "it would be hard to conceive of a greater waste for so small a benefit" than to require that an additional 3,080 cfs flow unused to the ocean to provide less than 1 cfs for municipal use.

This case raised two fundamental issues. First, could a water right be used to regulate streamflow for water quality, in this case salinity? The *Antioch* court recognized that "an appropriator of water from a stream for domestic and similar uses has the right to enjoin the pollution of the stream above him," but considered that diverting water, as opposed to discharging something noxious into a stream, could not be considered an action "that in the least affects the purity of the water." Thus, streamflow was not a water quality parameter. Second, should the interests of downstream water users prevent the construction of upstream storage? The court noted that, if it acceded to Antioch's request, this would set a precedent that could impede the construction of storage, an outcome "highly detrimental to the public interests."

Storage was of keen interest. The 1920 drought demonstrated that additional storage was needed to accommodate new applications for water rights. Also, since the early 1900s electric utilities had become interested in hydropower; these included Southern California Edison, which was looking to the headwaters of the San Joaquin River.

The 1913 act limited annual water use on uncultivated riparian land to 2.5 acre-feet per acre. This was at issue in *Herminghaus v. Southern California Edison* (1926). The Herminghaus family owned undeveloped riparian land along the San Joaquin River, used for pasture and, since 1896, leased to Henry Miller. For forage, they had the practice of temporarily damming the river during the spring runoff so that it overflowed the land and produced a crop of natural grasses. Their 1913 act limit was approximately 54,000 acre-feet, but they diverted the entire river flood flow, around 1.8 million acre-feet. Edison had riparian rights to the upper San Joaquin, and constructed its first hydropower facility in 1911. In the 1920s it was planning to expand that system. Its reservoirs would store water used to flood the Herminghaus lands, and Herminghaus sued. Edison claimed the right, as a riparian, to store water when the volume of flow far exceeded the needs of downstream *irrigators,* releasing it in the late summer when irrigation water was in short supply while generating electricity for public use. Herminghaus argued that it had a riparian right to divert the flood water.

The court ruled that the riparian right did include the use of flood waters, but it did *not* include storage, so Edison impounded water as a mere appropriator. The public benefit of Edison's reservoirs was irrelevant. At that time, California courts applied a standard of reasonable use in disputes among riparians, among appropriators, and between a riparian and an appropriator where the riparian claimed unreasonable use by the appropriator. But they refused to apply reasonable use to a claim by an appropriator against a riparian. Following that principle, the lower court ruled for Herminghaus. On appeal, the California Supreme Court upheld the ruling. There was a wave of public outrage. There was also "a rash of new cases" against hydropower projects (Miller 1989, 103). In response, the legislature placed a constitutional amendment on the ballot declaring that "the general welfare requires that the water resources of the State be put to beneficial use to the fullest extent of which they are capable," and required that all surface water use—riparian and appropriative alike—be reasonable and beneficial. The amendment passed in November 1928.

CONSTRUCTING A HYDRAULIC SOCIETY

By then, California was immersed in water projects. The notion of transferring surplus Sacramento River water to the drier San Joaquin Valley was first suggested in 1858 and was the subject of extensive investigations by the State Engineer, William Hammond Hall, in 1877–88. It resurfaced in 1919 in a proposal by Robert Marshall. Prompted by the 1920 drought, the legislature allocated funds for water resources investigations and, following the 1924 drought, it allocated additional funds. The report, in 1927, offered a coordinated plan for developing the state's water resources (Division of Engineering and Irrigation 1927). That was a banner year. Besides placing the constitutional amendment on the ballot, the legislature authorized the Department of Finance to file to reserve appropriative water rights that might be needed for a statewide water plan, with the notion that it would assign those rights only to users whose projects conformed to that plan.[15]

In 1929, water planning moved into high gear. The Division of Water Rights was combined with the Division of Engineering and Irrigation within the Department of Public Works to form a single Division of Water Resources. Funding was allocated for an expanded water planning effort. The product was a detailed proposal for a State Water Plan (Division of Water Resources 1931). The plan's main focus was to provide storage upstream in the Sacramento Valley and to transfer water to the San Joaquin Valley.

In 1933, the legislature endorsed the project and authorized a bond issue, which was narrowly approved by voters. With California then in the depths of the Great Depression, the state made no attempt to sell the bonds. Instead, it turned to the federal government for help, first seeking grant or loan assistance and then, with those not forthcoming, asking it to take over the project. In 1935, President Roosevelt released emergency relief funds so that work could begin. In 1937, Congress formally approved the Central Valley Project (CVP) as a Bureau of Reclamation project. Construction of Shasta Dam started in 1937 and was completed in 1945; the hydropower and other ancillary facilities were completed in 1950. Construction of Friant Dam also started in 1937; it was completed in 1942. The Madera Canal was completed in 1945, and the Friant-Kern and Delta-Mendota Canals in 1951.

The war led to a boom in California's agriculture and economy, and it continued afterwards. Irrigated acreage grew from 5.1 million acres in 1939 to 6.6 million acres in 1949. The population increased from 6.9 million in 1940 to 10.6 million in 1950. The growth in potential demand for water clearly outstripped the supply expansion from the CVP.

California, by then chafing at federal control of the CVP, created a water planning authority in 1945 and funded a state-wide water resources investigation in 1947. In 1951, a report was released proposing a new large dam in the Sacramento Valley and an aqueduct through the San Joaquin Valley to Southern California (California State Water Resources Board 1951). That year, the legislature authorized what became the State Water Project (SWP) and appropriated funds for detailed engineering studies. This gained further momentum from a massive flood in 1956 which the new dam could have prevented. To implement the project, the Department of Water Resources was created that year as a superagency vested with all the powers and responsibilities relating to water from the Department of Public Works and other state agencies. In 1957, the department issued the culminating product of the decade-long state water investigation, a comprehensive master plan for the SWP. In 1959, the legislature authorized bonds for the first stage of the SWP. This was the largest bond issue ever offered by any state, and it was made subject to voter approval. The bond was narrowly approved in an election in 1960; the northern counties, reluctant to send "their" water south, rejected the proposal; the southern counties, containing the majority of the beneficiaries, provided the margin of victory.

Meanwhile, the CVP was also being expanded. Folsom Dam was completed in 1956, and Trinity Dam in 1963. The CVP division delivering water to the west side of the Sacramento Valley was largely completed by 1965. The new CVP and SWP dams released water into the Sacramento River to flow into the Delta for pumping southward. This required additional pumping capacity in the Delta and additional conveyance capacity in the San Joaquin Valley. These were supplied through a federal–state partnership, which included a new aqueduct. The federal portion was known as the San Luis Aqueduct. The SWP portion, known as the California Aqueduct, carried water over into Southern California. Construction of the CVP portion was completed in 1968, and of the SWP portion in 1973. With that, California's modern hydraulic system was in place.

That system represented a strategy of supply expansion rather than more efficient management of existing resources. What was the role of water rights in the strategy?

The 1928 constitutional amendment was crucial. Without it, most of the new dams could not have been built. However, following a 1935 Supreme Court ruling, the reasonable use doctrine fell into dormancy, losing the effective ability to restrict wasteful water use. Also, when the Division of Water Rights was folded into the Division of Water Resources in 1929, this reduced the resources and attention devoted to water rights administration.

The projects' financing was based on water users' paying the cost of the water supplied by the project. For those receiving water from the new CVP canals, that amount could readily be quantified.[16] Not so for users diverting water from the San Joaquin or Sacramento River. With the San Joaquin River, which would be dried up downstream of Friant Dam, the diverters would receive a like amount of water from the Delta-Mendota Canal as a free replacement.[17] The Shasta and Trinity Dams changed the seasonal timing and volume of Sacramento River stream-flow, producing less flow in the spring but more in the late summer, and created the risk that existing diverters would grab the augmented late-season flow. These diverters could divert the amount of their pre-existing right to river water (their *base supply*) for free; if they wished, they could divert an extra quantity, for which they paid (*project water*).[18] Determining those quantities of water was problematic. For existing users with riparian rights, there was no quantity associated with their right. For users with pre-1914 appropriative rights, those rights remained unquantified because they were outside the Water Commission Act's purview. For users with post-1914 rights senior to the CVP, the loose administration of post-1914 water rights made those quantities uncertain, too.

The burden of negotiating quantities fell on the CVP and SWP. This was harder in the Sacramento Valley than in the San Joaquin Valley.

In the San Joaquin Valley, a key factor was Henry Miller's dominance. He had been exceptionally litigious in protecting his water rights.

A 1933 suit against the Madera Irrigation District had generated "nearly a complete adjudication of rights" to the San Joaquin River upstream of Mendota Pool (Graham 1950, 597). Moreover, his company was now in a period of decline and, for several reasons, "needed to sell its water rights before it lost them" (Garone 2011, 165). In 1939, it reached an agreement with Reclamation to sell the water that flooded its pasture lands, and it received an exchange contract to provide substitute water for its irrigated croplands in the San Joaquin Valley. The agreement became the model for contracts with other riparian landowners and ultimately fixed the level of their compensation (Miller 1992, 173).[19]

In the Sacramento Valley, there were many small landowners, with no dominant land owner like Miller and little prior litigation that might have quantified water rights. Impounded water flowed along the river channel to the Delta. The legal point for CVP and SWP diversions was near the Delta, several hundred miles downstream from the points of storage, with many intervening users in between. As the state's chief water lawyer warned in 1942, whether an adequate amount of project water would be available for export depended "upon the degree to which the rights of these intervening users are defined with exactitude, as well as the extent to which those users voluntarily confine themselves thereto. In the existing condition of human nature it may be confidently predicted that [they], finding an abnormal increment in the stream, will each for himself define and exercise their rights in their own favor with substantial elasticity" (Holsinger 1942, 13). Exactitude in the definition of those rights required a statutory adjudication. The Reclamation supervisor in Sacramento had recommended this in 1939, but his superiors in Washington rejected it. California now proposed this in December 1942, but Washington again rejected it. By 1951, with the CVP in full operation, the situation had worsened. Sacramento Valley diversions, which had averaged about 1 million acre-feet annually from 1924 to 1940, soared to 2 million acre-feet in 1951. More diversions meant less project exports. Reclamation leadership was concerned enough to consider requesting an adjudication. At that point, the water users in the Valley brought Congressional pressure on Reclamation to abandon that notion (Bain, Caves, and Margolis 1966, 477). For another decade, Reclamation and the Sacramento River water users continued to disagree about water amounts for settlement contracts. Finally, after 20 years of negotiations, and under pressure from the Secretary of the Interior, Reclamation began signing settlement contracts in 1964.

THE WATER RIGHTS BOARD ERA

To prevent further conflict of interest, the 1956 reorganization that created the Department of Water Resources to operate the SWP also established a separate entity, the State Water Rights Board (SWRB), which assumed the responsibilities of the Department of Public Works regarding surface water rights. The new SWRB was a diminished version of the 1920s Division of Water Rights, and had less ability to manage post-1914 rights. It lost the authority to provide water master service, which stayed with the Department of Water Resources. That department also retained many experienced staff, including engineers employed in field investigations to measure water diversions, which "severely handicapped" the SWRB and limited its effectiveness (California State Water Rights Board 1957, 34). It faced a "heavy backlog" of applications for new permits, accumulated "over a long period of years" (Holsinger 1957, 686).

With new applications, the SWRB tightened the information requirements to validate the amount of water claimed. But it was in a weak position with respect to existing rights, lacking authority over riparian and pre-1914 rights, and with few resources to monitor post-1914 rights.[20] There was no more collection of data on diversions.[21] In 1965, an ineffectual attempt was made to change this, motivated by concerns over uncapped riparian diversions. Riparians and appropriators with both pre- and post-1914 rights were required to file a statement every three years with the SWRB detailing their monthly diversions.[22] But the information was "for information purposes only," and failure to file lacked legal consequences. Smaller diverters in the Delta and others received exemptions. The result was little compliance: in 1978, only 10 percent of holders of riparian and pre-1914 rights filed statements.[23]

The SWRB was also in a weak position to address the water projects. Though the CVP's permits had not yet been issued—this was the immediate task for the SWRB—it was delivering water on a massive scale as a *fait accompli*. The SWRB could approve permits for the CVP (and SWP) only if (1) unappropriated water was available, (2) this was in the public interest, and (3) it did not impair existing vested rights. That depended on the amount of water controlled by existing vested rights, something not clearly known and beyond the authority of the SWRB to determine.[24] There had been a conscious decision in 1939–42 and in 1951 against holding an adjudication, and the SWRB was not going to touch that.

In its first CVP decision in 1958 (D-893), the SWRB considered applications to divert American River water by the CVP, the city of Sacramento, and various others, significantly exceeding the river flow. There also were state applications for future water developments in upstream counties, and flows for fish conservation requested by the Department of Fish and Game. There were multiple protests, including claims that the diversions would harm vested rights downstream by increasing salinity in the Delta. The SWRB rejected most of the applications but granted the CVP and Sacramento applications, subject to (1) future agreements among the parties to control Delta salinity, (2) future reductions for within-watershed development, and (3) compliance with recent Fish and Game agreements for fish flows.[25] This set the pattern for CVP and SWP applications: the SWRB granted the projects' applications on an interim basis subject to the resolution of ongoing negotiations about vested water rights, salinity levels, and/or fish conditions. At intervals, the SWRB reopened the decision process, heard evidence, and made a new interim decision on similar terms. This pattern of deference to negotiations among the interested parties continues to the present day, and the negotiations remain largely unresolved.[26]

What did change was the growing power of environmental concerns. In 1949, California adopted the first comprehensive water pollution control law in the United States, creating a statewide water pollution control agency. By 1961, interest in water quality had broadened beyond human health protection: enhancement of fish and wildlife resources was declared an official purpose of the SWP. The issue of salinity in the Delta refused to go away. In the 1940s, the CVP's strategy was to rely on releases from Shasta to control salinity in the Delta. In the 1950s, the CVP backed away from that commitment. The Delta was the hub for moving project water to agricultural and urban users in the San Joaquin Valley and Southern California. Salinity in the Delta mattered to users both there and in export areas. In 1959, the Delta Protection Act was passed, mandating that the Delta be kept fresh enough for these purposes. Freshwater releases not only reduced salinity but also protected fish and the Delta's aquatic ecosystem. The two concerns became mutually reinforcing. A 1966 legislative report asserted that downstream water quality was receiving inadequate attention from the SWRB: "The problem of resolving the protection which the Delta water users should receive, based on their vested rights, is beyond the ability of the presently organized SWRB to solve" (California Assembly Interim Committee on Water 1966, 28). The report called for the combination of water rights and water quality regulation in a single entity. In 1967, the State Water Resources Control Board (SWRCB) was created, combining the functions of the SWRB and the water pollution agency. In 1972, when the federal Clean Water Act mandated state water pollution regulation under EPA oversight, the SWRCB became the state's designated water pollution regulation agency and its water rights agency.

The SWRCB started off energetically, but then bogged down. Its first Delta decision, D-1379 in 1971, strengthened the conditions imposed earlier for salinity control and fish protection, and introduced new water quality standards to protect agricultural and urban uses in the Delta. But before the decision could be implemented, it was stayed by a suit challenging the SWRCB's authority to impose conditions on permits held by Reclamation as a federal agency. The U.S. Supreme Court resolved the legal issue in the SWRCB's favor in July 1978. In August 1978, the SWRCB issued D-1485 together with a Delta water quality control plan. These further strengthened water quality standards and introduced monitoring for compliance. To ensure that Delta water quality would not be impaired by the projects, D-1485 required them to release water and/or curtail diversions if the flow into the Delta would otherwise be insufficient.[27] It was greeted with a barrage of lawsuits from water users. In 1986, a California Court of Appeal ruling dismissed the challenges to the SWRCB's authority.

By then, compared to the SWRB in 1956, the SWRCB had powerful tools in its water rights arsenal. As Gray (chapter 4, this volume) shows, California courts used the reasonable use doctrine, revived by the 1967 *Joslin* decision, to enhance the SWRCB's regulatory jurisdiction with a reasonableness criterion responsive to changing circumstances. The 1986 Appeal Court ruling held that D-1485 was too narrow. While, under water law, the SWRCB had respected vested nonproject rights in conditioning project permits, under its water quality authority it could regulate *all* water users to ensure a reasonable level of water quality protection. Both sources of authority should be exploited. And the public trust doctrine, upheld in the 1983 *National Audubon* ruling, gave the SWRCB power to overturn settled water rights if subsequently found to violate the public trust, including protection of environmental resources.

Thus armed, a more ambitious SWRCB reopened its decision process. A draft report in November 1988 called for a "California Water Ethic" with more vigorous urban and agricultural conservation, a cap on Delta exports, and tighter water quality standards. The water users howled, and the governor pressured the SWRCB to withdraw the draft and let the parties negotiate among themselves. Following agreement between urban and environmental, but not agricultural, interests, the SWRCB issued a draft decision, D-1630, which limited exports in dry years and required additional fish flows. At the behest of agricultural users, the governor vetoed the decision in April 1993, leaving D-1485 still in place.

Now the EPA intervened, threatening to impose its own water quality standards for the Delta. An eleventh-hour agreement averted this in 1994, establishing CALFED, a collaborative planning process overseen by state and federal agencies and key stakeholders, largely sidelining the SWRCB.[28] CALFED produced some positive results, including scientific investigations, but no agreement. Native fish species continued a steep decline, with no simple remedy in sight. With leadership changes in Washington and Sacramento, the political support and funding that had sustained CALFED a decade earlier evaporated, and it was terminated in 2006 (Hanemann and Dyckman 2009).

CALFED was blamed for failing to reverse the Delta ecosystem's decline or to improve supply reliability for water users. Yet CALFED had no power to limit diversions or set water quality standards—those powers remained with the SWRCB. The SWRCB is tasked with "the orderly and efficient administration of the water resources of the state" (Water Code, Article 174). It has "primary responsibility" for implementing the reasonable use doctrine and for ensuring "meaningful implementation of the public trust" (Robie 2012a, 1175–76). It has not lived up to those obligations, instead displaying chronic passivity and regularly deferring to hoped-for stakeholder agreement.[29] Why? The water rights section was, and is, chronically understaffed for both water use monitoring and scientific analysis. The board is under the governor's thumb, whether indirectly or through his open intervention. Governors, kowtowing to water users, have controlled the board.[30] This is a political failure: a lack of political will to ensure that the board's regulatory functions are performed.

The SWRCB's weakness affects not only permits and water quality standards but also the enforcement of water rights generally.[31] Riparian and pre-1914 rights remain largely unverified and unquantified. Compliance inspections of diverters during 1998–2003 found that 38 percent were in violation of their water rights, and another 11 percent were subject to revocation for nonuse. In three watersheds, an inspection found 50 percent of small reservoirs diverting without a right (California State Water Resources Control Board 2008, 6). Besides personnel, enforcement authority was still inadequate. "Currently, [the SWRCB] does not possess sufficient authority to effectively monitor and enforce water right laws. . . . In particular, the law does not (1) provide clear authority for SWRCB to require monitoring by diverters, (2) authorize monetary penalties for monitoring and reporting violations, (3) have adequate penalties for unauthorized diversions and violations of cease and desist order, and (4) have provisions for interim relief" (8).

A NEW ERA?

Around 2005, new concerns arose that the Delta was in crisis due to ecosystem decline and levee vulnerability to seismic and erosion risks. Levee failure could permit massive saltwater intrusion, jeopardizing water exports to the south. In March 2007, Governor Schwarzenegger appointed a Delta Vision Blue Ribbon Task Force, mandating a report by November 2007 and a strategic plan by October 2008. Legislation implementing those recommendations was introduced in February 2009, but bogged down amid water user opposition. It emerged, somewhat shorn, in the last hours of the session in November 2009. It established "the two coequal goals of providing a more reliable water supply for California, and protecting, restoring and enhancing the Delta ecosystem." To that end, a Delta Stewardship Council was created, tasked with developing a comprehensive Delta Plan. It created a Watermaster for the Delta, appointed by the SWRCB and the Delta Stewardship Council, to "exercise the [SWRCB's] authority to provide timely monitoring and enforcement of [the SWRCB's] orders and . . . permit terms." It introduced penalties for failure to file diversion reports, removed the reporting exemption for in-Delta diverters, and authorized additional enforcement staff for the SWRCB. Missing were increased penalties for illegal water diversions, enhanced SWRCB enforcement authority, and independent power to initiate an adjudication.[32]

The Delta Watermaster has displayed vigor and independence in monitoring diversions and enforcing water rights in the Delta. Compliance with the diversion-reporting requirement is now 99 percent in the Delta, and 70 to 85 percent elsewhere. Overall, however, the SWRCB's monitoring and enforcement authority for water rights remain "weak" and "inconsistent with its broad enforcement authority over water quality" (Wilson 2012, 3, 10).

Yet, following the declaration of a drought emergency in January 2014, the SWRCB has shown unaccustomed forcefulness, more than in previous droughts.

In the 1977 drought, the SWRCB sent out 4,858 notices of shortage on various streams. Based on projections of summer demand and streamflow, the notices identified time periods when each broad user category (riparian, pre-1914 or post-1914) could take no water or had to reduce diversions by a given percentage. However, lacking the authority to supervise water distribution, the SWRCB relied for enforcement on complaints received and field visits. In May 2014, based on similar projections of demand and streamflow, the SWRCB issued 8,596 notices of curtailment for all post-1914 rights in the Sacramento–San Joaquin watershed, requiring submission of curtailment certification within seven days.[33] There was only a 29-percent response. Invoking its reasonable use authority, the SWRCB adopted emergency regulations in July to streamline and better enforce curtailment, subjecting noncertification of future curtailment to a penalty of $500 per day.[34] It also adopted emergency regulations mandating urban water suppliers to impose conservation measures equivalent to limiting outdoor use to two days per week, with monthly monitoring reports required.

These actions also exposed the SWRCB's weaknesses. Commenters complained that the proposed 2014 emergency regulations were illegal if applied to riparian or pre-1914 rights. By disregarding individual facts and circumstances and making a blanket determination of unreasonable use for a broad user category, the SWRCB had violated due process. The shortage projections were unreliable because the SWRCB's diversion data were incomplete (pre-1914 data lacks priority dates), unverified, and inaccurate. The curtailments shielded the projects from their existing responsibility to meet Delta water quality standards, and were an attempt to shift that responsibility to other users without a formal decision process. It is presently unknown whether those objections will be litigated and sustained. Also unknown is whether the SWRCB's new forcefulness will continue when the drought ends.

CONCLUSION

California started out with a surface water right that was extremely unsuited to its location in a semi-arid region with highly variable stream flow. While still keeping the riparian right, it invented another type of water right, the appropriative right, which became the standard in the other Western states. As originally implemented in California, the appropriative right lacked effective recording, quantification or enforcement. Other states discovered this once they emulated California's system, but they soon instituted reforms that made the appropriative right more functional. The reforms provided a usable record of who had rights and with what seniority, and an apparatus on the ground for monitoring diversions and enforcing seniority. Moreover, while appropriative rights obtained earlier were initially grandfathered, they were subsequently brought into compliance with the new administrative system. In California, by contrast, reform of appropriative rights was long blocked by water users and arrived relatively late, in 1914. When California acted, what it did was limited. There was recording of post-1914 rights but not supervision of the distribution of water to ensure conformity with those rights. Pre-1914 rights stayed outside the authority of the SWRCB and its predecessors; they still remain unquantified in many cases, and they require litigation to quantify or enforce them. Riparian rights are also outside the SWRCB's authority, and they are unenforceable and unquantifiable except through litigation. A systematic quantification of surface water rights in California's Central Valley would require a statutory adjudication or something equivalent. This was mooted several times, including in 1939, 1942, and 1951, but was rejected as too time-consuming. Seventy-five years later, it is still needed.

The system of water rights affects the allocation of surface water in several ways. It hinders the re-allocation of water through water marketing. Riparian rights can be transferred to nonriparian land only through guile. Without more formal verification, unquantified or poorly quantified appropriative rights can be leased short-term but not leased long-term or permanently transferred. Hence, the vast majority of water marketing in California has been restricted to short-term leases. When there is a drought, the system's weaknesses show up. In past droughts, water suppliers and users have worked things out informally among themselves, sidelining the SWRCB or its predecessors and largely bypassing seniority. When the SWRCB did attempt to enforce seniority, it acted in a simplistic manner: in the 1977 and 2014 droughts, it treated all post-1914 rights as the same without regard to seniority within the category. (It did the same in 1977 for pre-1914 rights, treating them as a homogeneous category.) The lack of data on quantities associated with water rights and on actual diversions left no alternative. The system performs worst with regard to the allocation of water between instream and off-stream uses. Large-scale diversions have been occurring from the Delta since 1949, but there is still no authoritative determination of responsibility for meeting Delta water quality objectives. The SWRCB has relied so far on restricting CVP and SWP diversions to meet those objectives, despite the 1986 Appeal Court ruling that it could regulate *all* water users to ensure a reasonable level of water quality protection in the Delta. As the demand for water in California grows, as the Delta ecosystem declines, and as drought becomes more common with climate change, the SWRCB's failure to exercise its full legal authority will become increasingly burdensome and costly.

NOTES

1. Space precludes discussion of groundwater rights in California. Put simply, that situation has been even worse, although legislation enacted in September 2014 may eventually lead to some effective regulation of groundwater overdraft.

2. That restriction was not initially enforced in California.

3. Some of these details emerged gradually through court rulings in the period 1897–1927.

4. Meade (1903, 199–202) provides several examples where, between 1860 and 1890, ditch companies lost their entire investments because they were unaware of superior appropriative rights elsewhere on the stream.

5. In order to secure a complete and consistent settlement of water rights to a stream, all potential claimants must be brought into the same suit, either through a suit to quiet title or through a procedure known as a stream adjudication.

6. The states adopting appropriative rights as the only form of surface water right were Alaska, Arizona, Idaho, Montana, Nevada, New Mexico, Utah, and Wyoming. The states maintaining a dual system of riparian and appropriative rights were Kansas, Nebraska, the Dakotas, Oklahoma, Oregon, Texas, and Washington (relatively humid, non-mining states).

7. Nebraska, Nevada, Utah, Idaho, the Dakotas, New Mexico, and Arizona.

8. The court ruled that under certain conditions that did not hold in this case—if the appropriator began using water from a stream *before* a riparian acquired his property—the appropriation doctrine would prevail.

9. United States v. State Water Resources Control Board, 182 Cal. App. 3d 82 (1986), p. 170 (italics in original).

10. Wiel (1914, 446), states that "there was a promise of opposition to the bill if such provisions went into it."

11. Wiel (1914, 446); Chandler (1913, 162, 168). Chandler comments, "There is little use in securing an adjudication unless properly authorized officials are charged with the regulation of headgates in accordance therewith."

12. The commission originally had authority over riparian rights in a statutory adjudication; that authority was removed in 1917. It had authority over riparian rights in a court reference if the court so decided.

13. Under the 1913 act, the commission (or division) was authorized to initiate an adjudication on its own authority. As of 1922, this authority had never been exercised (Division of Water Rights 1922, 8). It was never exercised subsequently, either.

14. Division of Water Rights (1924, 38). Another modification in 1923 changed the filing fee from a flat charge to a fee that varied according to the amount of water applied for. This "played an important role in eliminating many purely speculative filings" (Division of Water Rights 1926, 26).

15. The legislation waived the diligence requirement that these rights be exercised within a fixed period or be lost.

16. These users received what were called *water service contracts*.

17. These contractors are *exchange contractors*, because they exchanged their right to river water for CVP deliveries.

18. These contractors are *settlement contractors*. The SWP also has some settlement contractors on the Feather River.

19. Following litigation, owners of land along the dewatered San Joaquin River segment received compensation for the loss of their land value.

20. Authority to prevent unreasonable use of water remained exclusively with the Department of Water Resources until 1971.

21. Starting with the 1924 drought, there had been an annual inventory of individual diversions in the Central Valley. This ceased in 1956.

22. For post-1914 rights, the "face value" (maximum diversion) is known, but the amount *actually* diverted is "only a fraction of face value" and is "undetermined" (California State Water Resources Control Board 2008, 4).

23. Governor's Commission to Review California Water Rights Law (1978, 18). During the 1977 drought, the SWRCB received about 150 complaints of illegal diversions or violations of permit terms. Upon investigation, 30 cases were found to merit enforcement actions. This was "the first time since enactment of the Water Commission Act in 1914 that the State has enforced its jurisdiction to enjoin illegal diverters" (Division of Water Rights 1978, 18).

24. The Governor's Commission (1978, 26) suggested that, once signed, the settlement and exchange contracts provided de facto quantification of the riparian and pre-1914 rights of CVP contractors. But those contracts represented a judgment by an agency (Reclamation) that lacked the authority to determine water rights in California. Similarly with the SWP contracts signed by the Department of Water Resources. Olson and Mahaney (2005, 82), staff counsel to SWRB's successor, reject the commission's suggestion: "Various agencies conducted studies in order to make assumptions regarding the physical characteristics involved, including estimates of existing water rights. However, these studies did not determine actual water rights, and clearly state that assumptions may differ substantially from the actual rights as determined in a court or by the [SWRB]."

25. In 1959, the SWRB received formal authority to reserve jurisdiction to modify or delete terms when issuing CVP and SWP permits.

26. See Hanemann and Dyckman (2009), who argue that the negotiations are a zero-sum game, thus inherently incapable of yielding a stable bargaining outcome.

27. In 1981, the SWRCB added a condition, Term 91, prohibiting users junior to the projects from making diversions when stored project water was being released to meet Delta quality standards. First seen as an interim measure pending more comprehensive studies of water availability for all diverters, and then

made permanent when that approach was abandoned "due to lack of adequate data," this condition was applied to permits with post-1978 rights, and subsequently to some post-1965 permits. Invoked almost every year since 1984, it covers only 233 out of 5,500 diverters, and enforcement is limited.

28. In 1995 the SWRCB adopted a Delta Water Quality Plan with new standards for fish and wildlife, based on recommendations agreed to by the parties. It planned a water rights hearing to allocate responsibility for meeting flow-dependent objectives, but canceled this "to facilitate negotiations" that might lead to a settlement among the parties, meanwhile leaving the CVP and SWP with ultimate responsibility for those objectives (D-1641, WRO-2001-05). Despite legal challenges, the decision was largely upheld in 2006, except that the court ruled that the SWRCB could not substitute flow objectives agreed to by the parties for those in the 1995 plan.

29. See Robie (2012b, 9–11): "the Board has been too timid in its leadership, and overall has been a disappointment. . . . It remains critical for the [SWRCB] to take a more active role in applying the reasonable use and public trust doctrines . . . *of its own accord*. . . . I urge the Board to be more proactive, more bold . . . in fulfilling its adjudicatory and regulatory functions." Also see Nawi and MacMillan (2008, 4): "When [the SWRCB] has taken effective action, this has tended to be the result of consensus reached by parties outside the Board's process."

30. The formulation currently used by the governor's staff when telling you that you are off the board is: "The space is needed for someone else."

31. From 1983 to about 2000, Sawyer (2005, 36) notes, the SWRCB "was less interested in water right enforcement" due partly to gubernatorial directives and partly to staff shortages, as personnel were reallocated from enforcement to other tasks. (Staffing resources, of course, also reflect gubernatorial priorities.) Interest in enforcement revived somewhat thereafter.

32. The last was recommended, to no avail, by the Governor's Commission in 1978 as well as by the Delta Task Force in 2008.

33. The notice warned that, if current conditions continue, the SWRCB might also curtail pre-1914 and riparian diversions.

34. The draft regulations applied to all diverters, including pre-1914 and riparians; pre-1914 and riparian rights were omitted from the version adopted.

REFERENCES

Bain, Joe S., Richard E. Caves, and Julius Margolis. 1966. *Northern California's Water Industry.* Baltimore, MD: Johns Hopkins Press.
California Assembly Interim Committee on Water. 1966. *A Proposed Water Resources Control Board for California: A Staff Study.* Sacramento.
California State Water Commission. 1917. *Report of the State Water Commission of California.* Sacramento.
———. 1918. *Second Biennial Report of the State Water Commission of California, 1917–1918.* Sacramento.
———. 1921. *Third Biennial Report of the State Water Commission of California, 1919–1920.* Sacramento.

California State Water Resources Board. 1951. *Water Resources of California.* Bulletin No. 1. Sacramento.

California State Water Resources Control Board. 2008. Letter from Dorothy Rice, Executive Director, to John Kirlin, Executive Director, Delta Vision Task Force, May 12, 2008. Sacramento.

California State Water Rights Board. 1957. *Annual Report of the State Water Rights Board of the State of California, Covering Period July 5, 1956 to June 30, 1957.* Sacramento.

Chandler, Alfred E. 1913. "The 'Water Bill' Proposed by the Conservation Commission of California." *California Law Review* 1:148–68.

Clay, Karen, and Gavin Wright. 2005. "Order without Law? Property Rights during the California Gold Rush." *Explorations in Economic History* 42:155–83.

Division of Engineering and Irrigation, California Department of Public Works. 1927. *Summary Report on the Water Resources of California and a Coordinated Plan for their Development.* Bulletin No. 12. Sacramento.

Division of Resource Planning, California Department of Water Resources. 1957. *The California Water Plan.* Bulletin No. 3. Sacramento.

Division of Water Resources, California Department of Public Works. 1931. *Report to Legislature of 1931 on State Water Plan, 1930.* Bulletin No. 30. Sacramento.

Division of Water Rights, California Department of Public Works. 1922. *Report of the Division of Water Rights, November 1, 1922.* Sacramento.

———. 1924. *Biennial Report of the Division of Water Rights, November 1, 1924.* Sacramento.

———. 1926. *Biennial Report of the Division of Water Rights, November 1, 1926.* Sacramento.

Division of Water Rights, California State Water Resources Control Board. 1978. *Drought 77 Dry Year Program.* Sacramento.

Dunbar, Robert G. 1983. *Forging New Rights in Western Waters.* Lincoln: University of Nebraska Press.

Garone, Philip. 2011. *The Fall and Rise of the Wetlands of California's Great Central Valley.* Berkeley: University of California Press.

Governor's Commission to Review California Water Rights Law. 1978. *Final Report.* Sacramento.

Graham, Leland O. 1950. "The Central Valley Project: Resource Development of a Natural Basin." *California Law Review* 38:588–637.

Hanemann, Michael, and Caitlin Dyckman. 2009. "The San Francisco Bay-Delta: A Failure of Decision-Making Capacity." *Environmental Science and Policy* 12:710–25.

Hobbs, Gregory J. 1999. "Colorado's 1969 Adjudication and Administration Act: Settling In." *University of Denver Water Law Review* 3:1–19.

Holsinger, Henry. 1942. "Necessity for Comprehensive Adjudication of Water Rights on the Sacramento and San Joaquin Rivers in Aid of the Central Valley Project." Memorandum dated December 2. Sacramento: California State Water Resources Control Board archives.

———. 1957. "Procedures and Practice before the California State Water Rights Board." *California Law Review* 45:676–87.

Meade, Elwood. 1903. *Irrigation Institutions*. New York: Macmillan.

Miller, M. Catherine. 1985. "Riparian Rights and the Control of Water in California, 1879–1928: The Relationship between Agricultural Enterprise and Legal Change." *Agricultural History* 59:1–24.

———. 1989. "Water Rights and the Bankruptcy of Judicial Action: The Case of Herminghaus v. Southern California Edison." *Pacific Historical Review* 58:83–107.

———. 1992. *Flooding the Courtrooms*. Lincoln: University of Nebraska Press.

Nawi, David, and Jeannette MacMillan. 2008. *Authority and Effectiveness of the State Water Resources Control Board*. Report to the Delta Vision Committee. Sacramento.

Olson, Samantha K., and Erin K.L. Mahaney. 2005. "Searching for Certainty in a State of Flux: How Administrative Procedures Help Provide Stability in Water Rights Law." *McGeorge Law Review* 36:73–115.

Pisani, Donald J. 1984. *From the Family Farm to Agribusiness*. Berkeley: University of California Press.

———. 2002. *Water and American Government*. Berkeley: University of California Press.

Robie, Ronald B. 2012a. "Effective Implementation of the Public Trust Doctrine in California Water Resources Decision-Making: A View from the Bench." *UC Davis Law Review* 45:1155–76.

Robie, Ronald B. 2012b. *What Is, What Has Been, and What Ought To Be*. Ann J. Schneider Memorial Lecture, University of California, Davis, May 1.

Rhode, Paul W. 1995. "Learning, Capital Accumulation, and the Transformation of California Agriculture." *Journal of Economic History* 55:773–800.

Sawyer, Andrew H. 2005. "Improving Efficiency Incrementally: The Governor's Commission Attacks Waste and Unreasonable Use." *McGeorge Law Review* 36:209–52.

Squillace, Mark. 1989. "A Critical Look at Wyoming Water Law." *Land and Water Law Review* 24:307–46.

———. 1991. "One Hundred Years of Wyoming Water Law." *Land and Water Law Review* 26:93–101.

Tarlock, A. Dan. 2000. "Prior Appropriation: Rule, Principle, or Rhetoric?" *North Dakota Law Review* 76:881–910.

Wiel, Samuel C. 1914. "Determination of Water Titles and the Water Commission Bill." *California Law Review* 2:435–49.

Wilson, Craig M. 2012. *Improving Water Right Enforcement Authority*. Report to the State Water Resources Control Board and the Delta Stewardship Council, September 19. Sacramento, CA.

CHAPTER 4

The Reasonable Use Doctrine in California Water Law and Policy

BRIAN E. GRAY

The cardinal principle of California water law is that all water rights, and all uses of water, must be reasonable. This seemingly simple and innocuous sentence masks a world of meaning and complexity, however, because the requirement of reasonable use embraces at least four interrelated concepts. The determination of reasonable water use is *utilitarian*: the law seeks to encourage relatively efficient, economically and socially beneficial uses of the state's water resources. It is *situational*: the evaluation of individual reasonable use concerns not only the water right holder's own uses but also other competing demands (both consumptive and ecological) on the water resource. The reasonable use doctrine is also *dynamic*: the definition of reasonable use varies as the economy, technology, demographics, hydrologic conditions, environment, and societal needs evolve. And, because all uses of water must be consistent with this interdependent and variable definition of reasonable use, the law renders all water rights *fragile*. A water right that was reasonable when first recognized, and which may have been exercised reasonably for many years, may become unreasonable as hydrologic conditions change, as California's economy evolves, as population grows and new demands for water arise, as ecological needs are better understood, and as the environmental laws that protect the state's aquatic ecosystems and native species are applied in ways that limit the impoundment and diversion of water for consumptive uses.

The doctrine of reasonable use is thus both a policy mandate and a limitation on water rights. A part of California's Constitution since 1928, it applies to all branches of government, to all levels of governmental administration of the state's water resources, and to public and private uses of the state's waters. Its overarching directives, comprehensive reach, and infusion into the water rights system make it the most powerful of all of the laws that govern California's water resources.[1]

THE CONSTITUTIONAL DOCTRINE OF REASONABLE USE

Article X, section 2 of the California Constitution declares that, because of the state's hydrologic and economic conditions, the general welfare requires that its water resources "be put to beneficial use to the fullest extent of which they are capable, and that the waste or unreasonable use or unreasonable method of use of water be prevented." It also stipulates that "the conservation of such waters is to be exercised with a view to the reasonable and beneficial use thereof in the interest of the people and for the public welfare."[2]

Article X, section 2 then ties the reasonable use requirement to the water right itself: "The right to water or to the use or flow of water in or from any natural stream or water course in this State is and shall be limited to such water as shall be reasonably required for the beneficial use to be served, and such right does not and shall not extend to the waste or unreasonable use or unreasonable method of use or unreasonable method of diversion of water." It concludes with the statement that the amendment "shall be self-executing, and the Legislature may also enact laws in the furtherance of the policy in this section contained."

In one of its early interpretations of article X, section 2, the California Supreme Court emphasized the fundamental purposes of the doctrine of reasonable use, explaining in *Peabody v. City of Vallejo* (1935): "The waters of our streams are not like land which is static, can be measured and divided and the division remain the same. Water is constantly shifting, and the supply changes to some extent every day. A stream supply may be divided but the product of the division in nowise remains the same. When the supply is limited public interest requires that there be the greatest number of beneficial uses which the supply can yield."

Three decades later, in a case that presaged the modern era in California water law and policy, the court expressly linked these aspects of article X, section 2, to the definition of water rights.

REASONABLE USE AND THE PROPERTY RIGHT IN WATER

In *Joslin v. Marin Municipal Water District* (1967), the Supreme Court rejected the claims of riparian landowners who harvested sand and gravel deposited by the natural flow of Nicasio Creek for commercial sale. The landowners alleged that the district violated their water rights by constructing a dam that impaired the natural flow and thus deprived them of the suspended materials carried by the water.

The court observed that, to prevail on their damages claim, the Joslins must "first establish the legal existence of a compensable property interest" and that such an interest "consists in the right to a reasonable use of the flow of water." Although evaluation of reasonable use "depends on the circumstances of each case," the court reasoned, "such an inquiry cannot be resolved *in vacuo* isolated from statewide considerations of transcendent importance. Paramount among these [is] the increasing need for the conservation of water in this state, an inescapable reality of life quite apart from its express recognition in the 1928 amendment." The court then concluded that the Joslins' reliance on the unimpaired flow had become unreasonable in light of the new demands for municipal water supply. Moreover, "since there was and is no property right in an unreasonable use [of water] there has been no taking or damaging of property by the deprivation of such use and, accordingly, the deprivation is not compensable."

Although *Joslin* is sometimes read as a simple decision not to countenance a use of water that required an inordinate percentage of the flow of the stream, a closer reading reveals that the Supreme Court had broader purposes in mind. The opinion emphasized the utilitarian goals of the doctrine of reasonable use: to ensure that the state's water resources are used in ways that serve the public interest, not just to benefit senior water right holders. Equally importantly, the court focused on the dynamic nature of the reasonable use inquiry and the consequent fragility of the property right in water. Water rights are defined by reasonable use, and they are thereby limited by reasonable use. A use of water that may have been lawful when established—and that continued to be exercised lawfully for many years—may become unreasonable as conditions change. Moreover, because the property right in water is defined by contemporary standards of reasonable use, a water right (or certain aspects of the right) may cease to exist as a result of changes in hydrologic, economic, demographic, or environmental conditions that are well beyond the control of the water right holder.

REASONABLE USE AND WATER RIGHTS ADMINISTRATION

In the years following *Joslin*, the California courts applied the doctrine of reasonable use principally to enhance the regulatory jurisdiction of the State Water Resources Control Board (SWRCB). In *In re Waters of Long Valley Creek Stream System* (1979), the Supreme Court held that when the board conducts a statutory adjudication of all surface water rights, it has authority to relegate unexercised riparian rights to a priority below all active uses of water in the system—both riparian and appropriative.[3] In *People v. Shirokow* (1980), the court ruled that a nonriparian water user may not claim prescriptive rights outside of the SWRCB's permitting and licensing jurisdiction. Both decisions emphasized that the legislature had created the board for the express purpose of implementing and enforcing the constitutional reasonable use mandates.[4] Assertions of previously dormant riparian rights to preempt valid existing uses and claims to water rights based on prescriptive use were unreasonable, the court concluded, because they created uncertainty and undermined the SWRCB's ability comprehensively to administer California's surface water rights system.

In the wake of *Joslin*, the California Courts of Appeal also bolstered the SWRCB's power directly to enforce the reasonable use doctrine. In *People ex rel. State Water Resources Control Board v. Forni* (1976), the court held that the board had authority under Water Code section 275 to enjoin vineyards along the Napa River from diverting water to spray on their crops during periods of low temperatures to prevent wine grapes from freezing. The lawsuit was based on the board's determination that "direct diversion during the frost season may at times dry up the river and deprive many of the vineyardists of water which they need to protect their vines from frost" (Cal. Code Regs. Title 23, § 735). The court affirmed the board's decision to compel all vineyards—including those that diverted water pursuant to riparian rights—to construct storage to minimize aggregate demands on the available water. According to the court, the "overriding constitutional consideration is to put the water resources of the state to a reasonable use and make them available for the constantly increasing needs of all the people. In order to attain this objective, the riparian owners may properly be required to endure some inconvenience or to incur reasonable expenses."

In *Imperial Irrigation District v. State Water Resources Control Board* (1986, 1990), the court upheld the SWRCB's finding that the

Imperial Irrigation District's unlined canals and lack of regulating reservoirs were causing flooding and waste of water within its service area. It also rejected the irrigation district's argument that, as a pre-1914 appropriator, it was not subject to the SWRCB's regulatory authority. As in *Forni,* the court held that section 275 conferred independent jurisdiction on the board to enforce the reasonable use mandates of article X, section 2. Faced with the prospect of losing a substantial portion of its water rights, the district agreed to a 35-year transfer of more than 100,000 acre-feet of conserved water each year to the Metropolitan Water District of Southern California.[5]

The courts also have affirmed the SWRCB's assertion of its reasonable use powers to set water quality and flow standards for the Sacramento–San Joaquin Delta and to establish operational constraints on the Central Valley Project, the State Water Project, and other water right holders as required to protect water quality, fisheries, and other instream uses in the Delta ecosystem. In *United States v. State Water Resources Control Board* (1986), the Court of Appeal again emphasized the multifaceted and dynamic nature of article X, section 2:

> We perceive no legal obstacle to the Board's determination that particular methods of use have become unreasonable by their deleterious effects upon water quality. Obviously, some accommodation must be reached concerning the major public interests at stake: the quality of valuable water resources and transport of adequate supplies for needs southward. The decision is essentially a policy judgment requiring a balancing of the competing public interests, one the Board is uniquely qualified to make in view of its special knowledge and expertise and its combined statewide responsibility to allocate the rights to, and to control the quality of, state water resources.

The court concluded that the "power to prevent unreasonable methods of use should be broadly interpreted to enable the Board to strike the proper balance between the interests in water quality and project activities in order to objectively determine whether a reasonable method of use is manifested."

Finally, in something of a sequel to *Forni,* the Court of Appeal recently affirmed the SWRCB's power to regulate the diversion of water from the Russian River system—including most of its tributaries and all hydrologically connected groundwater—to protect coho salmon. The board determined that simultaneous diversions and pumping by vineyards for frost-prevention purposes were unreasonable because the aggregate withdrawal of water caused migrating juvenile salmon to become stranded in the river bed. In *Light v. State Water Resources*

Control Board (2014), the court held that the board could include riparians and pre-1914 appropriators (including those that extract hydrologically connected groundwater from the river system) within this regulatory scheme: "That the Board cannot require riparian users and pre-1914 appropriators to obtain a permit before making reasonable beneficial use of water does not mean the Board cannot prevent them from making unreasonable use. Any other rule would effectively read Article X, Section 2 out of the Constitution."

THE REASONABLE USE DOCTRINE AND ENVIRONMENTAL QUALITY

The California Supreme Court's most important applications of the reasonable use doctrine following *Joslin* came in two high-profile cases that pitted municipal water use against environmental protection. In *Environmental Defense Fund v. East Bay Municipal Utility District* (1980), the court held that environmental advocates may rely on article X, section 2, to claim that a proposed upstream point of diversion for water supplied by the CVP to the East Bay Municipal Utility District was unreasonable because of its adverse effects on water quality, fish and wildlife, and recreational uses in the lower American River. Three years later, in its landmark opinion in the Mono Lake litigation, the court held that the mandate of reasonable use also embraces the *public trust*, an ancient doctrine that protects recreational access, boating, fishing, and ecological uses of the state's navigable waters.

In *National Audubon Society v. Superior Court* (1983), the Supreme Court ruled that Los Angeles's long-standing rights to appropriate water from the streams that supply Mono Lake are subject to the public trust. Just as the doctrine of reasonable use serves as an inherent limitation on the exercise of all water rights, the court declared that the public trust doctrine "imposes a duty of continuing supervision over the taking and use of the appropriated water. In exercising its sovereign power to allocate water resources in the public interest, the state is not confined by past allocation decisions which may be incorrect in light of current knowledge or inconsistent with current needs."

Although *Audubon* is best known as the case in which the Supreme Court incorporated the public trust doctrine into California's water rights system, it is equally important as a reasonable use decision. First, the court explained that, although the public trust doctrine and the water rights laws "developed independently of each other," its integration of the two would bring both under the umbrella of article X, section 2. The constitutional amendment "establishes state water policy. All uses of water, including public trust uses, must now conform to the standard of reasonable use."

Second, the court recognized that water rights do not exist in isolation from the broader society and environment. In its earlier reasonable use decisions, the paramount goal was conservation of scarce water resources to accommodate new consumptive demands as California's population and economy continued to grow. In *Audubon,* the court built on its *Environmental Defense Fund* holding and emphasized that fish, wildlife, recreation, and other in-stream uses that depend on that same water are also important societal interests that must be taken into account.

Third, the *Audubon* court's articulation of the evolving nature of the public trust was consonant with its dynamic conception of the doctrine of reasonable use. Both laws recognize that "the state is not confined by past allocation decisions which may be incorrect in light of current knowledge or inconsistent with current needs."

Fourth, this dynamic feature means that water rights are mutable under both the reasonable use and public trust doctrines. *Joslin* held that there "is no property right in an unreasonable use of water" and that the definition of "what constitutes reasonable water use is dependent upon not only the entire circumstances presented but varies as the current situation changes." *Audubon* embellished these principles, emphasizing that the law "prevents any party from acquiring a vested right to appropriate water in a manner harmful to the interests protected by the public trust."

LEGISLATIVE DECLARATIONS OF REASONABLE USE

The legislature also has exercised its constitutional authority under article X, section 2, to declare that certain environmental uses of California's water resources are reasonable. For example, section 1243 of the Water Code states that "the use of water for recreation and preservation and enhancement of fish and wildlife resources is a beneficial use of water." Similarly, the California Wild and Scenic Rivers Act provides: "It is the policy of the State of California that certain rivers which possess extraordinary scenic, recreational, fishery, or wildlife values shall be preserved in their free-flowing state, together with their immediate environments, for the benefit and enjoyment of the people of the state. The Legislature declares that such use of these rivers is the highest and most beneficial use and is a reasonable and beneficial use of water within the meaning of Section 2 of Article X of the California Constitution" (Cal. Pub. Res. Code § 5093.50).

More recently, the legislature exercised its reasonable use authority to enact the Delta Reform Act of 2009, which established a Delta Stewardship Council to formulate a Delta Plan and to oversee actions that may affect the waters and resources of the Delta ecosystem. The act declares that waters of the Sacramento–San Joaquin River and Delta system shall be administered to achieve the "co-equal goals" of "providing a more reliable water supply for California and protecting, restoring, and enhancing the Delta ecosystem" (Cal. Water Code §§ 85020, 85054). The legislature also stated that the "longstanding constitutional principle of reasonable use and the public trust doctrine shall be the foundation of state water management policy and are particularly important and applicable to the Delta" (§ 85023).

In *California Trout v. SWRCB* (1989), the Court of Appeal upheld the legislature's authority to make these types of categorical declarations of reasonable use. One of California's oldest environmental protection statutes directs that the "owner of any dam shall allow sufficient water at all times to pass through a fishway, or in the absence of a fishway, allow sufficient water to pass over, around or through the dam, to keep in good condition any fish that may be planted or exist below the dam" (Fish and Game Code § 5937). The court rejected the claim that this statute violates article X, section 2. It emphasized that the Constitution expressly authorizes the legislature to enact laws in furtherance of the policy of reasonable use and held that where "various alternative policy views reasonably might be held whether the use of water is reasonable within the meaning of article X, section 2, the view enacted by the Legislature is entitled to deference by the judiciary."

THE REASONABLE USE DOCTRINE AND WATER ALLOCATION

Although the SWRCB and the courts have authority under the reasonable use doctrine to reallocate water (out of priority) between water right holders, and to require consumptive users to provide more water to environmental uses, the relationship between water rights priorities and reasonable use continues to raise questions.

In its most recent groundwater rights case, *City of Barstow v. Mojave Water Agency* (2000), the California Supreme Court overturned a decision

Statutory Encouragement of Efficient Water Use

The legislature has enacted a variety of statutes that are designed to encourage more efficient use of the state's waters. For example:

- It has declared that the cessation or reduction in the extraction of groundwater—either as a result of the use of alternative sources or to allow for the replenishment of the aquifer—is a reasonable and beneficial use (Water Code §§ 1005.1–1005.4).

- It has stated that the cessation or reduction in the use of water made possible by the substitution of recycled, desalinated, or treated polluted water is a reasonable beneficial use (§§ 1010(a)).

- It has mandated similar treatment for cessations or reductions in water use because of conservation or the use of groundwater that is managed as part of a conjunctive use program (§§ 1011(a), 1011.5(a)).

- It has authorized the transfer of water that is made available by conservation or the substitution of these alternative sources and has guaranteed that the transferor's rights to the transferred water will be protected and preserved during the term of the transfer agreement (§§ 1010(b), 1011(b), 1011.5(b), 1014–1017).

- It has required municipal water agencies to meter and report on water use, and it has required agricultural supply agencies to monitor and report on groundwater levels (§§ 500–535, 10920–10936).

- It has authorized counties and local agencies to conjunctively manage surface and groundwater supplies and has required urban and agricultural water agencies to adopt best management practices to promote conservation and efficient use (§§ 10608–10608.64, 10610–10656, 10750–10783.2, 10800–10853).

The legislature also has granted public water agencies authority to use "allocation-based conservation water pricing"—i.e., tiered water pricing—which it identified as "one effective means by which waste or unreasonable use of water can be prevented and water can be saved in the interest of the people and for the public welfare, within the contemplation of Section 2 of Article X of the California Constitution" (§§ 370–374).

that "equitably apportioned" the safe yield of an overdrafted groundwater basin among all water right holders regardless of the type or priority of water right. "We have never," the court stated, "endorsed a pure equitable apportionment that completely disregards overlying owners' existing legal rights."

Yet, the court also rejected the view that priority of water right alone should determine which users should curtail their pumping and which may continue. It confirmed the broad holding of *Joslin* that article X, section 2, "dictates the basic principles defining water rights: that no one can have a protectible interest in the unreasonable use of water, and that holders of water rights must use water reasonably and beneficially." Although "water right priority has long been the central principle in California water law, . . . the corollary of this rule is that an equitable physical solution must preserve water right priorities to the extent those priorities do not lead to unreasonable use." In crafting a "physical solution" to the problem of aggregate overdraft, the court concluded, a trial court "may neither change priorities among the water rights holders nor eliminate vested rights in applying the solution *without first considering them in relation to the reasonable use doctrine*" (emphasis added).

The two other courts that have confronted this question of the relationship between water rights priorities and reasonable use have followed this approach. In *El Dorado Irrigation District v. State Water Resources Control Board* (2006), the Court of Appeal overturned a decision by the board that required all permittees and licensees in the Sacramento River basin to cease diversions whenever the CVP or SWP are releasing stored water to meet Delta water quality standards—regardless of the appropriator's priority *vis-à-vis* the two projects. The court explained that "sometimes the use of water under a claim of prior right must yield to the need to preserve water quality to protect public trust interests, and continued use under those circumstances may be deemed unreasonable." If, for example, "El Dorado's diversions of natural flow contribute to the degradation of water quality in the Delta, the Board has a legitimate interest in requiring [the district] to reduce its diversions to contribute toward the maintenance and improvement of water quality in the Delta."

The court cautioned, however, that the board must respect the relative water rights priorities:

> When the Board seeks to ensure that water quality objectives are met in order to enforce the rule against unreasonable use and the public trust doctrine, the Board must attempt to preserve water right priorities to the extent those priorities do not lead to unreasonable use or violation of public trust values. In

other words, in such circumstances the subversion of a water right priority is justified only if enforcing that priority will in fact lead to the unreasonable use of water or result in harm to values protected by the public trust.

The Court of Appeal also recently grappled with the question of priority of water rights in the context of the Russian River adjudication. It acknowledged that when "the supply of water is insufficient to satisfy all persons and entities holding water rights, it is ordinarily the function of the rule of priority to determine the degree to which any particular use must be curtailed. Yet even in these circumstances, the Board has the ultimate authority to allocate water in a manner inconsistent with the rule of priority, when doing so is necessary to prevent the unreasonable use of water" (*Light v. SWRCB*, 2014). This is especially true, the court stated, when the board is acting to protect the public trust. It added, quoting *El Dorado*, that because "'no one can have a protectible interest in the unreasonable use of water' . . . when the rule of priority clashes with the rule against unreasonable use of water, the latter must prevail."

These cases add an important caveat to the law of reasonable use. Although all water rights are defined by reasonable use and must be exercised reasonably in light of contemporary conditions and standards, the doctrine does not apply *carte blanche*. Aggregate unreasonable use (such as groundwater overdraft or harm to water quality or fisheries) does not necessarily mean that every water user is acting unreasonably. Nor does it mean that the SWRCB and the courts may necessarily require *pro rata* reductions in water use or impose equal conditions on all water users regardless of their relative priority of right. While the reasonable use doctrine grants the board and the courts broad powers to correct the overall problem, the remedies applied to each water user must be more nuanced. The trier of fact must make individualized determinations of reasonable use and must be guided by the rule of priority. As all three of these cases make clear, the SWRCB and the courts may depart from the underlying water rights priorities only if they justify their decisions on findings of individual unreasonable use *vis-à-vis* the other potentially affected water right holders. The reasonable use doctrine is powerful, but it is not a magic wand.

THE FUTURE OF REASONABLE USE

Although there is no single response to the challenges of California water management in the twenty-first century, the reasonable use doctrine will play an important role in helping to effectuate a variety of necessary improvements in California water policy.

Prevention of Waste and Improvements in the Efficiency of Water Use

The state has estimated that available improvements in the efficiency of water use could conserve between 180,000 and 1.1 million acre-feet per year in the agricultural sector and 1.2 million to 2.1 million acre-feet per year currently used for municipal and industrial purposes (California Bay-Delta Authority 2005). Although this analysis probably underestimates the statewide water conservation potential, it does suggest that many water-use practices may be unreasonable in light of the existing strains on the state's developed water supplies and the future diminution in useable supplies that is a predicted consequence of climate change.

The state—acting principally through the SWRCB or the courts—has authority to investigate individual cases of unreasonable use and to declare unreasonable a variety of water practices that may have been acceptable in the past, but which are no longer tolerable in the face of contemporary and future water supply challenges. These may include excessive evaporative and conveyance losses, inefficient irrigation techniques, failure to adopt or to implement best management practices, and perhaps other profligate uses such as the irrigation of water-intensive crops and landscaping. Future unreasonable use also may include excessive reliance on imported water instead of shifting to a more varied water portfolio that incudes cost-effective alternatives such as demand reduction, use of recharged groundwater, and recycling of reclaimed wastewater.

The Delta Watermaster has issued a report to the SWRCB and the Delta Stewardship Council advocating greater enforcement of the reasonable use doctrine to address wasteful water practices and to create incentives to achieve more efficient water use (Wilson 2010). "The underlying premise of this report," he stated, "is that the inefficient use of water is an unreasonable use of water." The Watermaster then provided examples of a variety of currently available agricultural water management practices that could be required to promote the reasonable use of water, including "weather-based and deficit irrigation scheduling, water distribution systems that can supply water to farmers 'on-demand,' and improved irrigation methods, such as substituting drip and sprinkler irrigation for flood irrigation."

The report provides a template for focused and proactive application of the reasonable use doctrine to promote greater efficiency in water use. Although the report addressed only agricultural water practices, there is a significant role for reasonable use investigations of water use in California's urban and suburban areas as well. There exists the potential for significant water savings, especially in irrigation of landscaping and other outdoor uses (Hanak et al. 2011, 171–73).

Regional Water Management, Water Pricing, and Water Use Efficiency

In the Delta Reform Act of 2009, the legislature declared a state policy "to reduce reliance on the Delta in meeting California's future water supply needs through a statewide strategy of investing in improved regional supplies, conservation, and water use efficiency" (Cal. Water Code § 85021). It then directed that "each region that depends on water from the Delta watershed shall improve its regional self-reliance for water through investment in water use efficiency, water recycling, advanced water technologies, local and regional water supply projects, and improved regional coordination of local and regional water supply efforts." A variety of statutes empower local and regional agencies to promote greater efficiency in water use. These include the requirements that municipal water agencies meter and report on water use (Cal. Water Code §§ 500–535), that agricultural water agencies monitor and report on groundwater levels (§§ 10920–10936), and that agencies adopt urban and agricultural water management plans that include best practices to promote conservation and efficient use (§§ 10608–10608.64, 10610–10656, 10750–10783.2, 10800–10853). The legislature also has authorized (with voter approval) more than $2 billion in bond funding to support forty-six integrated regional water management programs that allow cities, counties, and other agencies to coordinate their water supply, water management, and flood control efforts (Hanak et al. 2011, 365–68). All of these laws build on the constitutional mandate that California's water resources be administered to promote reasonable, and reasonably efficient, water use.

Water pricing also plays an important role in encouraging reasonable use. The legislature has authorized public water agencies to adopt "allocation-based conservation water pricing" (Cal. Water Code §§ 370–374). The agency may set a base rate that is designed to cover its fixed costs and then one or more higher rates that increase with volume of water use. Tiered rate structures create incentives for conservation and more efficient use, because the cost per unit of water rises as each customer's demands increase (Hanak et al. 2011, 270–73). Agencies with allocation-based tiers typically use revenues from the upper tiers to fund conservation programs within their service area. Since the legislature's express authorization of allocation-based conservation pricing in 2008, a number of water supply agencies have adopted tiered rates (Hanak et al. 2011, 270–73).

Although enacted pursuant to article X, section 2, tiered rate pricing has raised questions under article XIIID of the California Constitution, which was passed by the voters in 1996 as Proposition 218. This law provides *inter alia* that water rates "shall not exceed the proportional cost of the service" attributable to each parcel of land that receives water service (Cal. Const. art. XIIID, § 6(b)(3)).

In *City of Palmdale v. Palmdale Water District* (2011), the California Court of Appeal invalidated a tiered-rate structure that set different rates (and different percentage increases between tiers) for residential, commercial, and irrigation customers. The court recognized that the district had adopted the tiered rates for the purpose of encouraging conservation and efficient use, consistent with the constitutional reasonable use mandate as well as the legislature's authorization of allocation-based conservation pricing. It explained, however, that "article X, section 2 is not at odds with Article IIID so long as . . . conservation is attained in a manner that 'shall not exceed the proportional cost of the service attributable to the parcel.'" The court concluded that the district had not explained "why [these other laws] cannot be harmonized with Proposition 218 and its mandate for proportionality. PWD [Palmdale Water District] fails to identify any support in the record for the inequality *between* tiers, depending on the category of user."

The court's insistence on a cost-based justification of water rates and rate differentials may well be required by Proposition 218, but overly strict judicial interpretations of the law will present challenges for contemporary water administration. Allocation-based conservation pricing is one of the most direct and proactive means of implementing article X, section 2's goals of conservation and efficient use. It accomplishes these goals through price incentives, rather than government fiat; and it fairly distributes the costs of water service by requiring those who use the most to pay the most. The question in these types of cases should not be how to ensure that article X, section 2, "is not at odds with" Proposition 218, but how to ensure that the ratemaking strictures of Proposition 218 do not undermine the most important principles of California water law and policy (Gray et al. 2014).

Integrated Management of Groundwater and Surface Water Resources

One of the most vexing problems in California water law is the antiquated separation between the law of surface water rights and the law governing groundwater (Sax 2003; Hanak et al. 2011, 322–28). Although there is now integrated management of surface and groundwater resources in twenty-two adjudicated groundwater basins and special groundwater management districts, many problems remain (see chapter 8 in this volume).

The most logical and direct response to these problems would be for the legislature to enact a statute empowering the SWRCB to exercise integrated permitting and regulatory authority over surface and groundwater rights. There is no question of the legislature's power to do this under article X, section 2. With limited exceptions, however, it is unlikely to occur.[6]

Yet, the courts have their own constitutional reasonable use authority to address problems of groundwater overdraft and conflict between surface and groundwater uses. For example, the physical solution and final judgment in the *Mojave* adjudication included rights to both groundwater and surface water, based on the trial court's determination that the two are hydrologically connected: diversions from the river reduce groundwater recharge, and groundwater pumping reduces the volume and flow of water in the river. Indeed, the water management system that the judgment created places a water replacement charge on all water extraction that exceeds each user's "free production allowance." The revenues from these charges are used to fund the acquisition of imported surface water to augment and replenish the native groundwater supplies (Littleworth and Garner 2007).

This integrated management of surface water and groundwater supplies established an important precedent: that the courts have authority under article X, section 2, to unify the law of surface and groundwater rights situationally where unintegrated management and regulation would result in unreasonable use. And the courts have this constitutional power despite the general legal distinction between the surface water and groundwater systems. Indeed, as a result of earlier groundwater adjudications and special legislation, integrated surface and groundwater management is now an important feature of regional water administration in several parts of California. These include the Orange County Water District, the Water Replenishment District of Southern California, the Santa Clara Valley Water Agency, and a number of smaller districts that manage adjudicated groundwater basins (Blomquist 1992).

The SWRCB's assertion of its article X, section 2, authority to limit surface water diversions and groundwater withdrawals from the Russian River system to protect coho salmon is another example of how the reasonable use doctrine can facilitate integrated water management. If the board were confined to its direct authority over surface water permittees and licensees, its efforts to protect the salmon would be frustrated. Not only would riparians and groundwater right holders be exempt, but those appropriators who *are* subject to the regulation could simply shift from surface water diversions to groundwater pumping and evade the restrictions. The aggregate effect would be to place an already endangered species in further jeopardy of extinction, which would unquestionably be an unreasonable exercise of water rights. As the Court of Appeal recognized in *Light v. SWRCB*, integrated surface and groundwater regulation in this context is therefore an appropriate and necessary exercise of the SWRCB's reasonable use powers.[7]

Incentives for Water Conservation and Transfer

Many of the important reasonable use cases have involved reallocations of water from senior water right holders whose existing uses or methods of use had become unreasonable in light of new consumptive demands on the resource or new environmental requirements. The courts in these cases have consistently held that an unreasonable use of water—unreasonable under contemporary standards—may not be asserted to block the new use or to obtain compensation from the new user. Although these principles are a *sine qua non* of reasonable use, some water users, economists, and policymakers have criticized the doctrine for rendering water rights uncertain. This uncertainty is harmful, they argue, because it may deter investment and marketability: "If current owners of water rights do not have secure rights—even if the lack of security serves perfectly valid public purposes—they will have a difficult time finding buyers for those rights" (Haddad 2000, 41).

Properly administered, however, the reasonable use doctrine can place constructive pressure on existing water users not to waste water and to encourage the profitable transfer of water from potentially unreasonable uses. Indeed, California's two most prominent water transfers resulted from this interplay between reasonable use and the market.

As described above, in 1984 the SWRCB made a determination of unreasonable use against the Imperial Irrigation District (IID), finding that the district's unlined canals and lack of regulating reservoirs in its water distribution system were causing both waste of water and flooding of land adjacent to the Salton Sea. The board ordered the district to correct these problems and to conserve a minimum of 100,000 acre-feet per year. With its water rights in jeopardy of reduction, the district agreed to line its canals, construct regulating reservoirs, and make operational improvements to its distribution system. These conservation actions would be funded, however, by the Metropolitan Water District (MWD) as payment for a 35-year transfer of 106,110 acre-feet per year from the IID to the MWD (Gray 1994). The SWRCB could have simply divested the IID of its water rights to the extent of unreasonable use, but it chose not to do so in favor of the more constructive solution presented in the IID–MWD transfer.[8]

The waste and unreasonable-use laws also served as a catalyst for a subsequent transfer of conserved water from IID to the San Diego County Water Authority. Following years of focus on California's excessive use of water from the Colorado River, the U.S. Bureau of Reclamation made a formal finding that farmers within the IID were wasting water. Based on this finding, the bureau determined that the district was in violation of the beneficial use requirement of federal reclamation law, which includes a reasonable use standard, and it ordered a reduction in water deliveries to the IID of approximately 8 percent. This decision broke a decade-long deadlock in negotiations among the Department of the Interior, the IID, the MWD, and the SDCWA. Two months later, the Southern California water agencies pledged to reduce their use of Colorado River water by 800,000 acre-feet per year over the next 14 years. This Quantification Settlement Agreement (QSA) brought California into compliance with the 4.4 million acre-feet per year limit of the Boulder Canyon Project Act, which governs the allocation of Colorado River water among Arizona, California, and Nevada. In the QSA, IID also agreed to conserve and transfer 277,000 acre-feet per year to SDCWA for a period of 35 years (Gray 2005).[9]

The IID transfers were the product of the state and federal governments' enforcement of the mandate of reasonable use. The SWRCB and the Department of the Interior applied the doctrine of reasonable use aggressively, but also flexibly, to give IID and its members a choice:

forfeit their water rights to the extent of unreasonable use, or correct the problem and benefit economically from the conservation and transfer of their previously wasteful practices. Application of the reasonable use doctrine in this context thus served three salutary purposes. It induced the conservation and more efficient conveyance and use of water within IID. It led to the transfer of the conserved water to higher-valued uses within MWD and SDCWA. And it reduced MWD's and SDCWA's long-term demands for water from both the Colorado River and other sources (such as the Delta or new water projects in the Sierra Nevada).

One of the goals of the modern water transfer statutes is to create economic incentives for water right holders and their derivative users to conserve water and to transfer that water to higher-valued uses by presenting them with the opportunity costs of their existing uses—that is, by showing them that they may earn more revenue from selling water than they can through their own uses. Enforcement of the reasonable use mandate to induce these types of transfers is an important means of effectuating these statutory policies and should become a more prominent feature of California's efforts to foster greater efficiency in water use and water allocation. The two IID transfers are a model for this vital synergy between water transfers and reasonable use.

Compliance with Environmental Standards and Protection of the Public Trust

The reasonable use doctrine also serves the important purpose of helping to implement and enforce the public trust and the other environmental laws that protect water quality, endangered species, aquatic habitat, and other *in situ* uses. These laws establish fundamental limitations on the amount of water that water right holders may impound and divert from California's rivers, lakes, and estuaries.

Federal statutes, such as the Clean Water Act (33 U.S.C. §§ 1251 *et seq.*) and the Endangered Species Act (16 U.S.C. § 1531 *et seq.*), are preemptive of California water rights law, as are the water quality standards, biological opinions, and other regulations and administrative actions that implement them. California's environmental laws also take precedence over water rights in the event of conflict. The environmental baselines these laws establish define the quantity of water available for impoundment and diversion (*In re Bay-Delta Programmatic Environmental Impact Report Coordinated Proceedings*, 2008).

Yet, some courts have struggled to understand the relationship between environmental mandates and the reasonable use doctrine as a limit on the exercise of water rights. For example, in *Tulare Lake Water Storage District v. United States* (2001), the U.S. Court of Federal Claims ruled that restrictions on SWP operations required by biological opinions issued under the federal Endangered Species Act to protect winter-run Chinook salmon and Delta smelt were a taking of property, because the operational constraints caused water shortages for some SWP contractors. The court ordered the United States to pay the contractors approximately $26 million in damages. Although the court noted that the reasonable use and public trust doctrines might preclude the appropriation of water under conditions that would imperil endangered species of fish, it declined to consider either aspect of California water rights law as part of its analysis.

Similarly, in *Casitas Municipal Water District v. United States* (2008), the U.S. Court of Appeals for the Federal Circuit held that the United States's directive that a local water district allow water to pass through a fish ladder was a taking of property. The releases of water were needed to support migration of steelhead, which are also protected under the federal ESA. The court reasoned that the federal government had physically diverted the plaintiff's water for its own purposes—protection of the endangered fish. The court never addressed the question of whether California's reasonable use and public trust doctrines might limit the plaintiff's exercise of its water rights in a manner that could harm the protected fish.[10]

Yet, analysis of the reasonable use doctrine in these settings should be straightforward, both for the advocates of environmental protection and for the courts. Article X, section 2, declares as a matter of California constitutional and property rights law that existing uses of water are unlawful if they cause unreasonable harm to water quality, fish, aquatic ecosystems, or other in-stream beneficial uses. Not only does the state have a duty to enforce the reasonable use mandate, but it may do so without violating the water rights of those users who must reduce their impoundment and diversion of water, limit discharges, or otherwise alter their water use practices to comply with this supervening law. As *Joslin* and its progeny make clear, because there is no valid property right in an unreasonable use, when the state acts to abate water practices that unreasonably harm the environment it may do so without compensation.

Although an adjudicatory reasonable use determination always requires an assessment of the competing interests, it is difficult to imagine a case in which a court would find reasonable a use of water that violates water quality standards or jeopardizes the continued existence of endangered or threatened species. As described above, the California Court of Appeal has held that statutes that allocate water to environmental uses, or which place limits on the impoundment and diversion of water to protect against environmental harm, are presumptively reasonable and are entitled to significant deference (*California Trout v. SWRCB*, 1989). Conversely, uses of water that violate state and federal environmental laws—or the water quality and streamflow standards, effluent limits, biological opinions, incidental take limits, and other regulations that implement those laws—should be presumptively unreasonable and a substantial burden placed on the water right holder to prove otherwise.

Constructive Pressure to Reform

The reasonable use doctrine will continue to be a vital component of California water policy. It is the foundation of the state's water rights system and applies to all water rights. It confers authority on all branches and levels of government to ensure that water is used reasonably (and reasonably efficiently) to maximize the general welfare of Californians. This includes drinking-water supplies and economic uses, as well as the environment. The reasonable use doctrine (sometimes working in tandem with the public trust) therefore serves to ensure that the impoundment and diversion of the state's waters for consumptive uses do not degrade aquatic ecosystems or harm the aquatic and terrestrial species that also depend on these waters.

Some future applications of the reasonable use mandate will be obvious. As discussed in the preceding subsection, assertion of the doctrine both to protect endangered fish and to limit the exercise of water rights that threaten to jeopardize such species is one example. Another obvious (and perhaps easy) application of the reasonable use mandate would be to restrict the groundwater pumping on the west side of the San Joaquin Valley that has caused overdraft and compaction of local aquifers, with attendant land subsidence of almost 30 feet. This overdraft and subsidence now threaten the geologic stability and flow capabilities of the Delta-Mendota Canal and the California Aqueduct, which deliver irrigation water to approximately 3 million acres of farmland in the San Joaquin Valley and Tulare Basin and to more than 16 million residential, commercial, and industrial customers in Southern California (U.S.

Geological Survey 2013). This is a compelling example of the state's obligation to enforce the doctrine of reasonable use for the benefit of California's people and economy.

In less egregious situations, the reasonable use doctrine is likely to play a more indirect role in improving water management and water use. The Delta Watermaster's call for an investigation of irrigation practices is an excellent beginning. Coupled with the statutes that protect existing users' rights to conserved water and allow them to transfer water made available through voluntary conservation, such investigations could induce some farmers and irrigation-water managers to correct wasteful practices and perhaps even profit from their reforms. Presented with both the opportunity costs of their existing uses and the threat of loss of water rights for failure to act, these users may choose to do the right thing. The two IID transfers of conserved water are useful templates for this type of interplay between reasonable use and market incentives.

Moreover, some environmental groups have suggested that the state should assert its reasonable use authority against those who irrigate water-intensive crops, such as alfalfa and pasture (Natural Resources Defense Council 1996). The doctrine also could be used to put pressure on municipal water agencies and their customers to minimize their use of water for landscaping and other outdoor uses. Whether state and local regulators would have the will to take such actions is an open question. As the foregoing demonstrates, however, the reasonable use mandates clearly apply to uses that demand an inordinate share of the available water in light of contemporary competing demands. It should not matter in this context whether the excessive demands are the result of an unreasonable point of diversion, method of conveyance, or place of use, or are caused by the type of use to which the water is put.

The doctrine of reasonable use may therefore be best understood as a source of pressure on all water users to exercise their rights in a manner that accounts for the effects of their water-use practices on other existing and potential uses—both consumptive and environmental—and that keeps pace with the times. Although the law may not necessarily require that individual water uses be as efficient as technology permits, or that water uses be changed to ensure optimal allocation (however that might be determined), the doctrine does set an enforceable standard of reasonably efficient use and reasonably efficient allocation as current conditions warrant. A consistent and palpable threat of regulatory enforcement of reasonable use may serve as a constructive inducement to better water use and more optimal allocation.

CONCLUSION

Toward the end of his life, Wallace Stegner looked back over a century of water resources policy and wrote: "The West cannot carry what it has lifted. It will make heroic efforts, always in the direction of more grandiose engineering works, and in the end it will subside back to what it was meant to be, an oasis civilization with one great deficiency—water." We need, he concluded, "a Redeemer" (Stegner 1986).

The twin problems that Stegner identified—overdevelopment of an arid environment and an unrequited faith that we can somehow engineer a solution to water scarcity—are even more palpable 30 years on. Yet, as Stegner knew full well, there will be no redeemer. There is only our capacity to learn from the past, to repair the problems that we have created, and to place ourselves on a more sustainable future path. As California moves forward in the twenty-first century to confront the challenges posed by overuse and misallocation, groundwater overdraft, ecological degradation, continued population growth, and the predicted effects of global warming and climate change, the responsive and dynamic mandates of the reasonable use doctrine will be an essential guide.

NOTES

1. The common law, statutory law, and constitutional law of water rights also contain a "beneficial use" requirement, which means that all uses of water must be for a socially beneficial use (Cal. Const. art. X, § 2; Cal. Water Code § 1240). This chapter focuses on the reasonable use requirement because, both as a water policy directive and as a limitation on water rights, it is the more significant of the two.

2. Although the doctrine of reasonable use was part of the common law of riparian and appropriative rights, the voters placed it in the Constitution in 1928 to overturn a series of California Supreme Court decisions that prevented appropriators from alleging unreasonable use against riparians. The consequence was to allow riparians to enjoin any nonriparian use of water that diminished the natural flow of California's rivers, regardless of the unreasonableness of the riparian's claims. This in turn threatened the development of the state's economy, which was increasingly dependent on water exported from the Sierra Nevada to the Bay Area, the Tulare Basin, and Southern California (Gray 1989; Hundley 2001).

3. In *City of Barstow v. Mojave Water Agency* (2000), the Supreme Court suggested that this same principle may apply to unexercised groundwater rights held by property owners whose lands overlie the aquifer (and hence have first priority to its safe yield).

4. The court relied in both cases on the Water Code, § 1050, which declares that the SWRCB's regulatory authority is "in furtherance of the policy con-

tained in Section 2 of Article X of the California Constitution and in all respects for the welfare and benefit of the people of the state, for the improvement of their prosperity and their living conditions."

5. This transfer, as well as the interplay between the reasonable use doctrine and water transfers, will be discussed in the final section of this chapter.

6. The legislature has granted the board authority to engage in integrated surface and groundwater rights administration in the Scott River system, where the board has the power to conduct a statutory adjudication of all water rights, including "ground water supplies which are interconnected with the Scott River" (Cal. Water Code § 2500.5).

In September 2014, California governor Jerry Brown signed into law three bills that empower local agencies to regulate groundwater pumping: AB 1739, SB 1168, and SB 1319. The legislation also authorizes the SWRCB to regulate groundwater pumping if the board determines that (1) the local groundwater sustainability plan is inadequate and "is not being implemented in a manner that will likely achieve the sustainability goal," and (2) "the basin is in a condition where groundwater extractions result in significant depletions of interconnected surface waters" (Cal. Water Code § 10735.2(a)(5)(B)(i), (ii)). The new law stipulates that before January 1, 2025, however, "the state board shall not establish an interim plan under this section to remedy a condition where the groundwater extractions result in significant depletions of interconnected surface waters" (§ 10735.8(h)). Although this legislation is an important first step toward integrated regulation of ground and surface water resources, it does not alter the long-standing general legal divide between the two.

7. In May 2014, a Superior Court applied this reasoning to hold that the public trust doctrine may limit groundwater pumping that lowers surface flows in the Scott River and thereby harms fish and recreational uses (*Environmental Law Foundation v. SWRCB*, 2014). This decision is consonant with the integrative and comprehensive interpretation of article X, section 2, described in the text.

8. The board's forbearance of its power to divest the IID of a portion of its water rights was supported by the legislature's general declaration that water conservation, as well as the transfer of conserved water, is a reasonable and beneficial use (Water Code §§ 1011(a), (b)). The legislature also enacted a special law to protect the IID against forfeiture or diminution of its water rights as a result of water conservation and to insulate the district from liability for any adverse effects on the Salton Sea that might result (§§ 1012, 1013(a)).

9. As with the IID–MWD transfer, the legislature enacted special legislation to facilitate the IID–SDCWA transfer (Cal. Water Code § 1013(b)–(h)).

10. In a later opinion in the case, the Federal Circuit came closer to the reasonable use question, recognizing that article X, section 2, defines the property right in water. It held that because Casitas had not proved that the loss of the water that the government required to pass though the fish ladder had reduced the amount that the district could apply to beneficial use, the district had failed to establish an interference with its water rights. Under California law, the court concluded, "the concept of beneficial use provides an 'overriding constitutional limitation' on a party's water rights" (*Casitas Municipal Water District v. United States*, 2013).

REFERENCES

Blomquist, William. 1992. *Dividing the Waters: Governing Groundwater in Southern California.* San Francisco: ICS Press.

California Court of Appeal cases

People ex rel. SWRCB v. Forni, 54 Cal. App. 3d 743, 126 Cal. Rptr. 851 (1976).

United States v. SWRCB, 182 Cal. App. 3d 82, 227 Cal. Rptr. 161 (1986).

Imperial Irrigation District v. SWRCB, 186 Cal. App. 3d 1160 (1986).

California Trout v. SWRCB, 207 Cal. App. 3d 585 (1989).

Imperial Irrigation District v. SWRCB, 225 Cal. App. 3d 548, 275 Cal. Rptr. 250 (1990).

El Dorado Irrigation District v. SWRCB, 142 Cal. App. 4th 937; 48 Cal. Rptr. 3d 468 (2006).

City of Palmdale v. Palmdale Water District, 198 Cal. App. 4th 926, 131 Cal. Rptr. 3d 373 (2011).

Light v. SWRCB, 226 Cal. App. 4th 1463, 173 Cal. Rptr. 3d 200 (2014).

California legislation

AB 1739, SB 1168, and SB 1319, California Legislature, 2013–14 Regular Session.

California Fish and Game Code § 5937.

California Wild and Scenic Rivers Act, California Public Resources Code § 5093.5.

California Water Code §§ 275, 370–374, 500–535, 1011–1013, 1050, 1240, 2500.2, 10920–10936, 10608–10608.64, 10610–10656, 10735.2(a)(5)(B)(i), (ii), 10735.8(h), 10750–10783.2, 10800–10853, 85020–85023, 85054.

California regulations

23 California Code Regs. § 735.

California Superior Court cases

Environmental Law Foundation v. SWRCB, No. 34-2010-80000583 (Sacramento Supr. Ct. 2014).

California Supreme Court cases

Peabody v. City of Vallejo, 2 Cal. 2d 351, 383, 40 P.2d 486 (1935).

Joslin v. Marin Municipal Water District 67 Cal. 2d 132, 140, 429 P.2d 889, 60 Cal. Rptr. 377 (1967).

In re Waters of Long Valley Creek Stream System, 25 Cal. 3d 339, 358–59, 599 P.2d 656, 158 Cal. Rptr. 350 (1979).

Environmental Defense Fund v. East Bay Municipal Utility District, 26 Cal. 3d 183, 605 P.2d 1, 161 Cal. Rptr. 466 (1980).

People v. Shirokow, 26 Cal. 3d 301, 605 P.2d 859, 162 Cal. Rptr. 30 (1980).

National Audubon Society v. Superior Court, 33 Cal.3d 419, 443, 658 P.2d 709, 189 Cal. Rptr. 346 (1983).

City of Barstow v. Mojave Water Agency, 23 Cal. 4th 1224, 1242, 5 P.3d 853, 99 Cal. Rptr. 2d 294 (2000).

In re Bay-Delta Programmatic Environmental Impact Report Coordinated Proceedings, 43 Cal. 4th 1143, 184 P.3d 709, 77 Cal. Rptr. 578 (2008).

Federal Court cases

Tulare Lake Basin Water Storage District v. United States, 49 Fed. Cl. 313 (2001).

Casitas Municipal Water District v. United States, 543 F.3d 1276 (Fed. Cir. 2008), *petition for rehearing denied,* 556 F.3d 1329 (2009).

Casitas Municipal Water District v. United States, 708 F.3d 1340 (Fed. Cir. 2013).

Federal legislation

Clean Water Act, 33 U.S.C. §§ 1251 *et seq.*

Endangered Species Act, 16 U.S.C. § 1531 *et seq.*

California Bay-Delta Authority. 2005. *Final Draft Year-4 Comprehensive Evaluation of the CALFED Water Use Efficiency Element.* Sacramento: State of California.

Gray, Brian E. 1989. "In Search of Bigfoot: The Common Law Origins of Article X, Section 2 of the California Constitution." *Hastings Constitutional Law Quarterly* 17:225.

———. 1994. "The Modern Era in California Water Law." *Hastings Law Journal* 45:249.

———. 2005. "The Uncertain Future of Water Rights in California: Reflections on the Governor's Commission Report." *McGeorge Law Review* 36:43.

Gray, Brian E., Dean Misczynski, Ellen Hanak, Andrew Fahlund, Jay Lund, David Mitchell, and James Nachbaur. 2014. "Paying for Water in California: The Legal Framework." *Hastings L.J.* 65:1603.

Haddad, Brent M. 2000. *Rivers of Gold: Designing Markets to Allocate Water in California.* Washington, DC: Island Press.

Hanak, Ellen, Brian Gray, Jay Lund, David Mitchell, Caitrin Chappelle, Andrew Fahlund, Katrina Jessoe, Josué Medellín-Azuara, Dean Misczynski, James Nachbaur, and Robyn Suddeth. 2014. *Paying for Water in California.* San Francisco: Public Policy Institute of California.

Hanak, Ellen, Jay Lund, Ariel Dinar, Brian Gray, Richard Howitt, Jeffrey Mount, Peter Moyle, and Barton Thompson. 2011. *Managing California's Water: From Conflict to Reconciliation.* San Francisco: Public Policy Institute of California.

Hundley, Norris, Jr. 2001. *The Great Thirst: Californians and Water—A History.* Rev. ed. Berkeley: University of California Press.

Littleworth, Arthur L., and Eric L. Garner. 2007. *California Water II.* Point Arena, CA: Solano Press.

Natural Resources Defense Council. 1998. *Alfalfa Overview.* San Francisco, CA.

Sax, Joseph L. 2003. "We Don't Do Groundwater: A Morsel of California History." *University of Denver Water Law Review* 6: 269.

Stegner, Wallace. 1986. "Water in the West: Growing beyond Nature's Limits." *Los Angeles Times*, December 29.

United States Geological Survey. 2013. *Land Subsidence along the Delta-Mendota Canal in the Northern Part of the San Joaquin Valley, California, 2003–10.* Reston, VA: U.S. Government Printing Office.

Wilson, Craig M. [Delta Watermaster]. 2010. *The Reasonable Use Doctrine & Agricultural Water Use Efficiency: A Report to the State Water Resources Control Board and the Delta Stewardship Council.* Sacramento: State of California.

CHAPTER 5

Urban Water Demand and Pricing in a Changing Climate

JULIET CHRISTIAN-SMITH AND MATTHEW HEBERGER

The population of California is increasingly urban, with 98 percent of the state's 38 million residents living in cities and suburbs. California's urban water demand is likely to increase in coming years due to a variety of factors including population growth, development patterns, and climate change. When it comes to water, it doesn't just matter how much we grow, it also matters where and how we grow. While most of California's current population is clustered around the coastal cities of San Francisco, Los Angeles, and San Diego, much of the state's future growth is expected to occur in hotter, drier inland areas, where the average household uses much more water. In addition, warming due to climate change is contributing to increased water demand for landscapes, evaporative cooling, and energy production. In this chapter, we describe patterns of urban water use in California, discussing where and how water is used in California's cities, and examine how urban use has changed over time and how it is likely to change in the future. We describe how water managers forecast future water demands, and how demand forecasts can contribute to better water management. Ultimately, water use is influenced by many factors, and the price that users pay for water is among the most important of these factors. We conclude with a discussion of "sustainable water pricing," or how water service providers can set customers' water rates to achieve the goals of improving water-use efficiency and maintaining a reliable revenue stream to finance continued utility operations.

California's water utilities face two, often conflicting challenges: how to encourage water conservation and increase water supply reliability, while also maintaining fiscal solvency. Many water managers are calling this "the new normal": an era marked by decreasing water demands and increasing water prices that has led to a growing revenue gap between water sales and utility costs, making it difficult for water service providers to cover their costs. In order to address this new normal, many water service providers will need to consider improved water forecasts and more innovative pricing practices. Indeed, we use the term *water service provider* rather than *water supplier* throughout this chapter to indicate that there are a variety of water-related services that can be provided beyond just water supply. These include water conservation and efficiency programs, watershed protection and restoration programs, water reuse programs, and increased long-term water supply reliability.

Urban Water Use in California

How much water do Californians use? And how has water use changed over time? Given the central importance of water to California's economy and lifestyle, these are surprisingly difficult questions to answer. There is no comprehensive statewide program in place to measure and report water use, so the information presented here relies on estimates and modeling studies performed by various state and federal agencies. A first estimate comes from the United States Geological Survey, which estimates freshwater use every five years. According to their latest report, in 2005, California withdrew 7,550 million gallons of freshwater per day for public supply, industrial, and domestic use (Kenny et al. 2009). This is equivalent to 8.5 million acre-feet (maf), or 210 gallons per capita per day (gcd).

The California Department of Water Resources produced a slightly higher estimate for urban water use in the same year: 9 maf, or about 227 gcd. In figure 5.1, we show historical estimates of urban water use published by the department for 1972–2005. According to these official data, urban water use is strongly correlated with population, with one notable exception: water use declined in the early 1990s, during and immediately after the 1987–91 drought, a period of intense interest in water conservation. However, average rates of water use rebounded to pre-drought levels after the mid-1990s.

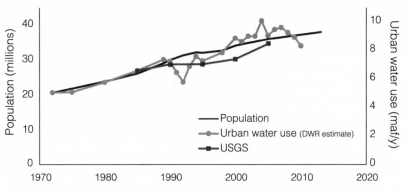

FIGURE 5.1. Historical population and urban water use in California, 1972–2005.

Droughts can have two opposing effects on urban water use. Dry conditions lead to increased demand for landscape irrigation. The state Legislative Analyst's Office notes that during dry years, urban use can actually increase by up to 10 percent due to increased water use for landscaping (Freeman 2008). On the other hand, prolonged drought can lead state and local authorities to call for water conservation and efficiency efforts, reducing water use.

For example, California experienced lower-than-average rainfall from 2007 to 2009. During that period, water suppliers launched a number of efforts to reduce demand. These included mandatory prohibitions on certain outdoor uses of water, increased rates, appliance rebates, and giveaways of efficient fixtures. There is evidence from several areas that per capita water use decreased in response to these efforts. In Long Beach, for example, per capita consumption was the lowest since the city began keeping records (Veeh 2010). A number of water service providers then raised water prices, after their customers' cutbacks led to less revenue from water sales. For example, the Metropolitan Water District, Southern California's biggest water wholesaler, saw sales drop 20 percent from 2007 to 2009, leading them to raise rates by 12 percent. Similar situations have been reported throughout the state (Fikes 2010).

Aside from the early-1990s drought years, per capita urban water use does not appear to have changed significantly over time in California. From 2000 to 2005, per capita water use averaged 229 gcd. This is somewhat higher than the average per capita water use in 1972 of 219 gcd; however, there is not enough evidence to state that there has been a statistically significant increase. Water-use data from the United States Geological Survey covering 1980–2005 show a similar pattern: average

TABLE 5.1   WATER USE BY SELECTED AGENCY
SERVICE AREA, 2006 *(in gallons per capita per day)*

San Francisco	95
Santa Barbara	127
Marin (MWD)	136
Los Angeles (LADWP)	142
Contra Costa (CCWD)	157
San Diego	157
East Bay (EBMUD)	166
Victorville (VVCWD)	246
Bakersfield	279
Sacramento	279
San Bernardino	296
Fresno	354

NOTE: This information was compiled by California Department of Water Resources staff from the Public Water Supply System database, and originally reported in the California Water Plan Update 2009 (4–49).

per capita water use does not appear to have changed significantly over the last four decades.

Patterns of Water Use in California

Per capita water use in California varies greatly from one part of the state to another. Climate can explain some of these differences, but history, culture, and water prices are also important. For example, some hot Southern California cities, such as Los Angeles, have managed to hold water use steady despite a growing population by significantly reducing per capita use (LA DWP 2012).

The most reliable estimates of individuals' water use come from water service providers, because these are based on measurements made by the providers at the local scale. Table 5.1 shows per capita water use for selected water service providers in 2006. These figures demonstrate the variability of urban water use in the state. Low consumption in San Francisco is usually attributed to the city's density, minimal outdoor water use, and cool coastal climate. High water use in the state's capital, Sacramento, can be explained by the hot climate and increased water-intensive landscaping (such as lawns), but also by the fact that 55 percent of residents do not have water meters, and thus pay a flat rate regardless of how much water they use (Khokha 2009b).

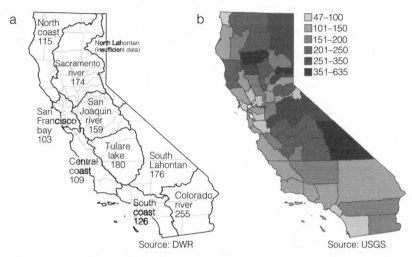

FIGURE 5.2. Per capita urban water use (in gallons per person per day), (a) by hydrologic region, from the Department of Water Resources, and (b) by county, from the U.S. Geological Survey.

The state's "20x2020" planning document, which provides a roadmap for reducing urban water use by 20 percent by the year 2020, reports per capita urban water use by hydrologic region (SWRCB 2010). California's ten hydrologic regions are planning areas developed by the state to manage watersheds and water supply. In the map in figure 5.2, county boundaries are shown by light-gray lines. Note that hydrologic region boundaries do not coincide with political divisions; some counties lie in two or three different hydrologic regions.

More than half of urban water use in California is in detached homes, or single-family residences. Table 5.2 reports the amount of water used in each major sector in California in the year 2000. According to state data, outdoor water use exceeds that used indoors (3.3 maf versus 2.3 maf, respectively). This is consistent with previous studies, including the national Residential End Uses of Water Study (Mayer et al. 1999), which reported outdoor water use as accounting for 58 percent of total water use (averaging 232 gpd outdoors, versus 168 gpd indoors).

FACTORS LIKELY TO AFFECT FUTURE URBAN WATER USE IN CALIFORNIA

Future water use depends on a number of factors, including social, cultural, and economic forces, which are notoriously difficult to predict.

TABLE 5.2   ESTIMATED URBAN WATER USE IN CALIFORNIA IN 2000, BY
SECTOR *(in million acre-feet per year)*

	Outdoor (maf/y)	Indoor (maf/y)	Total (maf/y)	Share of total urban use
Single-family residences	1.9	2.5	4.4	52%
Multi-family residences	0.36	0.8	1.2	14%
Commercial, industrial, institutional	0.63	1.6	2.2	26%
Large landscape	0.68	–	0.68	8%
Total urban use	3.6	4.9	8.5	100%

NOTE: These estimates are based on data in the California Department of Water Resources
Water Plan Update 2005 and personal communication with department staff.

We saw previously that San Francisco has among the lowest rates of per capita water use in the country, because many people live in apartment buildings and few have lawns. In contrast, suburban and rural residential areas tend to have larger lots and higher outdoor water use. While most new development is currently occurring in rural residential areas in the Central Valley and Inland Empire (Hanak and Davis 2006), some studies suggest that younger Californians may elect to live in denser, urban areas (Western 2013). Although it is difficult to predict what types of homes Californians will choose in the next few decades, it is possible to consider how policies and management choices will influence water use. Below, we consider how a number of key federal and state policies are shaping urban water use in California.

National Water- and Energy-Efficiency Legislation

The Energy Policy Act, or EPAct, passed by Congress in 1992, mandated water-efficiency standards for plumbing fixtures, as shown in table 5.3. These standards have led to reductions in household water use as more efficient fixtures replace older fixtures or are installed in new homes and businesses.

Building Codes and Plumbing Standards

Other planned changes to building and plumbing codes may further reduce per capita water demand in the coming decades. For example, the International Association of Plumbing and Mechanical Officials

TABLE 5.3   WATER EFFICIENCY STANDARDS FOR FIXTURES
REQUIRED BY THE 1992 U.S. ENERGY POLICY ACT

Fixture	Standard
Water closets (toilets)	1.6 gallons per flush
Showerheads	2.5 gallons per minute
Faucets	2.2 gallons per minute
Urinals	1 gallon per flush

(IAPMO) released its *Green Plumbing Mechanical Code and Supplement* in 2010 and issued a revised version in 2012. These codes are widely used by state and local governments in North America and thus have a major influence on the built environment. The new *Green Supplement* is meant both as "a resource for progressive jurisdictions that are implementing green building and water efficiency programs" and "a repository for provisions that ultimately will be integrated into the Uniform Codes." As water-efficient features move from voluntary to mandatory, it is likely to have effects on water use across the nation.

In addition to the standards and codes discussed above, there are voluntary programs that are already having impacts on water use. For example, the International Green Construction Code and the Leadership in Energy and Environmental Design (LEED) certification program establish rigorous efficiency standards that exceed current building codes. Likewise, the Environmental Protection Agency has developed the WaterSense program, a voluntary labeling program modeled after the EnergyStar program, to help customers identify and purchase efficient appliances. Products bearing the WaterSense label use less water than required under the national plumbing standards; the program is designed to help consumers choose efficient products and reward manufacturers for developing new water-efficient products.

Statewide Water Conservation and Efficiency Policies

For the past three decades, the state of California has provided various mandates and incentives for water service providers to promote water conservation or "demand management." For example, since 1983, state law has required every water agency that serves over 3,000 customers to prepare an Urban Water Management Plan every five years and submit it to the Department of Water Resources. The plan must include a description of the provider's demand management measures. In addition, many urban water service providers have voluntarily implemented a set of conservation and efficiency best management practices (BMPs) promoted by the California Urban Water Conservation Council (CUWCC 2011). However, a 2006 audit of the BMP program found that most water agencies, including most of the largest water suppliers, have not implemented all of the conservation practices, nor have they offered the requisite documentation explaining why the programs are not needed (CALFED 2006). In addition, it is unclear whether the BMP approach has resulted in meaningful water savings. The audit suggests that over 13 years the BMP approach may have been responsible for a reduction in per capita urban water use of only 2 percent.

The CUWCC's best practices–based approach to urban water conservation has always been a voluntary program; however, this is changing. A law signed in 2007 ties receipt of water-related state grant funding to BMP implementation. In effect, participation in the program will remain voluntary, but this may provide a stronger incentive for agencies to be fully compliant.

In addition, new legislation passed in 2009 (SB X7-7) requires that urban water agencies work to reduce per capita urban water use by 20 percent by 2020 in California. The intent behind the 20x2020 program is to prompt water service providers to expand water-conservation programs. A separate law passed in 2009 (AB 715) requires that only high-efficiency toilets and urinals be sold or installed after 2014. This law amends the 2007 California Plumbing Code and is stricter than the U.S. Energy Policy Act requirements described above. SB 407 of 2009 requires efficient toilets, faucets, and showerheads in all buildings. The law covers remodeled properties by 2014, all single-family homes by 2017, and multi-family and commercial buildings by 2019. While the law lacks an effective means of enforcement (such as requiring a retrofit as a condition of resale), it does require sellers to disclose whether the property is in compliance with the law.

The Model Water Efficient Landscape Ordinance

California water managers recognized the importance of addressing outdoor water use as more than half of households' water is used outdoors, mostly to water lawns and gardens. In 1990, the state legislature passed AB 325, which set water-efficiency requirements for landscapes at large commercial and public properties and for residences with professionally installed landscaping. Ten years later, an independent review of the program found several shortcomings in its implementation: "the legislation neither prescribed clear conservation goals, nor did it require meaningful levels of compliance" (Bamezai, Perry, and Pryor 2001). Simply put, many agencies did not have the resources or the will to monitor or enforce the law.

State legislators decided to try the approach again, passing AB 1881 in 2006. The new law required cities and counties to adopt local ordinances mandating minimum standards for landscape water efficiency. The law mostly covers commercial landscapes above a certain size (2,500 square feet), or large residential landscapes (over 5,000 square feet). Full implementation of this law is estimated to cover about 30 percent of California single-family homes (DeOreo et al. 2011). The ordinance requires landscapes to follow a water budget, for instance limiting the area planted as grass or other high-water-use plants and requiring efficient irrigation.

Local governments can use both carrots and sticks to promote efficient landscapes. In a number of areas, agencies have launched campaigns focused on outreach and education rather than regulation. For example, in the San Francisco Bay area, initiatives promoting low-water-use landscapes go by the name of Bay-Friendly Landscaping. These programs tout the many benefits of planting native vegetation: lower use of water, fertilizer, and chemicals; less waste and polluted runoff; and better habitat for birds and insects. Similar programs have been launched in Santa Rosa and Sacramento.

Residential Water Metering

It is often stated that "you can't manage what you don't measure." As of 2012, about 250,000 water customers in California were unmetered (DWR 2014). Research by the Public Policy Institute of California has found that, in cities with water meters, household water use is about 15 percent less than in unmetered cities (Khokha 2009a). Since 1992, state law has required the installation of water meters on all new construction. California has passed laws that will eventually result in universal metering, where every household has a water meter, by 2025 (AB 2572 of 2004; AB 975 of 2009). The move toward universal metering mostly affects residents of single-family homes, but some water municipalities have begun to require "sub-metering" of multi-family homes (SDCAA 2010). This is important, because nearly 30 percent of Californians live in multi-family homes. A 2004 study sponsored by the East Bay Municipal Utility District demonstrated water savings of 15.3 percent associated with sub-metering (Mayer et al. 2004). In the past few years, legislators have introduced bills to require greater sub-metering statewide; however, none of these bills has garnered enough support to be enacted into law.

The Graywater Law

In the summer of 2008, after two dry years, the California Senate passed SB 1258 requiring the state to revise building codes "to conserve water by facilitating greater reuse of gray water in California." Prior to August 2009, reuse of residential graywater from sinks, showers, and washing machines was limited. Although such systems were legal, they required detailed designs and often expensive permits. As of 2009, revised rules allow small-scale, laundry-to-landscape irrigation systems to be installed without a permit (California Plumbing Code, ch. 16). The ability to reuse graywater could have a significant impact on household water and energy use in the future.

"Show Me the Water" Laws and the Vineyard Decision

Many California historians have noted that the state's tremendous population growth would not have been possible without imported water. More recently, the attitude of planning for continued growth regardless of natural limits or the availability of natural resources has come under fire. This lack of coordination between land-use planning and water availability was addressed to an extent by the California legislature in 2001, when it passed SB 610 and SB 221, the so-called "show me the water" laws. Under these laws, developers of large projects (usually more than 500 housing units) must demonstrate that a 20-year water supply is available.

In a related California Supreme Court decision, *Vineyard Area Citizens for Responsible Growth v. City of Rancho Cordova* (2007), the court laid out general principles for dealing with water supply under the California Environmental Quality Act. The court stated that an applicant for a large project must do a thorough analysis of long-term water supply for the project. The court also stated that a developer must go beyond SB 221 and SB 610 in simply demonstrating the availability of water—the applicant must also anticipate the "reasonably foreseeable" environmental impacts of supplying water, and identify measures to mitigate any adverse impacts.

FORECASTING FUTURE WATER DEMAND

Water managers forecast future water demand for a variety of purposes, from planning for future water purchases or infrastructure investments to setting water prices. Several mathematical methods are in use for estimating future demand; these include extrapolating historic trends, correlating demand with socioeconomic variables, and more detailed scenario development. The simplest and most common means of forecasting future water demand is to multiply current per capita water use by the expected future population (McMahon 1993). While simple and easy to understand, this approach has several major shortcomings. In particular, it does not incorporate many important variables that we now know affect demand, such as efficiency standards, technological developments, and changing consumer preferences. Despite these limitations, a 2008 survey of water systems in the United States by the American Water Works Association found that most utilities are using this approach to forecast demand (AWWA and Raftelis Consultants 2009).

The result has been that, in most cases, demand forecasts overestimate future water demand. For example, figure 5.3 shows that the city of Seattle has consistently overestimated water demand from 1965 to the present day. Early forecasts were largely straight-line extrapolations, while more recent analyses have forecast slightly decreasing water use over the short term. But in each forecast, analysts failed to predict large drops in per capita demand that drove down aggregate water use even while the region's population grew. This can lead to problems, including the growing "revenue gap" facing many water service providers, discussed in more detail later.

Here, we describe a scenario-based planning tool developed to incorporate some of the physical, demographic, and economic forces that affect urban water demand, along with a series of possible management options, which are almost never included in water forecasts. The scenarios described below were modeled by the authors in *Urban Water Demand in California to 2100: Incorporating Climate Change* (Christian-Smith, Heberger, and Allen 2012). Using Microsoft Excel spreadsheets as the platform, the tool simulates future urban water demand in California to the year 2100. The tool is useful for constructing scenarios based on a series of user-defined inputs on urban water demand. It is important to emphasize that the tool does not make predictions but simulates future water demand based on user-defined scenarios.

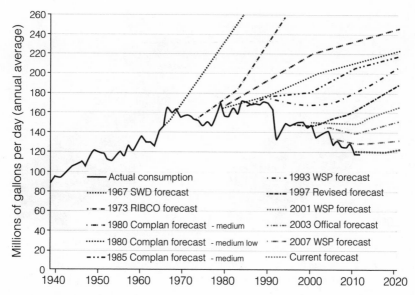

FIGURE 5.3. Actual water demand and past forecasts (from a Seattle Public Utilities presentation).

The scenario-based planning tool allows the user to choose among various inputs, including four global climate models; three climate scenarios; two potential evapotranspiration estimation methods; eight population projections; and the choice of running the model once, with fixed parameters entered by the user, or many times, in Monte Carlo mode, where each parameter is resampled and reported within 90-percent confidence intervals, capturing some of the uncertainty.

Several factors are considered *demand drivers*, or the primary forces that will determine water demand into the future:

- population growth
- development patterns, including where population growth occurs, what types of homes and landscaping people choose, and how and where commercial and industrial sectors grow or decline
- climate change and greenhouse gas emissions—we have included projections of the IPCC's B1 (low) and A2 (medium-high) scenarios, which forecast an average warming in California of 1.8 °C and 3.6 °C, respectively (Nakicenovic and Swart 2000; Maurer et al. 2010)
- management responses, such as increased water conservation and efficiency efforts and changes to water prices

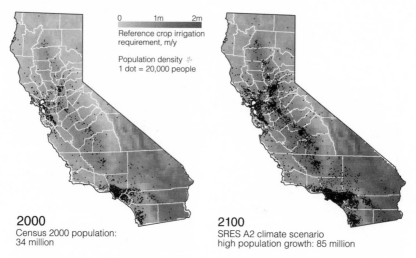

FIGURE 5.4. California's past and future population as they relate to water use. Source: Heberger, Christian-Smith, and Allen (2010).

This model was among the first to explicitly consider climate change impacts on water demand in California. Figure 5.4 is a map showing projected future conditions of two quantities that are expected to have an effect on urban water use. Due to climate change, temperatures are rising across the state, increasing evaporation and water use by plants. This is represented in the maps by "reference crop irrigation requirement" computed under a medium-high climate change scenario. The darker shades of gray indicate that plants will consume more water on average toward the end of this century than they did at the end of the last century. Dots indicate forecast population growth. Note that much of the future growth is projected to occur in inland valleys, which are hotter and drier than the coasts.

The tool can be used at various scales, from a customizable geographic scope (designed to allow individual water suppliers to isolate their service areas), to the county scale, to the hydrologic region, or statewide. Results are displayed as decadal averages from 2000–09 to 2090–99 and are always shown in comparison to a "static climate" (or the average of the climate from 1960 to 1999). This allows the user to quickly discern the impact of climate change, alone, on urban water demand. The results also describe the energy intensity of future water demand, allowing the user to understand how much additional energy— and greenhouse gas emissions—are associated with different scenarios.

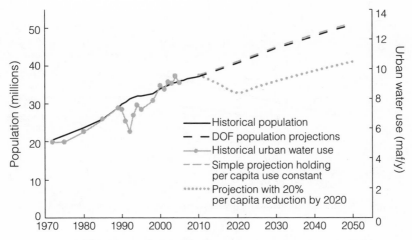

FIGURE 5.5. California's population and urban water use, 1972 to present (solid lines), with simple linear forecasts. The dashed line assumes that current patterns of water use hold steady, and the dotted line shows the effect of a 20-percent reduction by 2020 with no further improvements thereafter.

Results

Data on past water use show a strong correlation between urban water use and population over the past three decades. Figure 5.5 plots urban water use, 1972–2005, using data compiled by Department of Water Resources staff from various editions of the *California Water Plan Update* (DWR 2011). Population data come from the U.S. Census Bureau, which estimates population every 10 years, and the California Department of Finance, which provides estimates for the years in between.

We can use this relationship to make simple, first-order estimates of future water use. Population growth appears to follow a simple linear trend for the past 40 years, which we extrapolate into the future. Water use in 2005 was 225 gcd. Projecting both time series into the future, in 2050 there will be 60 million people consuming 15.5 maf of water per year. If the state is able to achieve the goal of a 20-percent per capita reduction in water use by the year 2020 (in accordance with the Water Conservation Act of 2009), water use will decline in the short term, because per capita water use is declining faster than population is growing. After 2020, if per capita water use remains at a reduced level, population growth still leads to an increase in urban water use over the next three decades; however, that rate of increase (dotted gray line) is not as great as in the business-as-usual scenario (dashed gray line).

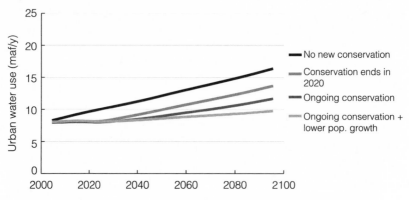

FIGURE 5.6. Simulated urban water use under four scenarios of conservation and population.

The rates at which conservation and efficiency improvements are put in place, and the degree to which they are continued in the future, will have major effects on urban water use. As shown in figure 5.6, when conservation efforts stop, water use grows linearly, in proportion with population, reaching 16.5 maf per year by the end of the century. For the baseline scenario, we assumed that California would meet the 20x2020 water conservation targets. If conservation efforts end in 2020, per capita use will begin to creep upward again after midcentury. This is a result of a warming climate and growth in inland valleys, both of which cause higher outdoor water use.

If conservation programs continue at the same rate necessary to achieve the 20x2020 targets until the end of the century, total urban water use would actually decline slightly by the end of the century, assuming historical climate. Yet, under a medium-high climate emissions scenario (which is most similar to current emissions levels), the gains from efficiency improvements are eventually overcome by the impact of warmer climate conditions. For example, after decreasing steadily to 158 gcd, urban demand begins to creep up over the last two decades of the century, reaching 162 gcd by 2100 despite continued water conservation and efficiency improvements. Thus, climate change is an important factor to consider when assessing future water demand.

The simulations of future water use model only "technologically feasible" conservation—the savings that could be achieved through greater market penetration of present-day, off-the-shelf appliances and fixtures. Much greater savings are possible when one considers changes in attitudes and behavior related to water, for example more people choosing smaller lawns or landscapes with low-water-use plants, or the future invention of additional water-saving devices.

A CHALLENGE TO CONTINUED WATER DEMAND REDUCTION: THE REVENUE GAP

Nationally, over the last decade, the costs for water service providers have risen more quickly than the costs to other sectors, and this is particularly true in California. There are several reasons behind the rapid cost increases, including a history of deferred maintenance, stricter water-quality standards, and climate change. Water utilities pass these increased costs on to customers, and the result has been that water prices have increased more rapidly than inflation (Beecher 2010). Maintaining adequate revenue continues to be a challenge for many water service providers in California (Black & Veatch 2012). Water managers rank deferred maintenance—necessary upgrades and repairs that have not been made in a timely or prudent manner—among the key drivers of increased costs.

As water infrastructure deteriorates, the pressure to keep water prices low has meant that there is little money to finance necessary upgrades. As these investments are continuously delayed, costs increase. The 2012 American Society of Civil Engineers *Infrastructure Report Card* estimates that, over the next 20 years, California will need to invest $39 billion in drinking water and nearly $30 billion in wastewater infrastructure; in both categories, California represents more than 10 percent of the needs estimated for the entire United States (ASCE 2013).

Part of the increased cost for financing new infrastructure projects is a result of constrained access to capital. In the wake of the financial crisis, the state and federal governments have reduced much of the grant and loan money that used to help pay for costly infrastructure (Emerson 2011). Historically, municipal bonds were a relatively inexpensive way to finance new infrastructure; today, credit rating agencies are downgrading some municipal water systems. According to a 2012 report by Ceres, the most common reason for these downgrades was that water price increases have not keep pace with spending on system maintenance or debt service coverage (Leurig 2012).

In addition, stricter water-quality standards mean that existing and new infrastructure projects tend to require additional levels of water treatment. There are two primary federal laws that govern water quality:

the Clean Water Act and the Safe Drinking Water Act. Together, they regulate water quality, in terms of the amount of pollution entering waterways as well as the maximum levels of pollutants in drinking water. Both are continually updated to ensure public safety. New chemicals, emerging contaminants, and their combined impacts are of particular concern to the water industry, and have resulted in stricter treatment standards. For example, a deeper understanding of the behavior of disinfection chemicals has resulted in stricter national standards for certain disinfection by-products that can be harmful to human and ecosystem health.

Finally, climate change is altering the timing, volume, and distribution of water supply through changes to precipitation and runoff. In addition, rising temperatures are raising outdoor water demand. Moreover, increasing frequency and severity of droughts, floods, and other extreme weather events mean new, climate-resilient infrastructure may be required (or existing infrastructure may need to be retrofitted) to accommodate increased climatic uncertainty.

The Revenue Gap

Water rate structures describe the way that total system costs are allocated among different customers. No matter which water rate structure is used, it should be effective at balancing total costs against total revenues. In other words, in order to maintain fiscal solvency, the total cost of providing water should be recovered through the prices customers pay to use that water. However, matching the price of using water with the cost of providing water can be difficult, because the costs are estimates and the price is set before the water is used. Any change in water demand or system operation can create an unexpected revenue loss or gain. For example, revenue losses can occur if more expensive water is needed to meet high demand during a drought, or if temporary drought conservation programs reduce water use below what was forecasted.

Increasing water costs, alongside decreasing water demand, lead to an ever-widening revenue gap, necessitating higher water prices. Between 1991 and 2006, California's average monthly charge for 1,500 cubic feet of water increased by more than $8, to $42 (in inflation-adjusted dollars; see table 5.4). As of 2006, the highest water charges in California have increased more significantly, by almost $50 for 1,500 cubic feet.

TABLE 5.4   THE RISING COST OF WATER IN CALIFORNIA

Year	Average water cha rge for 1500 cf (2013 dollars)	Highest water charge for 1500 cf (2013 dollars)
1991	$34	$103
2006	$42	$151

SOURCE: Black & Veatch (1995, 2006).

IMPROVED WATER FORECASTS AND PRICING PRACTICES ARE REQUIRED

The revenue gap can begin to be addressed by improved water forecasting and the application of more innovative pricing practices. Simplistic water forecasts that rely on past water usage and future population projections have historically overestimated water demand. Such forecasts, if used for water pricing, can lead water service providers to under-price water per unit and therefore under-collect revenue. Below, we describe a number of ways to improve water forecasting in California. In addition, we describe some of the innovative pricing practices adopted in California's energy sector to balance conservation with adequate revenue collection that are applicable to California's water sector.

Consider a Range of Factors That Are Likely to Affect Future Use

As previously noted, most agencies that forecast demand have used the simplest of all methods: multiplying future population estimates by historical per capita water use. This is a relatively straightforward way to conduct demand forecasting; however, forecasts that rely on historic conditions may not accurately reflect future patterns of water use, when one considers the factors described in the sections above.

While some changes are impossible to predict, others can be anticipated and fairly accurately estimated. For instance, the impact of new efficiency standards can be evaluated using a number of widely available tools. Water service providers may consult their local planning and finance departments to understand how drivers of water demand are expected to change in the future; for example, new industries and residential housing developments will increase water use in those sectors. Demand reductions from conservation mandates should also be included in the forecasts. At a minimum, California water service providers should consider the effects of water conservation mandates and climate change on future demand.

Include Price Effects

Most demand forecasts do not take into account the impact of price on demand. The assumption that per capita water demand will remain constant in the future implies zero price response, which is hardly ever the case (Chesnutt et al. 1997). Indoor water demand is generally considered less elastic (Olmstead and Stavins 2009), yet it has decreased over time in California for a number of reasons (Christian-Smith, Heberger, and Allen 2012). Outdoor water use, by contrast, is more elastic because it contains a larger "discretionary" component, such as landscape watering and pools, which customers are more likely to cut down in response to higher prices (Olmstead and Stavins 2009). Thus, it is critical to consider the demand response to changes in water prices over time.

Adopt Innovative Pricing Practices

California's energy sector has successfully implemented many pricing practices that seek to balance a commitment to energy conservation with utility financial health. While there are important differences between the water and energy sectors, there are several promising pricing practices that could be implemented, or further implemented, in the water sector. These include time-variant pricing, such as seasonal rates, along with innovative tools such as demand-response contracts; and lost-revenue-recovery mechanisms such as rate-stabilization funds. Some of these practices are already common in the water sector (e.g. tiered rates); some are becoming more widespread (e.g. seasonal rates, rate-stabilization funds), while others have not been widely applied to water (e.g. demand-response contracts and the calculation of an inherent commodity cost for water in the utility's revenue requirement). These practices are described in more detail by Donnelly, Christian-Smith, and Cooley in their 2013 white paper, *Pricing Practices in the Electricity Sector to Promote Conservation and Efficiency: Lessons for the Water Sector.*

Develop or Expand Affordability Programs

Water affordability is central to water access. When water prices make water unaffordable to particular communities, it can pose a health and safety issue and a myriad of administrative and political problems. Water affordability programs help customers who cannot afford increasing water prices. A number of water service providers in California currently provide some type of assistance to low-income customers. However, these programs tend to have very low enrollment. On the other hand, affordability programs in the energy and telecommunications sectors have relatively high participation rates. These programs can be characterized by their consistency: they often use the same or similar standards for eligibility, have stable sources of funding, and routinely release data about participation in the program in order to track progress. Water service providers could develop or expand affordability programs by using existing eligibility requirements from other sectors' affordability programs to enroll customers automatically.

CONCLUSIONS

In this chapter, we have looked at the portion of water in California that is used in our cities and suburbs. While this currently represents only about 20 percent of the state's water use, with the larger share going toward irrigated agriculture, total urban use is likely to continue growing along with the state's population, even if per capita demand continues to fall. While certain regions have held water use constant despite a growing population (Los Angeles and San Francisco are notable examples), on a statewide basis, urban water use has grown proportionally with population. Population growth is likely to be the most important driver of future water use, especially growth that is concentrated in the warm and dry inland valleys.

It is only recently that water planners have begun to incorporate information on climate change into forecasts of future water demand. Climate scientists are not able to predict with confidence whether California will receive more or less precipitation in the future. However, there is a consensus that the state will be warmer, which means increased evaporation and water use by plants, leading to possible increased demand for outdoor water use for lawns and landscaping. Analytical work has shown that climate change could contribute to an increase in future urban water demand in California of up to 15 percent by the year

2100. In fact, there is evidence that the effects of climate change are already being felt. Regional climate models show that temperature, evapotranspiration, and irrigation demand are slightly higher today than they have been over the preceding decades.

Aggressive conservation efforts can hold the state's urban water use steady, and may contribute to slight reductions. If successful, California's stated target of reducing per capita urban water demand by 20 percent by the year 2020 is likely to keep overall urban water use at or near current levels over the next decade. However, if efforts at improving water conservation and efficiency end in 2020, average per capita use will quickly begin to rise, driven largely by rising population growth in some of the driest parts of the state, where outdoor water demand is high.

Continued implementation of new water conservation and efficiency efforts can offset population growth. However, increased conservation and efficiency can also lower water service providers' revenue if not planned for properly. To successfully navigate the "new normal," an era of increasing water service costs and decreasing water sales, water service providers would be wise to improve water demand forecasting capabilities and consider more innovative water pricing policies. Other sectors, such as energy, have reduced consumption while maintaining utility financial solvency, and have instituted a variety of pricing tools and programs that could be much more widely adopted in the water sector. As climate change and other forces continue to challenge business-as-usual approaches to understanding water demand and pricing, many water service providers are becoming more nimble and forward-looking. It will be increasingly important for water service providers to be able to explain the services that they provide to their customers, beyond simply water supply, and to assign appropriate values to those services.

REFERENCES

ASCE. 2013. *2013 Report Card for America's Infrastructure: California Key Facts*. American Society of Civil Engineers. http://www.infrastructure reportcard.org/a/#p/state-facts/california.

AWWA and Raftelis Consultants. 2009. *Water and Wastewater Rate Survey 2008*. Denver, CO: American Water Works Association and Raftelis Financial Consultants, Inc. www.awwa.org/store/productdetail.aspx?ProductId=20743.

Bamezai, Anil, Robert Perry, and Carrie Pryor. 2001. *Water Efficient Landscape Ordinance (AB 325): A Statewide Implementation Review*. Sacramento: California Urban Water Agencies.

Beecher, Janice A. 2010. *Water Pricing Primer for the Great Lakes Region.* Great Lakes Commission. http://www.allianceforwaterefficiency.org/AWE-GLPF-value-water-project.aspx.

Black & Veatch. 1995. *California Water Charge Survey.* Irvine, CA: Black & Veatch.

―――. 2006. *2006 California Water Rate Survey.* www.kqed.org/assets/pdf /news/2006_water.pdf.

California Department of Water Resources. 2010. *California Water Plan Update 2009.* Sacramento: California Department of Water Resources.

CALFED. 2006. *Water Use Efficiency Comprehensive Evaluation.* Sacramento, CA: CALFED Bay-Delta Program. http://calwater.ca.gov/content/Documents /library/WUE/2006_WUE_Public_Final.pdf.

Chesnutt, T. W., J. A. Beecher, P. C. Mann, D. M. Clark, W. M. Hanemann, G. A. Raftelis, C. N. McSpadden, D. M. Pekelney, J. Christianson, and R. Krop. 1997. *Designing, Evaluating, and Implementing Conservation Rate Structures: A Handbook for the California Urban Water Conservation Council.* Santa Monica: California Urban Water Conservation Council.

Christian-Smith, Juliet, Matthew Heberger, and Lucy Allen. 2012. *Urban Water Demand in California to 2100: Incorporating Climate Change.* Oakland, CA: Pacific Institute. www.pacinst.org/reports/urban_water_demand _2100/.

CUWCC. 1991. *CUWCC Memorandum of Understanding (MOU).* California Urban Water Conservation Council. http://www.cuwcc.org/About-Us /Memorandum-of-Understanding.

DeOreo, William B., Peter W. Mayer, Leslie Martien, Matthew Hayden, Andrew Funk, Michael Kramer-Duffield, and Renee Davis. 2011. *California Single-Family Water Use Efficiency Study.* Sacramento: California Department of Water Resources, U.S. Bureau of Reclamation, and CALFED Bay-Delta Program. www.aquacraft.com/sites/default/files/pub/DeOreo-%282011%29-California-Single-Family-Water-Use-Efficiency-Study.pdf.

Donnelly, Kristina, Juliet Christian-Smith, and Heather Cooley. *Pricing Practices in the Electricity Sector to Promote Conservation and Efficiency: Lessons for the Water Sector.* Oakland: Pacific Institute, 2013. http://www. pacinst.org/publication/water-rates-pricing-practices/.

DWR. 2011. *Statewide Water Balance (1998–2005)* [Excel spreadsheet]. Sacramento: California Department of Water Resources. www.waterplan.water. ca.gov/docs/technical/cwpu2009/statewide_water_balance(1998–2005)04–28–11.xlsx.

―――. 2014. *California Urban Water Suppliers with Unmetered Connections.* Sacramento: California Department of Water Resources. www.water.ca.gov /urbanwatermanagement/2010_Urban_Water_Management_Plan_Data.cfm.

Emerson, Sandra. 2011. "Water Rate Hikes Tied to Aging Infrastructure." *San Jose Mercury News,* December 10. www.mercurynews.com/california/ci _19516462.

Fikes, Bradley J. 2010. "Water: Conservation, Recession Cause Wave of Rate Hikes." *San Diego Union-Tribune,* April 10. www.utsandiego.com/news/2010 /apr/10/water-conservation-recession-cause-wave-of-rate/.

Freeman, Catherine B. 2008. *California's Water: An LAO Primer.* Sacramento: Legislative Analyst's Office. www.lao.ca.gov/laoapp/pubdetails.aspx?id=1889.

Hanak, E., and M. Davis. 2006. "Lawns and Water Demand in California." *California Economic Policy* 2(2). San Francisco: Public Policy Institute of California. www.ppic.org/content/pubs/cep/ep_706ehep.pdf.

Heberger, Matthew, Juliet Christian-Smith, and Lucy Allen. 2010. *How Much Will We Use? Forecasting Urban Water Use in California with Changing Climate, Demographics, and Technology.* Oakland, CA: Pacific Institute. www.pacinst.org/publication/how-much-water-will-we-use/.

IAPMO. 2012. *2012 Green Plumbing and Mechanical Code Supplement for Use with All Codes.* Ontario, CA: International Association of Plumbing and Mechanical Officials. www.iapmo.org/Documents/2012GreenPlumbing MechanicalCodeSupplement.pdf.

Kenny, Joan F., Nancy L. Barber, Susan S. Hutson, Kristin S. Linsey, John K. Lovelace, and Molly A. Maupin. 2009. *Estimated Use of Water in the United States in 2005.* Circular 1344. Reston, VA: U.S. Geological Survey. http://pubs.usgs.gov/circ/1344/.

Khokha, Sasha. 2009a. "California's Water Meter Rebellion Withers." KQED's Climate Watch, May 17, 2009. http://blogs.kqed.org/climatewatch/2009/05/17/californias-water-meter-rebellion-withers/

———. 2009b. "Without Meters, Fresno Water beyond Measure." *National Public Radio.* www.npr.org/templates/story/story.php?storyId=104466681.

LA DWP. 2012. "LADWP Reminds Customers to Conserve Water: Reduced Snowpack & Increased Water Use Call for Increased Conservation." Los Angeles Department of Water and Power. www.ladwpnews.com/go/doc/1475/1426279/.

Leurig, Sharlene. 2012. *Water Ripples: Expanding Risks for U.S. Water Providers.* Boston, MA: Ceres. www.ceres.org/resources/reports/water-ripples-expanding-risks-for-u.s.-water-providers/view.

Maurer, Edwin P., Hugo G. Hidalgo, Tapash Das, Michael D. Dettinger, and Daniel R. Cayan. 2010. "The Utility of Daily Large-Scale Climate Data in the Assessment of Climate Change Impacts on Daily Streamflow in California." *Hydrology and Earth System Sciences* 14(6):1125–38.

Mayer, Peter, William B. DeOreo, Eva M. Opitz, Jack C. Kiefer, William Y. Davis, Benedykt Dziegielewski, and John Olaf Nelson. 1999. *Residential End Uses of Water Study.* Denver, CO: AWWA Research Foundation and American Water Works Association.

Mayer, Peter W., Erin Towler, William B. DeOreo, Erin Caldwell, Tom Miller, Edward R. Osann, Elizabeth Brown, Peter J. Bickel, and Steven B. Fisher. 2004. *National Multiple Family Submetering and Allocation Billing Program Study.* Boulder, CO: Aquacraft Inc. and East Bay Municipal Utility District. www.allianceforwaterefficiency.org/WorkArea/DownloadAsset.aspx?id=704.

McMahon, Thomas. A. 1993. "Hydrologic Design for Water Use." In *Handbook of Hydrology*, edited by David R. Maidment. New York: McGraw-Hill Professional.

Nakicenovic, Nebojsa, and Robert Swart, eds. 2000. *Emissions Scenarios: A Special Report of Working Group III of the Intergovernmental Panel on Climate Change.* Cambridge: Cambridge University Press. http://www.ipcc .ch/ipccreports/sres/emission/.

Olmstead, Sheila M., and Robert N. Stavins. 2009. "Comparing Price and Nonprice Approaches to Urban Water Conservation." *Water Resources Research* 45(4):W04301.

SDCAA. 2010. *San Diego Water Submetering Ordinance.* San Diego County Apartment Association. www.sdcaa.com/legislationandpublicaffairs/local-issue/229-san-diego-water-submetering.

Veeh, Matthew. 2010. "Long Beach Sets Another Water Conservation Record." Press release. City of Long Beach, CA. www.longbeach.gov/news/displaynews .asp?NewsID=4561&TargetID=55.

Western, Samuel. 2013. "A Demographer Predicts Big Changes for the West's Housing Landscape." *High Country News*, November 29. www.hcn.org /issues/45.20/a-demographer-predicts-big-changes-for-the-wests-housing-landscape.

CHAPTER 6

Coping with Delta Floods and Protecting California's Water Supply in a Regional Flood Management System

HOWARD FOSTER AND JOHN RADKE

THE DELTA, A NEXUS OF CALIFORNIA INFRASTRUCTURE

The Delta is the critical link conveying water from north to south for California's two major water distribution systems—the federal Central Valley Project (CVP), started in the 1940s and primarily serving central and southern valley agriculture, and the State Water Project (SWP), started in the late 1950s and serving urban water users in both the Bay Area and Southern California. For both these systems northern water, mostly carried by the Sacramento River, is discharged into the north Delta and pumped out from the southern Delta (table 6.1). Protecting the waterway link is a levee system, widely acknowledged to be fragile—it has been characterized by some levee engineers as "piles of dirt" (Bea et al. 2009, Breitler 2009)—and threatened by both earthquake-induced liquefaction failure and flood failures from large, "atmospheric river" storm events. Widespread failure of the levee system would result in a "Big Gulp," as the flooding of deeply subsided Delta lands alters river flows and allows an influx of salt water from San Francisco Bay. Saltwater introduction into the central Delta would force the shutdown of the CVP and SWP and have dramatic impacts on urban and agricul-

This work was supported in part by the National Science Foundation (EFR/RESIN grant no. EFR-0836047).

TABLE 6.1   PRINCIPAL TRANS-DELTA INFRASTRUCTURE SYSTEMS

Component	Properties
Central Valley Project (Central and Southern California deliveries)	Water delivered to the San Joaquin Valley and Southern California is released into the Delta at the Sacramento River where it flows southward, via the Delta channel network, and is pumped out at the C. W. Bill Jones Pumping Plant (formerly the Tracy Pumping Plant).
State Water Project (Central and Southern California deliveries)	Shares the Delta as the transport connection of water from the Sacramento River to the south. Water is pumped out of the Delta from the Clifton Court Forebay at the Harvey O. Banks Pumping Plant. About 70 percent of SWP water is used for urban areas and industry.
State Water Project: South Bay Aqueduct	Delta water provided by the SWP system, pumped out at Bethany Reservoir, serves eastern Alameda County and Santa Clara Valley.
Central Valley Project: Contra Costa Canal	Serves northern Contra Costa County; draws Delta water from south Hotchkiss Tract at Rock Slough.
State Water Project: North Bay Aqueduct	Serves Solano and Napa County water districts; draws Delta water from west of Hastings Tract at Baker Slough.
Mokelumne Aqueduct	Principal water supply for the East Bay.
PG&E electrical transmission	500 kV, 230 kV, and 115 kV transmission lines at Sherman Island and elsewhere.
Western Area Power Administration electrical transmission	500 kV lines, eastern and western edges of the Delta.
PG&E gas transmission	Connection of gas storage on Sherman Island with links throughout the state.
Kinder Morgan SFPP, IPP	Petroleum product pipelines.
BNSF Railroad	Bisects the Delta through Jones Tract.
Sacramento Deep Water Ship Channel	Serves the Port of Sacramento
Stockton Deep Water Ship Channel	Serves the Port of Stockton.
California Highways 4, 12, 160	Major thoroughfares.
AT&T and others, telecom/fiber optic	Proprietary, not mapped.

SOURCES: URS Inc. and Jack R Benjamin & Associates (2007), Tompos (2007).

tural water supplies. Other urban-serving infrastructures cross the Delta and are also vulnerable to levee failure. In 2005 the California Department of Water Resources predicted and analyzed the consequences of a magnitude-6.5 near-Delta earthquake (Jack R Benjamin & Associates 2005): thirty to fifty levee breaches on twenty one islands; major power and gas transmission lines damaged; inundation of State Highways 4, 12, and 160; and interruption of water deliveries in the two water projects for over a year, with 16 months to full recovery. Irrespective of earthquakes and climate change, everyday events also threaten levees. In 2004 the Jones Tract levee failed in June, in mild weather—absent typical storm-induced flood threats—due to levee seepage, perhaps from burrowing animals. The river flowing through the break immediately eroded the soft levee soils into a breach 300 feet wide and 55 feet deep (California Department of Water Resources 2009). Because it was summer, river and flood elevations were relatively low, and all available resources could be brought in to close the breach (after complete island inundation and isostatic balance) and to pump the island dry over a five-month period. Had it been winter, with higher water levels, more difficult working conditions, and competition for human and material resources, the cross–Jones Tract infrastructure could have been exposed to flood waters for a much longer period, and the BNSF rail lines, Mokelumne Aqueduct, and State Highway 4 could have been damaged, with dramatic consequences for urban California (Breitler 2009).

California's Sacramento–San Joaquin Delta is a challenging environment for infrastructure protection. It is an inland river delta, only one of two in the world; essentially, it is a drowned valley flooded by rising sea levels since the last ice age. It lies behind the Coast Range, in the center of the Sacramento-San Joaquin Valley, a depression into which fresh water from the Sacramento, Mokelumne, Consumnes, and San Joaquin Rivers and smaller tributaries mix with salt water flowing in from the Pacific Ocean through the Golden Gate and Carquinez Straits. European settlers found the Delta a constantly changing watery world of meandering rivers and moving islands, many covered in thickets of riparian vegetation. Navigation, however, was possible, with ships able to call at the ports of Sacramento and Stockton. The first attempts at stabilizing and reclaiming islands for agriculture began in the late 1860s, with Chinese labor constructing the first levees by hand. As the silty-organic soils are very productive and with easy access to fresh water, agriculture was and has remained a valuable Delta land use. In many places, the levees have been maintained with clam-shell dredges and drag

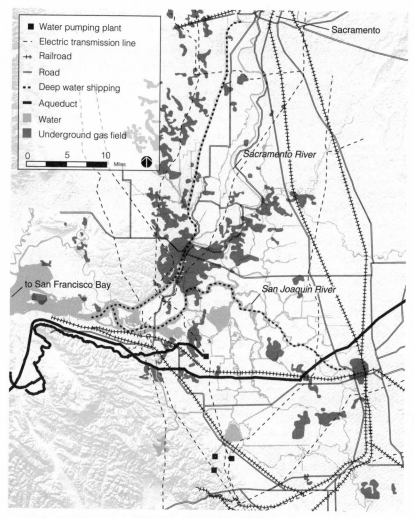

FIGURE 6.1. Infrastructure in the Delta, as portrayed by the California Department of Water Resources Delta Atlas, with details of the water management system (not shown: gas transmission and petroleum products pipelines). Redrawn from DWR (1995).

lines which scoop soil from the river bottom and deposit it on the levee top, and it is not compacted or otherwise engineered to fulfill contemporary levee standards. Because island stabilization has essentially locked this dynamic landscape into the present arrangement of rivers and fifty-seven reclaimed islands, the islands are no longer replenished with new soil from meandering rivers as in the past. Consequently, island soils are oxidizing and subsiding, with island elevations now well below sea level—greater than 16 feet below in some places (Mount and Twiss 2004). In addition, nineteenth-century placer mining throughout the Delta's watershed poured rock and soil debris into the Delta's rivers, raising river beds, reducing freeboard, and necessitating even greater levee heights. There have been approximately 160 levee failures since 1900 (Bates and Lund 2013).

PLANNING FOR POLICY FAILURE

In 2007 the Public Policy Institute of California estimated a $4 billion price tag for the Fortress Delta alternative plan to bring some Delta levees—only those required to protect freshwater transport—up to contemporary engineering standards (Lund et al. 2007, 2010). Unfortunately, such a proposal would compete with other proposed Delta construction projects, such as management of saltwater with seawater barriers, aqueduct construction, bypass schemes (e.g. the newly proposed Delta Tunnels), and environmental restoration projects. Michael Hanemann and Caitlin Dyckman, in their 2009 paper, "The San Francisco Bay-Delta: a Failure of Decision-Making Capacity," offer a compelling theory as to why no big Delta projects, with big price tags, are likely anytime soon. They describe the history of failure of Bay-Delta decision-making as due to an essentially perfect zero-sum game among participants with alliances unlikely from the "fundamental opposition of interests among stakeholders"—environmentalists, Delta farmers, San Joaquin Valley farmers, urban Northern California water users, and urban Southern California water users. For the past 60 years, no alliance could be made among any of these participants that could break the decision-making stalemate; thus, no Peripheral Canal, and thus the failure of the latest CALFED Bay-Delta Program management and restoration initiative. By Hanemann and Dyckman's argument, no big, expensive Delta-fix project is likely soon, because these interests have not significantly changed. For the foreseeable future, we must manage Delta flood risk for the system as it exists today.

FORESEEING ALTERED INFRASTRUCTURE

In 2011, University of California, Berkeley, professor John Radke and his colleagues performed a number of flood simulation studies for the Delta and San Francisco Bay Area to assess the alterations in travel time

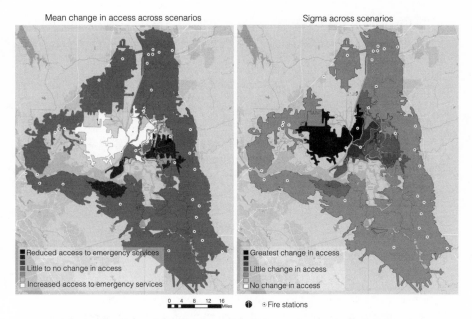

FIGURE 6.2. Flooding and the alteration of the Delta road network changes the availability of emergency services. Some areas are cut off from services, while other areas gain increased access. Source: Radke and Beach (2013).

and emergency service access that would be caused by island flooding and the piecemeal loss of the road transportation system (Radke and Beach 2013). The Delta simulations produced some surprising results. In some areas, as expected, access to emergency services was lost as flooding cut off low-lying roads and severed road access to the outside world. But in other areas, access to emergency services actually improved as emergency facilities were stranded by road closures and nearby residents enjoyed reduced competition for help from nearby fire stations, EMT, police, and other emergency management personnel (figure 6.2). A takeaway lesson from these studies is that the flooding of infrastructure—be it roads, communications, electrical, natural gas, or other networks—can reconfigure these systems, sometimes in surprising ways, and change the geography of service availability. More important, as these systems are reconfigured by disasters, new links (perhaps previously unidentified) become critical for service availability. For flood planning, such as material stockpiling, studies like Radke's can be performed as a number of Monte Carlo–like simulations to identify critical links and overall patterns of infrastructure loss. In the future, as models improve, these studies could be performed in real time and communicated to flood managers for deployment of flood fight resources.

FLOOD FIGHTING IN REAL TIME

When the storm has arrived and wind-driven waves threaten to overtop the levee, the last line of defense in the Delta is most likely the island's reclamation district personnel. Sandbags piled on top of the levee, along with plastic sheeting and seep-containment structures, are among the measures that these crews can take to protect the levee. Are these measures too little and too late? We find that a number of levee engineers dismiss flood fighting as a distraction from the more important task of rebuilding the levees to contemporary standards; but in studies of Delta levee failure probabilities performed by our colleagues, preliminary indications are that the actual levee failure rate is lower than what is suggested by an analysis of levee physical conditions—levee construction materials, failure mechanisms, and flood levels—and that difference suggests a positive effect of levee maintenance and flood fighting (Bea 2013). But there is no room for complacency. Significant planning effort is needed for effective flood fighting, including the evaluation of levee weaknesses and likely failure points, stockpiling of construction materials, and efficient communication among flood fighters, flood managers, and infrastructure managers.

PLANNING FOR LEVEE FAILURE

When a levee fails, is all lost? For the Jones Tract, that was essentially the case. There was an active effort to plug the breach, but it was ineffective: deposited plug materials were quickly swept away. It was only when the island was completely inundated that successful action could be taken. In this case the effort was not disaster prevention but disaster recovery. However, on other islands with significant topographic variation and other landform characteristics, a secondary flood fight effort may be possible.

To that end, the head of the San Joaquin County Department of Emergency Services, Ron Baldwin, decried the "Maginot Line mentality" of flood fighting and instead advocated a defense-in-depth strategy, with a planning effort resulting in what he called "flood contingency maps" (Baldwin et al. 2011)—maps delineating a detailed backup plan in the case of levee failure (figure 6.3). He and his colleagues started

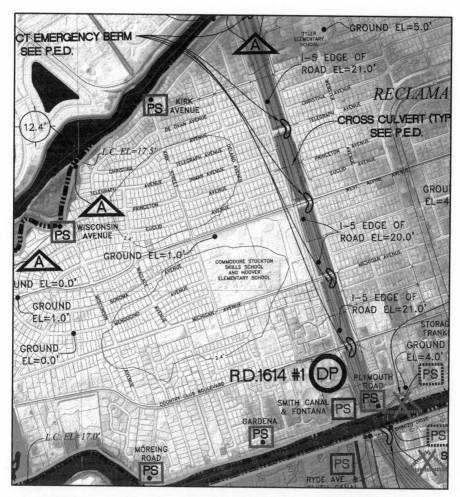

FIGURE 6.3. Portion of a flood contingency plan. These maps combine data from many sources: elevation differences (gray shaded backdrop); the location of emergency flood barrier berms and associated preliminary engineering studies (kidney-shaped graphics and PED labels); levee access points (triangle A); supply delivery points (circle DP); pump stations (square PS); and other facilities and elements critical to the flood fight. Sources: Baldwin et al. (2009), San Joaquin County Office of Emergency Services (2010).

with reclamation district staff and their engineering consultants, city and county emergency managers, and local residents to come up with a flood fighting plan that essentially identifies a series of fallback positions where the flood fight can continue after the levee is breached. The maps contain a number of location-specific flood-fighting actions—

some quite novel, such as damming freeway underpasses to make an interior flood barrier. Other prescriptions include the construction of temporary flood-control berms, levee relief cuts to return floodwater to the river, emergency pumps to dewater specified areas, and temporary ring dikes around critical infrastructure such as waste treatment plants. Many of the proposed actions are further developed by detailed preliminary engineering designs that specify, for example, the amount and type of material to be stockpiled for use in temporary dams. Even if a particular flood fight doesn't go exactly as anticipated, the flood contingency planning process has a significant benefit in creating a community of trust among the flood fight participants. By knowing whom to call on when disaster happens, these interpersonal relationships can greatly simplify and facilitate communication during an oftentimes chaotic real flood event.

PIECING IT TOGETHER: THE NEED FOR A NETWORK FOR SHARED SITUATION AWARENESS

An integrated approach to flood management would combine many activities: levee vulnerability assessments and monitoring by both people and instruments (smart levees), the marshaling of flood-fighting resources, flood fighting at the levee crest, the management of flooding after the levee breaks, real-time assessment of altered infrastructure networks, and updated identification of infrastructure-critical links. Along with the many flood management activities there are many actors. In California, as is typical elsewhere, we have FEMA, the governor's office, the California Office of Emergency Services, the Department of Water Resources, county offices of emergency services, and city agencies (e.g. fire and police); plus individual islands' reclamation districts and associated civil engineering consultants; plus, in the Delta, suppliers of critical materials such as the Dutra Group. In addition, there are the affected infrastructure operators (most of these privately owned), such as PG&E, other pipeline operators, public power agencies, water-supplying agencies, and telecom, rail, port, and other enterprises. For now and the foreseeable future, these are semiautonomous independent flood-fight participants, each of which represents a potentially competing demand upon flood response resources.

A necessary element in the coordination of all the flood actors is a technology to communicate the current status and conditions of the flood fight. This includes weather reports, likely levee breach points, current status of flooding, likely impacted infrastructures, status of flood-fighting resources (both material and human), and many other conditions which affect the flood fight. Equally important is that the communication not be one-way: the participants need to communicate their own response plans to all the other participants as the flood event unfolds. Geographic information system (GIS) mapping technology is evolving to fill this role (Falbo 2006), and information communicated on maps is part of a *technology delivery system*—defined by Bea (2011) and others as the combination of technologies and human management procedures necessary to manage a complex task.

For the management of Delta flooding, GIS technology is a technology delivery system element for the communication of spatial information pertaining to the flood-fighting actions—risk assessments, monitoring, marshaling, management, and updated progress and risk maps. GIS vendors, principally Esri, have developed a "common operating platform" (Esri Inc. 2008; National Alliance for Public Safety GIS Foundation 2012; Foster and Radke 2012) to portray a "common operating picture." Common operating pictures are typically composed of multiple data feeds, status maps, and other spatial information created by cooperating agencies to be viewed simultaneously on a computer desktop, either as overlays in a single map window or as multiple synchronized map windows. Typically, GIS-supported common operating platforms are Web applications with potentially an unlimited number of viewers. There are no hard technical problems standing in the way of such a system for flood management; there are a number of off-the-shelf technologies for implementing common operating platforms. The biggest impediments are social: there are significant security, coordination, and procedural hurdles that impede a shared GIS for Delta flood management.

THE PROBLEM WITH DATA

One difficulty is with data. A flood management mapping system requires access to many different datasets, including those describing the natural environment (high-resolution land surface elevation, land cover, river and stream data, current and predicted river stages and water surface elevations) and those describing the physical infrastructure (levees, roads, bridges, pipelines, power lines, emergency service locations, population and address information). These data already exist, but they have been created by and are maintained by different agencies, so one problem is

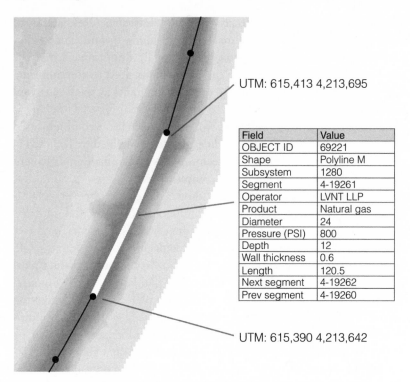

UTM: 615,413 4,213,695

Field	Value
OBJECT ID	69221
Shape	Polyline M
Subsystem	1280
Segment	4-19261
Operator	LVNT LLP
Product	Natural gas
Diameter	24
Pressure (PSI)	800
Depth	12
Wall thickness	0.6
Length	120.5
Next segment	4-19262
Prev segment	4-19260

UTM: 615,390 4,213,642

FIGURE 6.4. Hypothetical GIS data example: a line segment from a gas transmission infrastructure GIS dataset. Raw geographic data encodes exact spatial locations and can include sensitive attribute information.

providing access for flood managers. For example, the Department of Homeland Security has produced the Homeland Security Infrastructure Project (HSIP) dataset, a collection of 500 layers (HSIP level Gold) of detailed street location, addresses, building, infrastructure, and other data, developed by private vendors (Homeland Infrastructure Foundation Level Data Working Group 2011). Unfortunately, the contractors' licensing restrictions prevent these data from being widely distributed prior to disaster events, so they are unavailable for preparedness planning. When these data do show up in real events, data overload can interfere more than help with real-time disaster management. A higher barrier to data sharing exists when data are held by private infrastructure operators because sharing data raises security and market competition concerns. This is because geographic data are spatially exact and can contain overly revealing attribute information (figure 6.4). There are workarounds for these problems (Lawrence Livermore National Laboratory 2012), but these require the maintenance of additional datasets: one version for internal use and other versions for sharing that have been fuzzed up and stripped of potentially sensitive attribute information. Finally, there is the problem of interpretation. For flood management, infrastructure data must be made meaningful for planning and flood-fight coordination; specifically, facilities must be rated in terms of susceptibility to flooding and the consequences of such flooding must be understood. But who is to do these assessments? Typical flood planning GIS staffs do not have this expertise; it only resides with the engineering staffs of the infrastructure operators themselves. A way must be found by which infrastructure risk information can be shared without raising security concerns or giving away proprietary information.

RISK AS THE CONCEPTUAL GLUE TO BIND THE PARTICIPANTS TOGETHER

A risk assessment is a question-asking, analysis-forcing activity that can guide each participant's flood planning. There are many tools for risk analysis; a compendium is beyond the scope of this article. Conceptually, risk is the product of a probability of failure and the consequences of that failure, resulting in an expected loss in dollars or lives. But to be useful, risk need not be precisely quantified. It can be ranked (higher to lower), described in terms of combinations of probability and consequences (e.g., high probability + serious consequence = high risk), or described categorically. Useful risk categories can be distinctive to individual enterprises; for example, probabilities can described as "credible events" and outcomes can be described as "unacceptable time for service outage" or, more seriously, "precluded events"—those which all resources must be dedicated to avoid, such as the loss of electrical power at a nuclear reactor facility. In planning, risk is a commensurating function, by which disparate activities, costs, management procedures, and so on can be equated and combined, either quantitatively or categorically. In risk management, risk can be considered as managed *independently*, that is under the complete control of an enterprise (for example flood-proofing), *dependently*, when the management of risk depends partly upon services provided by others such as telecommunications or the Internet, or *interactively*, in which risk management (in real time) is part of a system whereby the actions of every participant affect all the others, for example the shared dependence upon a common pool of material stockpiles (Roe and Schulman 2010). In flood management, dependent and interactive risk management require a shared, interactive GIS.

When it comes to critical infrastructure, such as levees, roads, and power and gas transmission facilities, a risk approach offers the obvious directive for flood management: put resources toward defending those facilities whose damage would cause the most societal harm. There is a significant opportunity to develop risk assessments, and the technology exists to get around some of the obvious security problems with respect to sharing risk information. To perform risk assessments, we have the engineering staffs of many of the infrastructure providers, such as the Department of Water Resources, PG&E, Kinder Morgan, and BNSF Railways, who can assess their facilities' risk from flood events. Also, they typically have their own GIS staffs, who have the expertise to develop and post maps of risk information when time and circumstances demand. GIS technology has advanced to where it no longer requires access to raw data for map-making and sharing: risk assessments can be created by each participant in real time and shared as map pictures. In essence, the participants share geo-rectified/geo-located images—not data—and these identify only the facilities which must be protected in a particular flood. Contemporary enterprise GISs have the server tools to publish map images on the fly, and Web servers have graphic tools (transparency control, color-attribute filtering) to combine map images from many sources into single composite maps or linked multi-map displays.

PUTTING IT TOGETHER: SHARING FLOOD DATA, INFORMATION, AND RISK ASSESSMENTS—DELTA-WIDE MAPPING SYSTEM ALTERNATIVES

The Star

A centralized GIS for the management of flood information, with satellite consuming agencies, is one model for sharing and distributing flood-management information (figure 6.5). For the state of California, the examples we have seen are centralized systems with managers and staff to provide GIS services to a second tier of map product consumers. Such systems are often developed by a contractor who collects data from different providers, processes it to provide interpretive graphics, and develops a product-distribution architecture, and then either turns it over to a state agency or runs it as a service for that agency. The products of

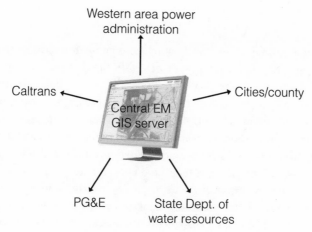

FIGURE 6.5. Star arrangement: centralized GIS server provides data and images to emergency management participants.

such a system can be in the form of finished maps, for example, as a digital atlas of traditional maps (a "map workbook"), or a data repository where data are provided for clients to download into their own GIS systems (Michael Baker Corporation 2012).

However, there are a number of problems with these centralized GISs: management of proprietary data, rigid design, and difficulties in funding long-term maintenance and upgrades. In the examples we have seen, the same sharing impediments dog these systems with respect to sensitive and proprietary data, resulting in the exclusion of important data. And there are other problems. Design parameters must be finalized before much developmental work can be undertaken, thus excluding knowledge gained from actual system use. Once specifications are written, the capacity for significant design change becomes constrained. Data timeliness is another issue. Data are downloaded from providers, but as providers update their data, the system managers must maintain data concurrency with additional maintenance to provide for the notification of changes from the provider, new downloads, and data verification. Centralized GIS systems must also be updated to adapt to new hardware and new client technologies. Finally, there is the problem of system financing. Because the costs of the entire system typically fall on one agency, a single department must commit to and budget for long-term system maintenance. Our experience with California state government agencies is that these issues are significant roadblocks to successful implementation of any new GIS systems.

The Ring

The conventional centralized, star arrangement overlooks one key opportunity for the development of a GIS flood management support system: virtually all the potential participants in a flood management system already have their own GIS systems and staffs. A ring, or federated, system attempts to utilize resources already in place to produce and share information relevant to the planning and management of flood-fighting resources (figure 6.6). There are potential advantages to such a system. One is that a federated system better accommodates the sharing of risk information. With their engineering and GIS staffs, the individual participants maintain security over their own data and can provide generalized risk assessments for planning purposes, or event-specific risk assessments for a particular flood fight in real time. Another advantage is that the data-concurrency problem goes away. In cases where raw data can be shared, for example via a local server, data that stay with their contributor are maintained by that agency, and the information that is shared is always the current version. Another advantage is that many design issues become more apparent early in the design process because the data providers are also the data consumers—that is, the participants already have an idea of what they want to learn, related to flood management, from other participants, and technology expectations are keyed realistically to the capabilities of the system participants. Another advantage is that a ring system can start small and be built incrementally: just two data providers need to agree to share (some) information, under (some) agreed-upon circumstances. This arrangement scales well to accommodate more providers and more functionality, always with the advantage of starting from an already functioning system. The approach avoids what has been called in the computer industry *software creationism*—the notion that complex systems can be successfully designed top-down, with all necessary knowledge already in hand at the start of the project, and without the need of knowledge gained from implementation (Raymond 1996). Finally, there is the issue of cost. The total cost does not have to be borne by one state agency; a distributed system distributes costs among the various engineering and GIS staffs of the participants.

However, not all is rosy with federated systems. One problem dominates all the other potential problems in a ring federation. The challenge is not technology but people and trust among participants. A great deal of face time is needed for liaison work by each participant to determine

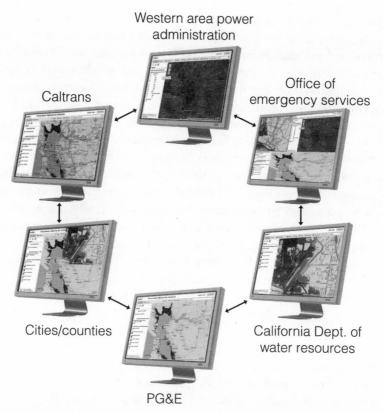

Western area power
administration

Caltrans

Office of
emergency services

Cities/counties

California Dept. of
water resources

PG&E

FIGURE 6.6. Ring of shared geographic data and images among flood
emergency management participants. Note that some of the displays
incorporate information from the other participants.

communication protocols such as when and how data are to be interpreted and displayed, with what graphics, shared with whom, and via
what distribution technology and safeguards. Distributed budgeting, an
apparent advantage, also has its own problems. Motivation will vary
among the participants, especially those with weak risk-assessment procedures, leading to uneven commitment of money and people. Counting
buy-in from potential participants, the cost of prototyping a federated
system might approach the cost of a single centralized system.

CONCLUSION

With new technologies to share geographic information, we now have
the basis for a shared virtual control room that keeps private data

private, yet provides for the sharing of important risk information for all participants to see, leading to enhanced situation awareness and improved flood management. Such a system has the potential to integrate much of the diverse, nontraditional and newly emerging geographic information for flood planning and flood fighting: dynamic real-time road and emergency service availability maps, for example those created by Radke and his associates (Radke and Beach 2013); island-specific flood-fighting plans such as those developed by Baldwin and the Delta reclamation districts (Baldwin et.al. 2012); real-time predicted inundation maps (Biging, Radke, and Lee 2012); dynamic infrastructure risk maps; and flood-fighting resource availability maps—all of which are under development by many individuals and agencies. FEMA and the Department of Water Resources are currently developing regional flood management plans, and the reclamation districts and cities and counties are developing flood safety plans. The technology already exists to integrate these and other planning efforts via GIS, and there are active projects underway to do so (U.S. Department of Homeland Security 2014). It remains a matter of applying human resources, which exist now. Considering all the flood-mitigation and levee-construction projects currently under consideration, the cost of a coordinated GIS risk and real-time management system is orders of magnitude lower, and offers perhaps the best promise in the near term of coping with critical urban infrastructure systems on top of sagging levees which are exposed to earthquakes, sea-level rise, and increasingly intense future storms.

REFERENCES

Baldwin, R., H. Foster, E. Roe, and L. Blake. 2012. *Guide to Flood Contingency Mapping.* 2nd ed. Federal Emergency Management Agency, Region IX. www.sjgov.org/oes/getplan/RDAB156/Flood_Map_Guide_Publication_Final_9-12.pdf.

Bates, Mathew, and Jay Lund. 2013. "Delta Subsidence Reversal, Levee Failure and Aquatic Habitat: A Cautionary Tale." *San Francisco Estuary and Watershed Science* 11(1):1–20.

Bea, Robert. 2011. *Technology Delivery Systems—TDS: You Can't Manage What You Can't Model.* White paper, UC Berkeley Center for Catastrophic Risk Management. http://ccrm.berkeley.edu/resin/pdfs_and_other_docs/TDS2.pdf.

Bea, Robert. 2013. *Final Report to National Science Foundation, Grant #EFR-0836047.* Resilient and Sustainable Infrastructure Networks (RESIN) project, University of California, Berkeley.

Bea, Robert, Ian Mitroff, Daniel Farber, Howard Foster, and Karlene Roberts. 2009. "A New Approach to Risk: The Implications of E3." *Risk Management* 11:30–43.

Biging, Greg, John Radke, and Jun Hak Lee. 2012. *Impacts of Sea-Level Rise and Extreme Storm Events on the Transportation Infrastructure in the San Francisco Bay Region*. CEC-500-2012-040. White paper, Climate Change Center, California Energy Commission. www.energy.ca.gov/2012publications /CEC-500–2012–040/CEC-500–2012–040.pdf.

Breitler, Alex. 2009. "Delta Debate Rages Five Years Later: Some Say Jones Tract Disaster Played Key Role." *The Record* [Stockton, CA], June 3.

California Department of Water Resources. 1995. *Sacramento-San Joaquin Delta Atlas*. Sacramento.

California Department of Water Resources. 2009. *Jones Tract Flood Water Quality Investigations*. Sacramento: Municipal Water Quality Investigations Program, Division of Environmental Services, California Department of Water Resources.

Esri Inc. 2008. *Geographic Information Systems: Providing the Platform for Comprehensive Emergency Management*. White paper. www.esri.com /library/whitepapers/pdfs/gis-platform-emergency-management.pdf.

Falbo, Dan. 2006. *A Geographic Information System's Role in Homeland Security: Supporting Strategic and Tactical Initiatives in Emergency Management and Public Safety*. Minneapolis, MN: Esri Inc. www.gis.state.mn.us/pdf /GCGI_EP12506.pdf.

Foster Howard, and John D. Radke 2012. *Elements and Applications of a GIS Technology Delivery System for the Sacramento-San Joaquin Delta*. Report to National Science Foundation, grant #EFR-0836047. Resilient and Sustainable Infrastructure Networks (RESIN) project, University of California, Berkeley.

Hanemann, Michael, and Caitlin Dyckman. 2009. "The San Francisco Bay-Delta: A Failure of Decision-Making Capacity." *Environmental Science and Policy* 12(6):710–25.

Homeland Infrastructure Foundation Level Data Working Group. 2011. *Presentation Agenda: HIFLD Overview, HIFLD to the Regions Overview, HSIP O verview*. ESRI HLS Summit Conference, July 10. http://proceedings.esri.com /library/userconf/hss11/papers/hifld_esri_hls_summit_10_jul201_final.pdf.

Jack R Benjamin & Associates. 2005. *Preliminary Seismic Risk Analysis Associated with Levee Failures in the Sacramento-San Joaquin Delta, Prepared for California Bay-Delta Authority and California Department of Water Resources*. Menlo Park, CA.

Lawrence Livermore National Laboratory. 2012. "New Tool Developed by Lawrence Livermore Team Aids Agencies in Preventing, Mitigating Terrorist Attacks." News release. www.llnl.gov/news/newsreleases/2002/NR-02–12–08 .html.

Lund, Jay, Ellen Hanak, William Fleenor, William Bennett, Richard Howitt, Jeffrey Mount, and Peter Moyle. 2010. *Comparing Futures for the Sacramento–San Joaquin Delta*. Berkeley: University of California Press and Public Policy Institute of California.

Lund, Jay, Ellen Hanak, William Fleenor, Richard Howitt, Jeffrey Mount, and Peter Moyle. 2007. *Envisioning Futures for the Sacramento-San Joaquin Delta.* San Francisco: Public Policy Institute of California.

Michael Baker Corporation. 2012. *California Delta Emergency Management Plan.* www.deltafloodemergencyplan.org/DEMP/Delta_Emergency_Management_Plan.html.

Mount, Jeffrey, and Robert Twiss. 2004. "Subsidence, Sea Level Risk, and Seismicity in the Sacramento-San Joaquin Delta." *San Francisco Estuary and Watershed Science*, 3(1).

National Alliance for Public Safety GIS Foundation. 2012. *GIS for Multi-Agency Coordination Centers (MACCs) including the Emergency Operations Centers (EOCs).* https://docs.google.com/a/coemergency.com/spreadsheet/viewform?hl=en_US&formkey=dG94TmM4Zl85bEotc1Z5eVd3Vy1FVmc6MQ#gid=0.

Radke, John D., and Tessa Beach, 2013. *Vulnerable by Design: Consequences of Infrastructure Failure in the Sacramento-San Joaquin Delta.* Conference of the Council of Educators in Landscape Architecture, University of Texas at Austin, March 27–30.

Raymond, Eric S., ed. 1996. *The New Hacker's Dictionary.* 3rd ed. Cambridge, MA: MIT Press. [Another edition of the dictionary is at www.outpost9.com/reference/jargon/jargon_toc.html.]

Roe, Emery, and Paul R. Schulman. 2010. *Assessing and Managing Failure Vulnerabilities of Interdependent Complex Infrastructure Systems: Resilience & Sustainability in the Interdependent, Interconnected and Interactive Critical Infrastructure Systems (I3CISs) of California's Sacramento Delta.* Unpublished white paper, Center for Catastrophic Risk Management, University of California, Berkeley.

San Joaquin County Office of Emergency Services. 2013. *Flood Contingency Map Index.* www.sjmap.org/oesfcm/FloodContingencyMapIndex.pdf.

Tompos, Eszter. 2007. *Where Does Our Water Come From?* [Map]. PG&E. www.pge.com/includes/docs/pdfs/shared/edusafety/training/pec/water/mapwaterdistricts_48x52_0307.pdf.

URS Inc. and Jack R Benjamin & Associates. 2007. *Technical Memorandum: Delta Risk Management Strategy (DRMS) Phase 1, Topical Area: Impact to Infrastructure, Final.* Sacramento: California Department of Water Resources.

U.S. Department of Homeland Security. 2014. *Unified Incident Command and Decision Support (UICDS), a Department of Homeland Security Initiative for Information Sharing Among Commercial, Government, Academic, and Volunteer Technology Providers to Support the National Incident Management System.* www.uicds.us/files/UICDS%20in%20Brief%20Gov.pdf.

CHAPTER 7

Portfolio Approaches to Reduce Costs and Improve Reliability of Water Supplies

ELLEN HANAK AND JAY LUND

Growing demands on California's finite and variable water supplies make scarcity a permanent consideration in water management. Managers will always be preparing for shortages, even in very wet years. This challenge will increase as California's population and economy continue to grow and as the climate changes; and it will become more severe and costly if water is not managed well. This chapter reviews institutions and options available to manage water scarcity. It begins with a brief discussion of the idea of *portfolio-based planning*, a useful way to think about how to combine water management actions for greater effect. It then examines California's use of the diverse set of tools available. The discussion illustrates how management will need to adapt to changing conditions in the natural and physical environment. In particular, new modeling results show how a dry form of climate change and a loss of water exports from the Sacramento–San Joaquin Delta (now a major hub for moving water from the state's wetter northern and eastern regions to demand centers to the west and south) may affect California's economy, and how aggressive increases in urban water conservation might help offset some of these costs. Institutional and legal barriers pose particular challenges for adopting promising actions.

This chapter draws on joint work with our colleagues Ariel Dinar, Brian Gray, Richard Howitt, Jeffrey Mount, Peter Moyle, and Barton "Buzz" Thompson for the book *Managing California's Water: From Conflict to Reconciliation* (Hanak et al. 2011).

ORCHESTRATING ACTIVITIES THROUGH
PORTFOLIO-BASED PLANNING

Most people are familiar with the use of portfolios in financial management to balance risks and returns through diversification. This concept also has become well accepted in many areas of infrastructure planning and operations, ranging from water to energy (Awerbuch 1993) to transportation (Johnston, Lund, and Craig 1995). The general notion is to employ a complementary mix of options—including supply-side, demand-side, and operational tools—to provide more cost-effective service that is reliable under a wide variety of conditions and able to serve multiple purposes.

Complementarities between some options can reduce costs and increase system reliability. For example, an inexpensive water conservation option may help avoid expensive expansions in supplies (sometimes called an "avoided cost"). But extreme levels of water conservation can be more costly than judicious use of other water management activities. Similarly, coordinated (or "conjunctive") use of surface and groundwater storage allows surface water purchased cheaply in wet years to be stored underground and retrieved for use in drier years, when surface water is more costly. In these cases, neither option would work as well alone. As with a financial portfolio, it is common for some components to do well when others do poorly. For instance, surface water storage does poorly during long droughts, whereas groundwater is more resilient to droughts. Recycled wastewater and desalinated seawater are relatively expensive options, but they are available even under extreme drought conditions.

Reliance on a variety of management techniques makes systems more stable when faced with such operational disturbances as droughts, adverse legal rulings, and mechanical breakdowns. It also makes them more resilient to longer-term planning and policy uncertainties from changing climatic, population, economic, and regulatory conditions.

Table 7.1 lists many of the options available to water managers seeking to balance supplies and demands. Options for expanding usable supplies include traditional methods, such as surface storage, conveyance, and water treatment, as well as more contemporary methods, such as improvements in operational efficiencies, conjunctive use of ground and surface waters, stormwater capture, and wastewater reuse. Water demand management options include improvements in water-use efficiency (e.g. low-flow plumbing fixtures and irrigation techniques to get "more crop per drop"), as well as reductions in water use below

TABLE 7.1 WATER SUPPLY SYSTEM PORTFOLIO OPTIONS

Demand and allocation options
- Urban water use efficiency (water conservation)*
- Urban water shortages (permanent or temporary water use below desired quantities)*
- Agricultural water-use efficiency*
- Agricultural water shortages*
- Ecosystem demand management (dedicated flow and nonflow options)
- Ecosystem water-use effectiveness (e.g. flows at specific times or with certain temperatures)
- Environmental water shortages
- Recreation water-use efficiency
- Recreation improvements
- Recreation water shortages

Supply management options
- Expanding supplies through operations (affecting water quantity or quality)
- Surface water storage reoperation (reduced losses and spills)*
- Conveyance facility reoperation*
- Cooperative operation of surface facilities*
- Conjunctive use of surface and groundwater*
- Groundwater storage, recharge, and pumping facilities*
- Blending of water qualities
- Changes in treatment plant operations
- Agricultural drainage management
- Expanding supplies through expanding infrastructure (affecting water quantity or quality)
- Expanded conveyance and storage facilities*
- Urban water reuse (treated)*
- New water treatment (surface water, groundwater, seawater, brackish water, contaminated water)*
- Urban runoff/stormwater collection and reuse (in some areas)
- Desalination (brackish and seawater)*
- Source protection

General policy tools
- Pricing*
- Subsidies, taxes
- Regulations (water management, water quality, contract authority, rationing, etc.)
- Water markets, transfers, and exchanges (within or between regions/sectors)*
- Insurance against drought
- Public education

*Options represented in the CALVIN model.

desired levels (called here "shortages"). Various general tools (pricing, water markets, taxes and subsidies, water markets, and public education) can motivate water users and water agencies to implement both supply- and demand-side options.

Each option provides different benefits, and each entails costs (table 7.2). The financial costs of most options vary considerably with location and water availability conditions. For instance, local water transfers in Northern California agricultural areas can make some water available for $50 per acre-foot or less, but farmers south of the Delta during the drought in the late 2000s were paying $500 or more for some water used by high-value crops, and during the critically dry 2014 season some of these farmers have paid more than $1,000 per acre-foot for water to keep their orchards alive. Similarly, cost ranges for new supply facilities, such as surface storage or recycled water, depend on the specific opportunities at different locations. Only a few options—such as low-cost water transfers, some agricultural efficiency measures, some conjunctive use, and conserving water by fallowing—are viable alternatives for most farming activities. Urban water agencies are more likely to employ a wider range of options, even though some options are costlier than many existing, but finite, supplies.

In some cases, it will be less expensive to endure temporary or even permanent shortages than to provide additional supplies. However, planned shortages can be controversial, particularly when water users had more abundant supplies in the past. The controversies are especially intense when agricultural or urban users' supplies are cut for reallocations of water to the environment. However, the environment has tended to face disproportionately high shortages during droughts (Null and Viers 2013).

Orchestration will often be more effective at the regional scale. When local agencies within a region coordinate their activities, they can benefit from economies of scale for some investments and create a more balanced portfolio. Coordination at the watershed and basin levels is required for some tools to be effective, such as groundwater basin recharge, water markets, source protection, and most large infrastructure projects.

Progress in Decentralized Portfolio Management

In recent decades, many urban water agencies have moved toward more diversified portfolio approaches, with greater emphasis on tools that

Method	Operational pros and cons	Illustrative cost range ($/acre-foot)
Demand and reallocation		
Water transfers	Pros: Flexible tool for lowering costs of dry-year shortages and enabling long-term reallocation of supplies as economy shifts Cons: Potential economic harm to selling regions	50–550
Agricultural water-use efficiency	Pros: Reduces total stream diversions and pumping; enables farmers to raise yields and limit polluted runoff Cons: May not generate net savings that make water available for other users; net use reductions often require fallowing	145–675 (per af of net use reduction)
Urban water-use efficiency	Pros: Savings can often occur without loss of quality of life; high net savings possible in coastal areas and with landscape changes; some actions also save energy Cons: Requires implementation by large numbers of consumers; can be especially difficult for outdoor water uses, which depend on behavior as well as technology	225–520 (per af of gross use reduction)
Supply management		
Conjunctive use and ground-water storage	Pros: Flexible source of storage, especially for dry years Cons: Slower to recharge and harder to monitor than surface storage	10–600
Recycled municipal water	Pros: Relatively reliable source in urban areas Cons: Public resistance can preclude potable reuse	300–1,300
Surface storage	Pros: Flexible tool for rapid storage and release Cons: Potential negative environmental impacts; small value of additional storage with a drier climate	340–820+ (state projects)
Desalination, brackish	Pros: Can reclaim contaminated groundwater for urban uses Cons: Brine disposal can be costly	500–900
Desalination, seawater	Pros: "Drought-proof" coastal urban supply tool, especially useful in areas with few alternatives Cons: Potential environmental costs at intakes and for brine disposal; sensitive to energy costs	1,000–2,500

NOTE: Costs are illustrative and vary widely with local conditions. For conjunctive use, the costs of water for banking may be additional. For most options other than water-use efficiency, cost estimates do not include delivery. For water transfers, conjunctive use, and surface storage, cost estimates do not include treatment. For agricultural water-use efficiency, cost estimates are for subsidies needed to implement measures that are not locally cost-effective and refer only to actions yielding net water savings. Many costs from DWR sources are from studies in the early-to-mid-2000s and may have increased with inflation. Some figures are rounded.

SOURCES: Hanak et al. (2011), using author estimates and data from the California Department of Water Resources and the U.S. Bureau of Reclamation.

stretch available water supplies to complement existing surface and groundwater sources. Pricing, subsidies, public education, and landscape-watering ordinances have been used to encourage urban demand reductions, and investments have been undertaken to augment usable supplies by desalting brackish groundwater, treating recycled wastewater, reducing operational losses, building interties (interconnections between water distribution systems, to allow utilities to manage their supplies jointly), and recharging groundwater basins with surface water and captured stormwater. In the agricultural sector, water-use efficiency techniques have become widespread in areas facing chronic shortages. In addition, an active water market has developed within the state, enabling temporary and longer-term reallocation of water from lower-value (mainly agricultural) activities to higher-value activities in farming and urban sectors and to the environment. This market has been combined, in some areas, with active groundwater recharge (or "banking") to balance supplies across wetter and drier years.

The state has promoted these shifts through legal reforms (e.g. to facilitate water marketing, to require low-flow plumbing fixtures), direct intervention (e.g. as a broker in the water market), and subsidies for some nontraditional activities (e.g. water-use efficiency investments and recycled wastewater plants). Often, these subsidies have sought to encourage collaboration among local agencies, most notably through the Integrated Regional Water Management program, which has allocated more than $2 billion in general obligation bond funds to these efforts since 2000.

Efforts to diversify water supply portfolios and increase coordination have helped improve California's ability to cope with scarcity. Nevertheless, major technical and institutional challenges remain to integrate these wide-ranging options into a coherent set of activities at local, regional, and state levels.

Technical Gaps in Portfolio Analysis

Determining how to combine options cost-effectively requires sophisticated analytical support and computer modeling.[1] Some local and regional agencies already employ decision support tools to develop their portfolios. The Metropolitan Water District of Southern California (2010), for example, uses a set of simulation models to develop a wide-ranging portfolio of water sources, storage facilities, and water conservation activities, as well as wastewater reuse, water marketing, and other options suitable for meeting regional demands over a wide range of wet and dry years. The San Diego County Water Authority (1997) has employed optimization modeling to identify and integrate a similarly wide range of water management actions. However, in many cases, investment choices are being made without the benefit of integrated decision support.

The technical gap may be most pronounced at the level of statewide planning. Although the last three issues of the *California Water Plan Update* (California Department of Water Resources 2005, 2009, 2013) have emphasized integrated portfolio approaches to water system planning, these exercises do not use portfolio modeling tools to quantify effective combinations of options. Instead, the plans discuss the potential water supply benefits of a range of options one by one, often without quantitative estimates of supply potential or costs. The plans acknowledge the complementarities among some options but make no attempt to quantify how they might interact and the relative roles each might have in cost-effective regional and statewide water management under different future scenarios.

The lack of integrated decision support will not stop innovation in water supply management, but it can lead to misjudgment of the actual savings potential from some options and a failure to recognize the benefits of others.

MODELING INSIGHTS ON THE BENEFITS OF INTEGRATION

To illustrate the value of integrating water supply management options statewide, this chapter provides some results from the CALVIN model. Computer models of water systems are commonly used in water management because they can explicitly represent what is known about complex systems, thereby providing a platform for exploring problems and solutions. The CALVIN model combines economic and engineering representations of most major elements of California's water supply system, such as water markets, pricing, reoperation and coordination of reservoir and aquifer operations, water conservation, water recycling, and desalination (these are identified by asterisks in table 7.1). CALVIN seeks least-cost ways to serve urban and agricultural water demands throughout most of the state while meeting environmental flow requirements.

CALVIN has provided insights into how California's water system can adapt to a wide variety of strategic opportunities and challenges (Jenkins et al. 2004; Pulido-Velázquez, Jenkins, and Lund 2004; Null and Lund 2006; Tanaka et al. 2006, 2011; Medellin-Azuara et al. 2008; Harou et al. 2010). In general, CALVIN has highlighted the value of tools that enhance the flexibility of the water system and make the most of existing system assets. Accurate price signals and a well-functioning water market are important for encouraging demand reduction and reallocation of water from lower- to higher-value uses. Integrated system operation—which treats all major groundwater basins, surface storage reservoirs, and conveyance facilities as part of a larger network—facilitates water marketing and makes it possible to better exploit the potential for conjunctive use of ground and surface water. Conveyance is generally the most valuable asset in this integrated system, in the sense that it is far more valuable to expand or enhance some interconnections (to facilitate conjunctive use and marketing) than to build new surface reservoirs.

All modeling has limitations. The CALVIN model idealizes water management in three important ways. First, it generally assumes that managers do not face institutional barriers to implementing the most cost-effective decisions. As a result, it can understate the costs of some adaptations—for example, if cumbersome administrative procedures or local political pressure in the source region prevents the use of water transfers, leading to greater shortages in other regions (Tanaka et al. 2011). Second, it assumes that managers have perfect foresight into hydrologic conditions. As a result, it somewhat understates the value of the higher-cost elements of a water supply portfolio as hedges against risk, and it overstates the benefits of reoperations, particularly for flood management (Draper 2001). Third, by representing water recycling and seawater desalination with average costs per acre-foot, when their initial investment costs are actually large and irreversible, the model often understates the costs of using these options.

Effects of Climate Change, Cutbacks in Delta Exports, and Urban Conservation

This section explores the implications of two major management challenges that California may well face by the mid-twenty-first century: (1) significant restrictions in water supply from the system's hub in the Sacramento–San Joaquin Delta and (2) drier overall conditions resulting from climate change. It considers how a major successful urban water conservation effort could help California cope with these challenges. The focus is on urban rather than agricultural conservation as an explicit policy tool because most agricultural water-use efficiency efforts do not result in net water savings.[2] The model does project large reductions in net agricultural water use from land fallowing under some conditions, because this is a relatively cost-effective way to respond to shortages.

To examine these changes, the analysis compares a base case with historical conditions for climate, Delta exports, and urban water use with scenarios where urban water use is cut by 30 percent and Delta exports are restricted (table 7.3).[3] The Delta currently serves as a conveyance hub for approximately 30 percent of water supplies for urban areas in the San Francisco Bay Area and Southern California and farms in the southern half of the Central Valley (Lund et al. 2010). Reductions in Delta exports reflect increasing restrictions on pumping operations arising from native-species declines as well as physical collapse of the system from widespread levee failure. Two climate scenarios are considered: historical climate conditions, and a warm-dry type of climate, as employed in the state's most recent biennial assessment of the potential effects of climate change (California Environmental Protection Agency 2010). Although most studies agree that temperatures in California will rise, there is no consensus on whether California's climate will be drier or wetter. This drier scenario provides a moderately extreme climate test of the state's water system.[4]

One important, and somewhat unrealistic, assumption is that this leap in urban water conservation is achieved for free. In reality, such conservation would incur significant up-front costs, at least in a transition period where existing water users change plumbing, appliances, and landscaping to technologies and plants that use less water (table 7.2; Hanak and Davis 2006). However, by allowing energy savings as well as water savings, many indoor conservation measures can actually save costs over the longer run.[5] Following a transition period, the assumption of no additional costs would be consistent with a shift in behaviors and tastes such that the new norms do not constitute a great overall hardship.

Key Findings

This modeling exercise yields important insights about the potential roles of conservation, infrastructure investments, and new water supply technologies in California's future.

1. *Urban conservation can significantly reduce pressure for Delta exports.* With a historical climate, the demand for Delta exports

TABLE 7.3 ASSUMPTIONS FOR 2050 WATER MANAGEMENT SCENARIOS *(gallons per capita per day)*

Scenario	Climate	Urban water use (gallons per capita per day)	Delta exports range	Costs per acre-foot of new supply technologies (2008 dollars)
Base case	Historical	2000 levels (221)	Full exports only (pre-2007 operating rules)	Desalination: $2,072 Recycled wastewater: $1,480
Policy changes with historical climate	Historical	30% reduction (154)	Full to zero exports	Desalination: $1,628 Recycled wastewater: $1,480 for new plants, $518 for existing plants
Policy changes with warm-dry climate	+8.1 °F and −26% streamflow	30% reduction (154)	Full to zero exports	Desalination: $1,628 Recycled wastewater: $1,480 for new plants, $518 for existing plants

NOTE: The model assumes 2050 land use and population, with 65 million residents (per Landis and Reilly 2003). Urban water use includes conveyance losses. The 30-percent reduction applies to residential and commercial uses but not to industrial uses, and the cuts are split proportionately between indoor and outdoor uses. The historical climate assumes conditions from 1922 to 1993. For other assumptions, see Ragatz (2013).

SOURCE: Ragatz (2013).

would drop from 5.7 million acre-feet (maf) (the base case) to 3.9 maf in response to 30-percent urban conservation. The savings are less with a warmer and drier climate: the base-case demand for Delta exports is higher (6 maf, essentially full pre-2007 capacity), and conservation reduces export demands only to 5.4 maf.

2. *Urban conservation can free up some supplies for agricultural uses.* This effect is particularly pronounced under a drier climate (figure 7.1). Given the high economic value of urban water use, which would probably increase following 30-percent urban water conservation, climate change and reductions in Delta exports have little if any effect on urban water deliveries. Almost all additional shortages from climate change and reductions in Delta exports are borne by agricultural water users, many of whom would still have incentives to sell water to urban users.

3. *Urban conservation can significantly reduce operating costs and generate energy savings.* Conservation reduces pumping for long-distance imports of water to Southern California from the Delta and the Colorado River. Reductions in Delta water export capacity further decrease water operation costs, mostly because less water is available to pump and treat. However, a drier climate increases use and costs for water reuse and seawater desalination. Although these results doubtless understate the initial costs to the urban sector of achieving conservation, the operational savings from conservation are likely to be durable.[6]

4. *A warmer, drier climate raises the costs of Delta pumping cutbacks substantially.* Drier conditions raise the costs of shortages by at least $1 billion per year for each scenario. (To see this, compare the pairs of bars in figure 7.2.) With a warmer, drier climate, the added costs of a complete shutdown of the pumps more than doubles, jumping to $2.8 billion per year, and more than wiping out the cost savings from the urban water conservation program (compare the orange bars in the base case and the no-export scenarios in figure 7.2). Increases in water shortages occur primarily in the agricultural sector, as water-short urban users purchase water from farmers with more secure rights (figure 7.3). The costs to the statewide economy would be even higher if these transfers were blocked.[7] These results highlight the value of building alternative conveyance—such as tunnels underneath the

Delta, now under consideration—to allow continued movement of some water to urban and agricultural water users.[8]

5. *Delta pumping cutbacks and a drier climate reduce the value of new surface storage.* Many policymakers and stakeholders have argued that new surface storage is a necessary part of any Delta-related water policy package. However, CALVIN model results show that Delta cutbacks substantially reduce the value of expanding Northern California surface storage, because it becomes increasingly difficult (and ultimately impossible) to move water to users south and west of the Delta. A warm-dry climate has a similar and more widespread effect, because most reservoirs rarely fill.[9] Even with a warmer-wetter climate, with more precipitation and earlier runoff, expanding conjunctive use and groundwater banking appears more cost-effective than expanding surface storage (Tanaka et al. 2006). Conjunctive use projects south of the Delta also become more difficult with Delta pumping cutbacks or a warm-dry form of climate change, because it is harder to obtain water for aquifer recharge. Delta pumping cutbacks also raise the value of new conveyance interties south and west of the Delta, to better employ available supplies.

6. *Delta pumping cutbacks and a drier climate make recycled water and desalination more valuable.* The 30-percent reduction in urban water demand, by itself, would lead water agencies to dramatically reduce new investments in these more expensive water sources (figure 7.3), although use of most existing water recycling plants would probably continue.[10] A drier, warmer climate plus an end of Delta exports encourage a significant increase in water reuse and desalination statewide, even with 30-percent urban water conservation. Nevertheless, water recycling and desalination would remain a small proportion of statewide water supplies.

In sum, these results suggest that a major effort in urban water conservation—along the lines now being sought under legislation passed in 2009—can lessen the brunt of Delta export cutbacks and the costs to the economy from a warmer, drier climate. However, even with substantial additional urban water conservation, a drier, warmer climate makes continued Delta water exports much more valuable, highlighting the value of new conveyance infrastructure to permit these exports to continue. Decisions on other major infrastructure investments also

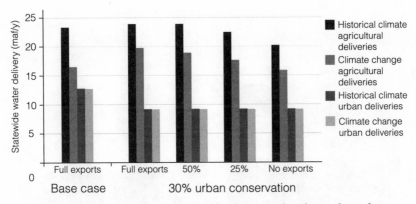

FIGURE 7.1. Urban water conservation would reduce agricultural water losses from reduced Delta water exports and a drier climate. Note that the figure shows conditions in 2050. See table 7.3 for scenario assumptions. Source: Ragatz (2013), as reproduced in Hanak et al. (2011).

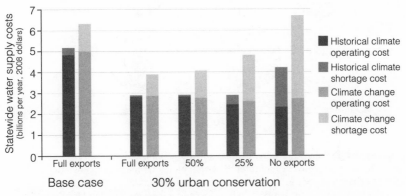

FIGURE 7.2. Ending Delta water exports would be particularly costly with a drier climate. Note that the figure shows conditions in 2050, in 2008 dollars. See table 7.3 for scenario assumptions. Source: Ragatz (2013), as reproduced in Hanak et al. (2011).

depend on these outcomes. In particular, it may be prudent to defer costly expansions of surface storage and focus on improving the ability of the existing system to work in an integrated manner, with the expansion of groundwater banking, selected interties, and water marketing institutions.

Of course, even if the state and federal governments succeed in implementing a long-term solution that allows substantial Delta exports from one or more tunnels under the Delta, it will be 10–25 years before such

Water portfolios (south of delta)

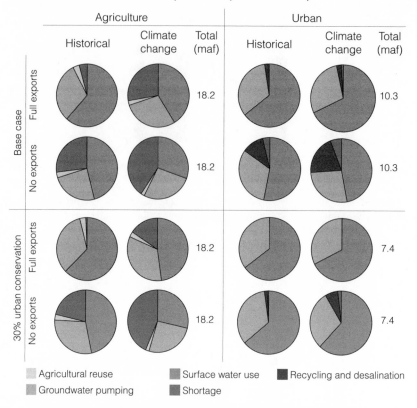

FIGURE 7.3. Ending Delta water exports and a drier climate would greatly reduce agricultural water deliveries south of the Delta. Note that the figure shows annual conditions in 2050. See table 7.3 for scenario assumptions. Sources: Ragatz (2013), Tanaka et al. (2011) (for base case without exports), as reproduced in Hanak et al. (2011).

facilities can be completed and operational. This implies a potentially long period of diminished water supplies for Bay Area and Southern California cities and southern Central Valley agriculture, with environmental pumping restrictions and the threat of a complete shutdown of the pumps from a major earthquake. Tools to enhance flexibility—such as infrastructure and institutions to facilitate water transfers—can help reduce agricultural and urban scarcity costs. In addition, early efforts to achieve conservation gains, along with other investments to stretch local resources (e.g. groundwater banking, stormwater capture, and waste-water reuse), can help build resiliency within urban areas.

OVERCOMING INSTITUTIONAL AND LEGAL HURDLES
TO PORTFOLIO MANAGEMENT

The modeling results presented above highlight the importance of linking management actions together, often over great distances, as part of a portfolio approach. To strengthen water supply portfolios, it will be necessary to overcome several important institutional and legal hurdles. This section highlights issues in three key areas: water pricing, water transfers, and groundwater management.

Water Pricing: An Underutilized Tool for Water Conservation

A variety of nonprice tools can encourage conservation: plumbing and appliance standards, landscaping ordinances and restrictions (e.g. limits on the planting of lawns and use of outdoor watering), rebates to encourage new-technology adoption, and public education (table 7.1). Water pricing should be an important part of any conservation effort, because it can reinforce the effectiveness of the many nonprice tools.[11]

Since the early-1990s drought, California's urban water agencies have made important advances in implementing conservation-oriented rate structures. In particular, many agencies have shifted from uniform to increasing block or tiered rates, which bill higher per-gallon charges when water use exceeds the threshold of one or more tiers (Hanak 2005b). Another reform—the switch to volumetric billing—has begun in the many Central Valley communities that traditionally did not bill by use, as a result of federal and state laws that require a phase-in of water meters. By 2006, roughly half of California's population lived in a service area with tiered rates, and fewer than 10 percent lived in communities with unmetered rates.[12] Over the past few years, there has been additional movement toward tiered rates, as urban utilities have sought to change consumer behavior in response to drought conditions and restrictions on Delta pumping (Baerenklau, Schwabe, and Dinar 2013).

In broad terms, tiered rate structures provide incentives to conserve (Hewitt and Hanemann 1995; Olmstead, Hanemann, and Stavins 2007). However, there have been debates about the extent to which different rate structures can meet a variety of potentially competing objectives: economic efficiency, revenue stability, political feasibility, and ability to cover utility costs (Hall 2009). From an efficiency perspective, water users should face a price signal corresponding to the marginal cost of new supplies, which typically exceeds the average cost of existing supplies.

Yet, if utilities charge everyone this long-run marginal cost, they raise too much revenue (Brown and Sibley 1986). From a political feasibility perspective, water rate structures need to be perceived as fair, which argues for transparency and simplicity. And from a revenue perspective, utilities need to be able to cover their fixed costs—typically a high component of overall costs—even if water use declines. (Structuring rates in this way is known as "decoupling," a standard feature of electricity rates in California for several decades.)

A particular type of tiered rate structure—often known in California as an "allocation-based" structure—can meet all these objectives. Allocation-based rates set tiers at different thresholds for different subgroups of ratepayers, so the volume in the base tier corresponds roughly to the amount of water an efficient household would need to use. Households using more face a higher price per gallon (reflecting the increasing marginal costs of new water for the system). The subgroups are defined based on readily observable factors that affect water use—household size, lot size, and climate zone—and the threshold can be adjusted across seasons to reflect the higher outdoor water requirements of plants in hotter, drier months. Utilities set the lower-tier price to recover fixed costs, and they can use additional revenues from the higher tiers to fund new supplies, including conservation programs. This system is transparent, and it sends a salient price signal to water users, because the conservation objectives embodied in the threshold are meaningful, and tailored to expectations of what water users with similar characteristics should be able to do. If the prices for the tiers are allowed to vary with drought conditions, this structure also allows utilities to meet their revenue requirement when water use declines (Hall 2009).

Allocation-based rate structures have been successful for several Southern California utilities since the early 1990s, including the City of Los Angeles and the Irvine Ranch Water District (Orange County), and in the past few years they have been adopted by several others, including the Eastern Municipal Water District and the Coachella Valley Water District (Riverside County) and the Rincon del Diablo Water District (San Diego County).[13]

In contrast, most tiered rate structures in California do not vary tiers by customer groups, making it harder to send salient price signals to most water users. In addition, with calls to restrict water use in the recent drought, many utilities found that they were unable to cover costs as water sales fell—evidence that they were relying on revenue from their upper tiers to cover fixed costs. The subsequent need to raise rates when customers have been reducing their water use raises political problems for utilities. Such problems could be avoided if utilities had the flexibility to implement a drought rate structure, whereby prices in the tiers are adjusted in advance to drought conditions. With an allocation-based structure, tiers also can be adjusted over time to encourage progressive conservation. For instance, Irvine Ranch recently reduced its base allocation to encourage higher outdoor water-use efficiency. Effective communication with the public is an important part of such programs. This includes information not only on why unit prices may need to rise when water use declines but also on which conservation actions can most effectively reduce water use. A recent survey for the Association of California Water Agencies found that a strong majority of the state's residents support the idea of reducing household water use (Fairbank et al. 2010). But this same survey found that most homeowners underestimated the dominant role of landscape irrigation in total water use.

As California moves to implement an aggressive urban water conservation program, more utilities should consider allocation-based rate structures. Opponents of this approach often voice concerns over the costs of implementation, given higher data needs. But advances in information technology have brought down the data costs of establishing allocations for different lot sizes. Digitized parcel maps are readily available for most counties, as are climate maps that reflect outdoor watering needs. And customers can have the option to declare household size. Another objection sometimes raised is that it is "unfair" to give larger base allocations to residents with larger lots (many of whom have higher incomes). Allocation-based rate structures are not fair in the sense of treating everyone exactly the same. But they end up being fair in a broader sense, because each group of customers ends up paying about the same average price per unit of water. By grouping customers more homogeneously by factors such as lot size and location, it is possible to send a meaningful price signal to all water users, to encourage efficient water use.

Water Marketing: Getting Past the Growing Pains of Adolescence

State and federal legislation passed in the 1980s and early 1990s paved the way for California's water market. New state laws clarified that transferring water is a beneficial use (to lessen sellers' fears that they might lose the rights to use water in subsequent years), extended "no injury" protections against negative "third-party" impacts on fish and wildlife (to ease concerns of environmental managers and stakeholders that water movements

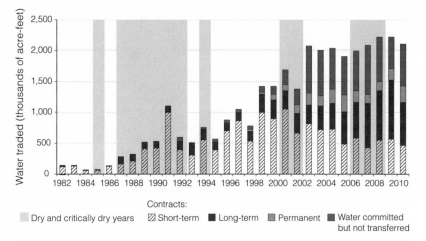

Contracts:

Dry and critically dry years   ⬚ Short-term   ■ Long-term   ■ Permanent   ■ Water committed but not transferred

FIGURE 7.4. California's water market grew in the 1990s but has flattened since the early 2000s. Note the figure shows actual flows under short-term and long-term lease contracts, estimated flows under permanent sale contracts, and the additional volumes committed under long-term and permanent contracts that were not transferred in those years. The database includes transactions between water districts, federal and state agencies, and private parties that are not members of the same water district or wholesale agency. Source: Hanak and Stryjewski (2012).

would negatively affect the quantity and quality of environmental flows), and required that owners of conveyance facilities lease space for transferred water if they had excess capacity. The federal Central Valley Project Improvement Act of 1992 also encouraged water marketing.

These legal changes, along with active participation in the market by both state and federal agencies, helped jump-start an active water market in the early 1990s, when California was in the midst of a major drought (Israel and Lund 1995; Gray 1996; Hanak 2003). The market continued to grow when the rains returned, and by the early 2000s, the annual volume of water committed for sale or lease was on the order of 2 maf, with roughly 1.3 maf moving between parties in any given year (figure 7.4).[14]

Consistent with the relative share of agricultural water use in the state's overall supply, farmers have always been the primary sellers in California's water market. But over time, there have been shifts in the nature of contracts and the uses of purchased water. During the 1990s, the market consisted primarily of short-term (single-year) transfers, with long-term contracts constituting only about 20 percent of total volumes. By the end of the 2000s, long-term and permanent sales accounted for most of the volume traded. Along with this transition, farmers have

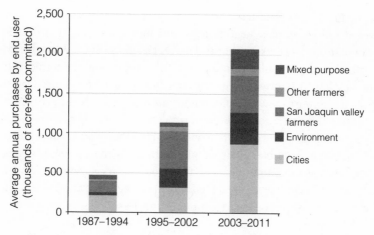

FIGURE 7.5. Urban water purchases now account for about half of the market. Note that the figure shows shares of all committed transfers (short-term flows and contract volumes for long-term and permanent sales). Mixed uses denote purchase by agencies with both urban and agricultural uses. Source: Hanak and Stryjewski (2012).

declined in importance as buyers, constituting only 27 percent of all contractual commitments in 2003–2011, versus 46 percent in the preceding 8 years (figure 7.5). Water purchases for environmental flows and wildlife refuges have remained important (one-fifth of the total), but the major increases have been by urban agencies, which now account for 42 percent of all commitments (more if one includes agencies with mixed water uses), compared to just 28 percent in 1995–2002.

Long-term contracts among water districts that use Colorado River water have accounted for a substantial share of this growth. With the conclusion of the Quantification Settlement Agreement in 2003, over 600,000 acre-feet of farm water transfers are now committed, mostly to urban users. Urban agencies within the Central Valley have also made local purchases from agricultural agencies, and some Southern California urban agencies have successfully purchased agricultural contract water from SWP contractors in the San Joaquin Valley. These long-term transfers have been made possible through a combination of system efficiency improvements (e.g. canal lining and operational improvements), agricultural land retirement, on-farm irrigation efficiency improvements (where improved efficiency generates net water savings, such as in the Imperial Irrigation District), and releases of water from surface and groundwater reservoirs (e.g. from Yuba County).

The growth of long-term and permanent transfers—which generally involve more complex negotiations and more in-depth environmental documentation—is a sign that the market is maturing. Long-term commitments are particularly important for supporting economic transitions. By law, urban water agencies need to demonstrate long-term supplies to support new development, and transfers can provide this assurance (Hanak 2005b). Long-term commitments for environmental flows provide flexibility for environmental managers and reduce the conflicts associated with regulatory alternatives to market-based transactions. Long-term commitments to make temporary supplies available—such as the recent 25-year transfer agreement between the Yuba County Water Agency and the Department of Water Resources—enhance operational flexibility.[15]

Despite these positive market developments, there is also evidence of overall weakening in market momentum. Overall trading volumes have leveled off since the early 2000s; excluding Colorado River transactions, both committed and actual flows have declined since 2001. During the recent drought (2007–2010), the market only made 500,000–600,000 acre-feet available of dry-year supplies, considerably less than in the previous drought.

A variety of impediments—some long-standing and some new—appear to be at work. One new problem relates to conveyance infrastructure. California's sophisticated supply infrastructure has made it possible to transfer water either directly or through exchanges throughout most of the state's demand and supply areas. However, the Delta is an important conveyance hub for north-to-south and east-to-west transfers, and new pumping restrictions since late 2007 have impeded both movements.

Other obstacles reflect legal and institutional impediments. Because California does not regulate groundwater at the state level, the no-injury protections for other legal surface water users (including fish and wildlife) do not extend to groundwater users. This omission has spurred the development of county ordinances restricting water exports in many rural counties that lack more comprehensive forms of groundwater management (Hanak and Dyckman 2003). Local groundwater ordinances have restricted direct sales of groundwater as well as transfers based on conjunctive use (selling surface water and pumping groundwater), and they have also restricted the development of groundwater banks in some places (Hanak 2003, 2005a). Although these ordinances were a useful stop-gap measure to prevent harm to local users, they are less efficient than comprehensive basin management schemes, which address locally generated overdraft as well as problems related to exports.

Another local concern in source regions has been the potential effects on the local economy of fallowing or land retirement. These "pecuniary" effects are not proscribed under state law, which generally views such changes as a natural consequence of shifts in the economy—much as a new freeway might affect local businesses for better or for worse.[16] However, fallowing conducted for sales to the drought water banks in the early 1990s generated local concerns, and many agricultural water districts disallow fallowing-related transfers unless the water is going to other lands leased or owned by the same farmer. Because fallowing of low-value crops is one of the most efficient and effective ways to make new net water available for other uses, continued local resistance will remain an obstacle to market development. In two long-term transfers of Colorado River water that involve fallowing (from Palo Verde Irrigation District and Imperial Irrigation District), buyers have supplied mitigation funds to address community effects (Hanak and Stryjewski 2012). Developing templates for such mitigation payments will be important for managing economic transitions. These programs should consider not only residents who may become unemployed as a result of fallowing but also the potential increase in social service costs and reduction in tax revenues for counties in the region where fallowing is occurring.

Another market obstacle relates to environmental protections. Over time, transfers have been subjected to additional environmental restrictions, beyond the requirement of no injury to environmental flow conditions. For instance, under the 2009 drought water bank program operated by the Department of Water Resources, fallowing of rice fields was restricted to protect the habitat of the giant garter snake, a listed species that now depends on artificial wetlands created by irrigation water. Use of diesel pumps for groundwater-substitution transfers was also restricted because it was deemed to violate Clean Air Act rules, which farmers are normally exempt from when they operate pumps for their own activities.

Uncertainties over the terms of these new restrictions, combined with the inability to move water through the Delta in the spring, depressed drought water bank activity: fewer than 80,000 acre-feet were acquired, though the goal was several hundred thousand acre-feet.

New mechanisms are needed to clarify and streamline environmental reviews for water transfers, particularly for medium-term agreements that create flexibility to transfer water quickly in the event of drought- or regulation-induced shortages. Water market development also will benefit from greater integration and more uniform treatment of the various types of water rights and contracts. Current rules heavily favor transfers between agencies within the same large project (Central Valley Project, State Water Project, Colorado River), resulting in less efficient reallocations for short-term water management and long-term economic shifts.

Filling the Gaps in Groundwater Management

Increased integration of surface water and groundwater is essential for portfolio management of California's water resources. Water banks use available space in aquifers to store imported surface water, recycled wastewater, and local stormwater, recharging the aquifer and making water available for subsequent pumping, and this type of storage is generally less costly than new surface storage (table 7.2).

To implement a groundwater banking project, it is necessary to have an effective means of measuring inflows (both imports and local recharge) and outflows (including pumping for local uses and for export). It is also necessary to anticipate possible effects of the project on local storage availability. Sometimes, importing water benefits local users by raising the level of the groundwater table, which reduces pumping costs. But in other cases, imported water may harm local users by displacing storage capacity in the aquifer that would have captured local recharge, to which they have superior rights. Water quality also may be an issue if the supplies available for recharge contain higher levels of salts or other pollutants that would contaminate water recharged from other sources.

The creation of water banks has been hampered by several lingering legal uncertainties. These include the artificial separation of surface water rights and groundwater rights systems in California water law, as well as questions about local landowners' rights to exclude others from using the aquifer space beneath their lands for storage of non-native water. These problems have largely been overcome in Southern California's adjudicated groundwater basins, where monitoring and accounting systems exist and there is clarity on who has rights to withdraw water from the aquifer. Banking is also relatively straightforward in the state's few special groundwater management districts, where a single agency is responsible for managing recharge and has authority to charge pump fees to cover the costs.

In some other areas—notably Kern County—active groundwater banking systems have been established based on looser arrangements, which include careful monitoring and an agreement with neighboring groundwater pumpers that withdrawals from the bank will not harm local parties (Hanak 2003). Such schemes worked effectively during the drought of the late 2000s—groundwater banks in Kern County and some arrangements organized by the Metropolitan Water District of Southern California made roughly 1.9 maf available from banks operated for off-site parties (Hanak and Stryjewski 2012). But generally, more comprehensive basin management mechanisms are needed to limit overdraft and increase conjunctive use operations in the state.

PRIORITIES FOR PORTFOLIOS IN WATER MANAGEMENT

California is not helpless in facing its chronic problems of water scarcity. More effective, robust, and cost-effective solutions to these problems are available by orchestrating a range of options at local, regional, and statewide levels. These "portfolio" solutions combine the strengths of individual options but require a higher level of analysis and integrated decision-making than is currently common in the state. California has already witnessed some progress in implementing portfolio approaches; numerous nontraditional tools have been tapped to cope with increasingly tight water supplies. Expanded efforts are especially needed in three areas: urban conservation, groundwater banking, and water marketing.

Urban conservation has the potential to play a major role in mitigating the effects of reduced export capabilities from the Delta and supply losses that may result from dry forms of climate change. Water-rate reform, using tiered rates with variable base allowances, can promote conservation in a flexible and fiscally responsible way.

The state should also work to loosen institutional barriers to groundwater banking and water marketing. For groundwater, one promising direction is for the state to establish criteria for integrating groundwater and surface water and for managing groundwater withdrawals, and to allow local entities to develop implementation plans (Hanak et al. 2011). To improve the functioning of the water market, the state needs to improve the predictability of the environmental review process, facilitate transfers across projects, and more generally ensure that market performance is a policy priority (Hanak and Stryjewski 2012). With better-functioning water markets, agricultural water conservation will increase in response to water scarcity and incentives to transfer water to agricultural, urban, and environmental activities in which water has a higher economic value.

Better information and stronger analytical tools will be needed to support modern portfolio management. The state has an interest in the collection and development of local, regional, and statewide information, as well as in regulations and incentives that foster the development of effective portfolios. Without such information and institutional prodding, water decision-making and conflicts will remain more difficult, expensive, and time-consuming to resolve.

NOTES

1. See Jenkins and Lund (2000) and Lund and Israel (1995) for some examples from the research literature.

2. This is because most excess irrigation water that is not consumed by crops is returned to the system as surface water runoff or groundwater basin recharge, and available for subsequent reuse (see Hanak et al. 2011, box 2.1).

3. More complete results are given by Ragatz (2013).

4. Other studies have shown that reduction in streamflow in this climate change scenario is more problematic for water management than increase in temperature, because existing surface reservoirs are able to absorb much of the additional early runoff associated with reduced snowpack and earlier snowmelt (Connell 2009). Given California's fairly large reservoir capacity, wetter climates tend to have much lower water supply costs but could easily have much greater flood management costs (Tanaka et al. 2006; Zhu et al. 2007).

5. See Cooley et al. (2010) for some examples, including low-flow showerhead replacement, more efficient front-loading clothes washers, faucet aerators, and a variety of commercial appliances.

6. With full exports, this conservation scenario reduces state and federal project energy use by 40 percent; if Delta exports are ended altogether, energy use goes down by more than two-thirds (Bates 2010).

7. In a scenario using historical hydrology and base-case demands, the loss of the ability to transfer water with a Delta shutdown increased costs by $700 million per year, or 47 percent (Lund et al. 2010; Tanaka et al. 2011).

8. These results are illustrative; they do not demonstrate that the administration's current proposal to build two new tunnels under the Delta as part of the Bay Delta Conservation Plan (spring 2013 administrative draft) would pass a benefit–cost test.

9. For results with Delta cutbacks on their own, see Ragatz (2013) and Tanaka et al. (2011). For a warm-dry climate on its own, see Ragatz (2013), Tanaka et al. (2006), Medellin-Azuara et al. (2008), and Harou et al. (2010).

10. Although model results show decreased use of water recycling from projected current levels with 30-percent urban water conservation, the sunk costs of existing recycling plants and other wastewater disposal and water supply

reliability considerations are likely to support continued use of existing water recycling plants.

11. The authors thank Michael Hanemann and Darwin Hall for discussion of many of the points raised here.

12. Authors' estimates, using rate structure information from the 2006 water-rate survey by Black and Veatch (2006).

13. For an analysis of the conservation effects of Eastern Municipal Water District's new tiered system, see Baerenklau, Schwabe, and Dinar (2013).

14. All data in this section are from Hanak and Stryjewski (2012).

15. In this agreement, supplies are made available annually to a pool of State Water Project and Central Valley Project contractors, who can bid on available volumes.

16. State law does require public hearings on transfers that will exceed 20 percent of local water use (§ 1745.05).

REFERENCES

Awerbuch, Shimon. 1993. "The Surprising Role of Risk in Utility Integrated Resource Planning." *Electricity Journal* 6(3):20–33.

Baerenklau, Kenneth A., Kurt A. Schwabe, and Ariel Dinar. 2013. *Do Increasing Block Rate Water Budgets Reduce Residential Water Demand? A Case Study in Southern California.* Working Paper 01–0913. Water Science and Policy Center, University of California, Riverside.

Bates, Matthew. 2010. *Energy Use in California Wholesale Water Operations: Development and Application of a General Energy Post-Processor for California Water Management Models.* Master's thesis, University of California, Davis.

Black & Veatch. 2006. *2006 California Water Rate Survey.* www.kqed.org /assets/pdf/news/2006_water.pdf.

Brown, Stephen J., and David S. Sibley. 1986. *The Theory of Public Utility Pricing.* Cambridge: Cambridge University Press.

California Environmental Protection Agency. 2010. *Climate Action Team Report to Governor Schwarzenegger and the California Legislature.* www.energy .ca.gov/2010publications/CAT-1000-2010-005/CAT-1000-2010-005.PDF.

Connell, Christina R., 2009. *Bring the Heat, but Hope for Rain: Adapting to Climate Warming in California.* Master's thesis, University of California, Davis.

Cooley, Heather, Juliet Christian-Smith, Peter. H. Gleick, Michael J. Cohen, and Matthew Heberger. 2010. *California's Next Million Acre-Feet: Saving Water, Energy, and Money.* Oakland, CA: Pacific Institute.

Draper, Andrew J. 2001. *Implicit Stochastic Optimization with Limited Foresight for Reservoir Systems.* Ph.D. dissertation, University of California, Davis.

Fairbank, Maslin, Maullin, Metz & Associates. 2010. *Key Findings from Recent Opinion Research on Attitudes toward Water Conservation in California.* Memo to the Association of California Water Agencies from Dave Metz and Shakari Byerly, June 1.

Gray, Brian E. 1996. "The Shape of Transfers to Come: A Model Water Transfer Act for California." *Hastings West-Northwest Journal of Environmental Law and Policy* 4:23–59.

Hall, Darwin C. 2009. "Politically Feasible, Revenue Sufficient, and Economically Efficient Municipal Water Rates." *Contemporary Economic Policy* 27(4):539–54.

Hanak, Ellen. 2003. *Who Should Be Allowed to Sell Water in California?* San Francisco: Public Policy Institute of California.

———. 2005a. "Stopping the Drain: Third-Party Responses to California's Water Market." *Contemporary Economic Policy* 23(1):59–77.

———. 2005b. *Water for Growth: California's New Frontier.* San Francisco: Public Policy Institute of California.

Hanak, Ellen, and Matthew Davis. 2006. "Lawns and Water Demand in California." *California Economic Policy* 2(2). San Francisco: Public Policy Institute of California.

Hanak, Ellen, and Caitlyn Dyckman. 2003. "Counties Wresting Control: Local Responses to California's Statewide Water Market." *University of Denver Water Law Review* 6(2):494.

Hanak, Ellen, Jay Lund, Ariel Dinar, Brian Gray, Richard Howitt, Jay Mount, Peter Moyle, and Barton Thompson. 2011. *Managing California's Water: From Conflict to Reconciliation.* San Francisco: Public Policy Institute of California.

Hanak, Ellen, and Elizabeth Stryjewski. 2012. *California's Water Market, by the Numbers: Update 2012.* San Francisco: Public Policy Institute of California.

Harou, Julien J., Josué Medellin-Azuara, Tingju Zhu, Stacy K. Tanaka, Jay R. Lund, Scott Stine, Marcelo A. Olivares, and Marion W. Jenkins. 2010. "Economic Consequences of Optimized Water Management for a Prolonged, Severe Drought in California." *Water Resources Research* 46:W05522. doi:10.1029/2008WR007681.

Hewitt, Julie, and Michael Hanemann. 1995. "A Discrete-Continuous Choice Approach to Residential Water Demand under Block Rate Pricing." *Land Economics* 71(2):173–92.

Israel, Morris, and Jay R. Lund. 1995. "Recent California Water Transfers: Implications for Water Management." *Natural Resources Journal* 35:1–32.

Jenkins, Marion W., and Jay R. Lund. 2000. "Integrated Yield and Shortage Management for Water Supply Planning." *Journal of Water Resources Planning and Management* 126(5):288–97.

Jenkins, Marion W., Jay R. Lund, Richard E. Howitt, Andrew J. Draper, Siwa M. Msangi, Stacy K. Tanaka, Randall S. Ritzema, and Guilherme F. Marques. 2004. "Optimization of California's Water System: Results and Insights." *Journal of Water Resources Planning and Management* 130(4):271–80.

Johnston, Robert, Jay R. Lund, and Paul Craig. 1995. "Capacity Allocation Methods for Reducing Traffic Congestion." *Journal of Transportation Engineering* 121(1):27–39.

Landis, John D., and Michael Reilly. 2003. "How Will We Grow: Baseline Projections of California's Urban Footprint through the Year 2100." In *Integrated Land Use and Environmental Models,* 55–98. Berlin: Springer.

Lund, Jay R., Ellen Hanak, William E. Fleenor, William Bennett, Richard

Howitt, Jeffrey F. Mount, and Peter B. Moyle. 2010. *Comparing Futures for the Sacramento San Joaquin Delta*. Berkeley: University of California Press and Public Policy Institute of California.

Lund, Jay R., and Morris Israel. 1995. "Optimization of Transfers in Urban Water Supply Planning." *Journal of Water Resources Planning and Management* 121(1):41–8.

Medellin-Azuara, Josué, Julien J. Harou, Marcelo A. Olivares, Kaveh Madani-Larijani, Jay R. Lund, Richard E. Howitt, Stacy K. Tanaka, Marion W. Jenkins, and Tingju Zhu. 2008. "Adaptability and Adaptations of California's Water Supply System to Dry Climate Warming." *Climatic Change* 87(1):S75–S90.

Metropolitan Water District of Southern California. 2010. *Integrated Resources Plan 2010 Update*. Report No. 1373 (October). Los Angeles, CA.

Null, Sarah E., and Jay R. Lund. 2006. "Re-Assembling Hetch Hetchy: Water Supply Implications of Removing O'Shaughnessy Dam." *Journal of the American Water Resources Association* 42(4):395–408.

Null, Sarah, and Joshua Viers. 2013. "The New 'Normal' Water Year in a Changing California Climate." Californiawaterblog.com, June 10. Center for Watershed Sciences, University of California, Davis.

Olmstead, Sheila M., Michael Hanemann, and Robert N. Stavins. 2007. "Water Demand under Alternative Price Structures." *Journal of Environmental Economics and Management* 54(2):181–98.

Pulido-Velázquez, Manuel, Marion W. Jenkins, and Jay R. Lund. 2004. "Economic Values for Conjunctive Use and Water Banking in Southern California." *Water Resources Research* 40(3):W03401.doi:10.1029/2003WR002626.

Ragatz, Rachel. 2013. *California's Water Futures: How Water Conservation and Varying Delta Exports Affect Water Supply in the Face of Climate Change*. Master's thesis, Department of Civil and Environmental Engineering, University of California, Davis.

San Diego County Water Authority. 1997. *Water Resources Plan*. San Diego, CA.

Tanaka, Stacy K., Chrtina Connel-Buck, Keveh Madani, Josué Medellin-Azuara, Jay Lund, and Ellen Hanak. 2011. "Economic Costs and Adaptations for Alternative Regulations of California's Sacramento–San Joaquin Delta." *San Francisco Estuary and Watershed Science* 9(2).

Tanaka, Stacy K., Tingju Zhu, Jay R. Lund, Richard E. Howitt, Marion W. Jenkins, Manuel A. Pulido, Mélanie Tauber, Randall S. Ritzema, and Inês C. Ferreira. 2006. "Climate Warming and Water Management Adaptation for California." *Climatic Change* 76(3–4):361–87.

Zhu, Tingju, Jay R. Lund, Marion W. Jenkins, Guilherme F. Marques, and Randall S. Ritzema. 2007. "Climate Change, Urbanization, and Optimal Long-term Floodplain Protection." *Water Resources Research* 43(6):W06421. doi:10.1029/2004WR003516.

CHAPTER 8

The Challenge of Sustainable Groundwater Management in California

DANIEL WENDELL AND MAURICE HALL

Unconstrained pumping of groundwater in California has led to widespread lowering of water tables, reduction or elimination of baseflow to streams, land subsidence, and impairment of existing surface water rights. Related environmental impacts include drastically decreased area of wetlands, shrinking riparian habitats along river corridors, and interruption of the late-summer and early-fall streamflows needed for passage of salmon and for the health of other aquatic species. Economic impacts include the need to deepen wells or install new wells, due to reduction in well yields from lowered water levels; higher energy costs due to increased pumping lifts; influx of poor-quality water that requires additional treatment; and repair of surficial infrastructure damaged by land subsidence.

How did we arrive at this situation, and what should we do about it? That is the focus of this chapter. We will discuss important physical concepts regarding how groundwater systems operate and demonstrate why these concepts need to be properly understood and applied in order to sustainably manage the resource. Most importantly, we will see how the failure to recognize the intimate connection between groundwater and surface water has led to significant environmental impacts and unintended consequences and the urgent need for groundwater management reform.

BACKGROUND

Groundwater refers to that portion of the subsurface environment where the openings in soil, sediment, and rocks are filled with water. The top of this saturated zone is referred to as the water table. Groundwater is an important source of supply for California, providing about 40 percent of the state's average water needs (Department of Water Resources 2013). About 75 percent of all groundwater pumped in the state is used by agriculture, with most of the remainder pumped to meet urban needs and a small amount used to sustain managed wetlands. Agricultural pumpage averages about 12.7 million acre-feet (maf) per year, but varies from 8 maf in a wet year like 2005 to as much as 16 maf in a dry year like 2009. Urban demands are relatively steady, at about 3.6 maf/y. Deliveries of surface water provide most of the rest of supplies in the state, averaging about 25 maf/y.

Confined and Unconfined Aquifers

The term *aquifer* is used to describe subsurface materials that are permeable enough to transmit water in sufficient quantities to be of interest to people. *Aquitards* are lenses and layers of subsurface materials that are relatively impermeable and therefore inhibit vertical groundwater flow. Aquifers that are near the surface and that have no overlaying aquitard layer are referred to as *unconfined*. Aquifers of this type are intimately connected to streams. Water pumped from unconfined aquifers comes from physical drainage of the pore space, which is referred to as *specific yield* and represents as much as 10 to 30 percent of the volume of the formation. *Confined* aquifers are overlain by confining units (aquitards) that impede hydrologic communication with overlying units. Water pumped from a confined aquifer comes from compression of the aquifer and expansion of water due to pressure release, not drainage of pores, and the amount of water produced in this fashion can be hundreds to thousands of times less than from an unconfined aquifer in similar circumstances. Figures 8.1 and 8.2 illustrate these and other key features of a groundwater flow system.

A pumping well causes a decline in water levels near the well, called *drawdown*—which, when looked at in three dimensions, results in a *cone of depression* (figure 8.3) around the well. As the well continues to be pumped, the cone of depression deepens and expands. Water-level drawdowns and the rate of expansion of the cone of depression at a given

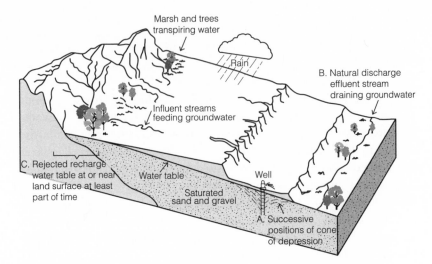

FIGURE 8.1. Sources of water for a pumping well. A. Groundwater storage from drawdown around the well, always the first source of water. B. Capture of natural discharge, the most common long-term (sustainable) source of water for a pumping well. C. Capture of rejected recharge, a relatively rare source. After Theis (1940).

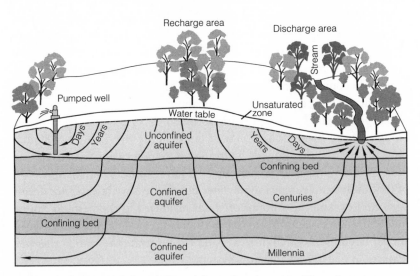

FIGURE 8.2. Movement of water through groundwater systems. Water that enters a groundwater system in recharge areas moves through the aquifers and confining beds comprising the system to discharge areas. After Heath (1983).

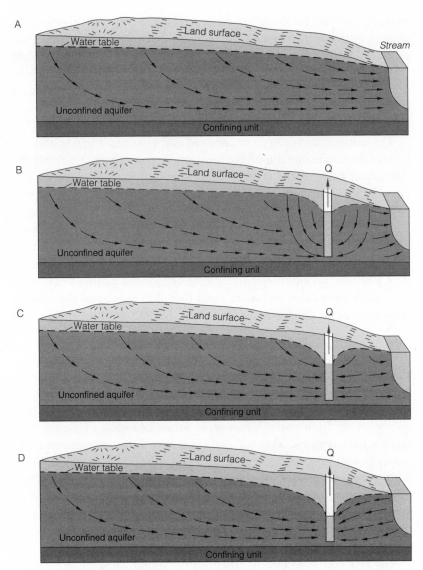

FIGURE 8.3. Streamflow depletion by a pumping well. A. Under natural conditions, recharge at the water table is equal to discharge at the stream. B. Soon after pumping begins, all of the water pumped by the well is derived from water released from groundwater storage. C. As the cone of depression expands outward from the well, the well begins to capture groundwater that would otherwise have discharged to the stream. D. In some circumstances, the pumping rate of the well may be large enough to cause water to flow from the stream to the aquifer, a process called induced infiltration. Streamflow depletion is the sum of captured groundwater discharge and induced infiltration. Modified from Barlow (2012).

pumping rate are far greater in a confined aquifer than in an otherwise equivalent unconfined system. No aquitard is completely impermeable, especially in the type of valley fill sediments present in California, and therefore the "leakiness" of an aquitard system plays an important role in controlling the shape and rate of growth of the cone of depression around pumping wells and of the timing of impacts to overlying aquifer units and surface water systems.

The distinctions between unconfined and confined aquifers and the cone of depression are important for a number of reasons. First, it is the interaction between the cone of depression and the surrounding environment that leads to the adverse impacts we ultimately experience. Second, there is a misconception that pumping from a deep, confined aquifer does not impact surface water systems because the overlying aquitard prevents flow between these two systems. This is not true. In fact, pumping from a confined aquifer can lead to greater impacts to distant streams than pumping an equivalent overlaying unconfined aquifer because of the greater drawdown and more rapid rate of expansion of the cone of depression in the confined aquifer (see e.g. Barlow and Leake 2012).

Sources of Water for Wells

Groundwater in an aquifer is in a continuous state of motion, moving from areas of recharge to areas of discharge. Recharge to aquifers commonly occurs from rainfall, deep percolation of excess irrigation, and seepage from streams, lakes and unlined canals. Areas of discharge commonly consist of streams, wetlands, lakes, springs, and estuaries. When a pumping well is introduced into this system it interrupts the natural flow pattern and becomes a new point of discharge. Water pumped from a well comes from somewhere, and understanding the ultimate source of this water is critical to determining the impacts of groundwater pumping on surface waters.

The first person to clearly articulate the source of pumped groundwater was C. V. Theis of the U.S. Geological Survey, who did so in his landmark 1940 paper, "The Source of Water Derived from Wells: Essential Factors Controlling the Response of an Aquifer to Development." In this paper, Theis demonstrated that there are three sources of water for a pumping well:

1. *Removal of water from storage.* Theis notes that this is always the initial source of water for a pumping well and is observed as

water-level drawdown around the well and formation of the cone of depression. However, this is not a sustainable source of water, and water levels would be in continuous decline if it were the only source.

2. *An increase in infiltration of water in the recharge area.* This occurs when a cone of depression reaches a natural recharge area and induces additional inflow. If this condition occurs it implies the presence of *rejected recharge.* In this case springs and wetlands are likely to be present, and to be adversely impacted by the loss of this water. It is important to note that recharge to an aquifer from rainfall and deep percolation of irrigation water, two major sources of recharge in California, are typically independent of water-table elevation and therefore independent of the effects of pumping.

3. *A decrease in discharge from the system.* This situation occurs when groundwater pumping captures water that otherwise would have flowed to streams, springs, wetlands, and lakes, or would have been lost by evapotranspiration from phreatophytes (plants with roots in the water table). In California, stream depletion is an important source of water for wells and can take the form of decreased discharge of groundwater to the stream or, if pumping stresses are great enough, reversal of flow directions, with resultant induced inflow from the stream.

These sources of water are illustrated in figures 8.1 and 8.2. The hydrogeologic principles presented in this section are discussed at more length in several well-written and very useful publications by the USGS: *Basic Groundwater Hydrology* (Heath 1983), *Sustainability of Ground-Water Resources* (Alley, Reilly, and Franke 1999), *Ground Water and Surface Water: A Single Resource* (Winter et al. 1998), and *Streamflow Depletion by Wells* (Barlow and Leake 2012).

BASIN MANAGEMENT CONCEPTS

Based on the discussion above, it should be apparent that groundwater pumping can be considered sustainable only to the degree that we accept associated impacts to surface water systems. Groundwater pumping is just another way of diverting surface water. One can think of it as "borrowed" water: water that will ultimately need to be "paid back" by streams and other surface water systems if groundwater levels are not to be in continuous decline. Deep wells in confined aquifers, and wells distant from streams, simply take longer to impact these surface water features.

Since the long-term yield of a groundwater basin can be viewed as the amount of depletion wells can impose on streams, it is important for stakeholders to decide how much of the resulting impacts they are willing to live with. Development and management of groundwater resources requires a trade-off decision, balancing use of pumped groundwater against use of local surface water systems by people and the environment. Unfortunately, these trade-offs typically go unacknowledged during development of groundwater supplies, with resultant unintended impacts to the environment and surface water rights holders. In this section we discuss important concepts regarding basin management, and in the subsequent section, methods of increasing yield.

Safe Yield

Safe yield is commonly stated as being a desirable goal for groundwater yields, and is commonly defined as "the amount of water which can be withdrawn from it annually without producing an undesirable result" (Todd 1959). This definition might seem adequate on the surface but it is too freighted with popular misconceptions, as well as law and the outcomes of legal proceedings that do not account for the physical reality of the connection between groundwater and surface water. One common problem with application of the safe-yield concept is that it is typically assumed to be equal to the average annual recharge rate of the groundwater basin. This is problematic for a number of reasons, including the fact that if all the recharge water is ultimately pumped out via wells, then no water would remain as natural discharge to meet nature's needs. In addition, the long-term yield of a basin is actually dependent on the amount of discharge the cone of depression can capture from surface water systems, not the amount of recharge water entering the system (Bredehoeft, Papadopulos, and Cooper 1982). For these and other reasons the classic safe-yield concept for groundwater basin management has been widely discredited (Bredehoeft, Papadopulos, and Cooper 1982; Sophocleous 1997; Alley and Leake 2004).

Sustainable Yield

The preferred management objective for groundwater basins should be *sustainable yield*, which we define as a yield that explicitly acknowledges the trade-offs between groundwater pumping and associated surface water impacts, and that is arrived at through a stakeholder-driven process. This process would account for the economic benefits of land cultivation, economic and cultural benefits of in-stream flow requirements for key fish, maintenance of important riparian habitat in select streams, and existing surface water rights. Strategies to maximize sustainable yield revolve around deliberately increasing recharge to the system (*managed recharge*), proactively managing the timing and location of pumping, and, ultimately, limiting the total amount of groundwater pumping from the basin. As we will discuss, this last conclusion brings with it a host of political, institutional, and legal issues, as well as questions regarding human nature.

Overdraft

Overdraft is a condition of continual water-level decline in a groundwater basin and is often referred to as *groundwater mining*. Overdraft is a consequence of pumping that exceeds the ability of the cone of depression to capture surface waters, including stream depletion. This condition is unsustainable in the long term and ultimately results in declining well yields and well failures. In some basins, severe water-level declines can also lead to unacceptable amounts of land subsidence. Management of land subsidence consists of maintaining groundwater levels above the value that triggers the subsidence—there is no other management option. Identifying this trigger value is complicated and associated with a high level of uncertainty. The safest management tactic is to operate basin water levels well above those known to trigger subsidence. In addition, it is wise to maintain additional water in storage to meet drought and other water-shortage emergency needs, which calls for even higher groundwater levels.

METHODS TO INCREASE THE YIELD OF A GROUNDWATER BASIN

The yield of a groundwater basin can only be increased by enhancing recharge, commonly referred to as *artificial recharge*, and by decreasing consumptive use of water, which represents water not returned to the system, such as evaporative loss. An overdrafted basin can only be brought back into balance by similar means. Some suggest that increasing agricultural water-use efficiency can increase basin yield, but, as discussed below, this is rarely the case.

Artificial Recharge

Artificial recharge refers to planned activities meant to increase the amount of water recharging a groundwater basin. This can take many forms, but the largest projects rely on recharge ponds. These ponds are most often constructed in and adjacent to ephemeral stream systems where favorable soil and aquifer characteristics exist. Only certain portions of the aquifer, referred to as *recharge areas*, are amenable to large-scale artificial-recharge projects. The limited extent of these areas underscores the importance of protecting them from development. Another important form of artificial recharge is called *in lieu recharge*, wherein surface supplies are provided to groundwater users "in lieu of" pumping groundwater. Other approaches to artificial recharge, such as injection wells and dual-use aquifer storage and recovery wells, can be also be useful in certain applications.

Artificial recharge has been used to successfully increase the yield of numerous groundwater basins in California (see e.g. Orange County Water District 2003). However, limited areas of suitable land, along with limited supplies of water available for recharge, ultimately limit the size of recharge projects. Although artificial recharge should generally be maximized to the extent economically feasible, the pump is mightier than the pond—it is simply not possible to recharge enough water to satisfy all potential groundwater demands. Therefore, groundwater pumping must ultimately be limited in some way.

Decreased Consumptive Use

Consumptive use refers to water that is actually removed from the system. Examples include water transpired by plants and water evaporated from the soil surface. In contrast, *nonconsumptive use* describes water uses in which the water is withdrawn and then returned to the system. Examples of nonconsumptive use include hydropower generation, instream flows, or the portion of irrigation that percolates back into the groundwater system.

It is the consumptive use of water that must be decreased in basins where overdraft is too great to be addressed by artificial recharge alone. The consumptive use of water in most basins currently in overdraft is predominantly for agricultural activities that rely heavily on groundwater pumping. Accordingly, the burden of curtailing consumptive use of groundwater rests heavily on the agricultural sector and will require hard choices such as switching to crops that consume less water and/or fallowing land. Economic impacts from these choices will be significant, both to farmers and to local communities. Possibly the best that can be said is that if these choices are made soon, the transition to sustainable groundwater supplies can be made in a planned and coordinated manner. The alternative is general failure of the groundwater supply to meet local needs, as experienced in the Paso Robles area, where overdraft has led to numerous well failures (Pierson 2013).

The Fallacy of Increasing Supply by Increasing Irrigation Efficiency

Increasing the efficiency of agricultural irrigation is frequently touted as an approach to increasing water supplies or reducing water use. This is largely a fallacy, because the vast majority of water consumptively used by agriculture is evapotranspiration losses from the crop, and this loss is not significantly reduced when irrigation efficiency is increased. Indeed, "excess" water applied in an "inefficient" manner, or even leaked from inefficient delivery systems, actually returns to the basin through runoff into streams or as groundwater recharge. In this way, excess (inefficient) irrigation water from one area becomes a supply for another area.

Failure to recognize this relationship has led to unintended consequences, one published example coming from the area of Walla Walla, Washington (Bower and Petrides 2009). Water managers and farmers in this area implemented numerous irrigation-efficiency and conservation projects, including changes in methods of applying water and shifting to lower-water-use crops. As noted in the article, "The seemingly successful restoration overlooked an important aspect of the surface-water management: the role of groundwater." The result was drying up of once-perennial springs, declining river flow, and ultimately adverse impacts to the very riparian and salmon habitats they were trying to restore. A key conclusion of the article supports our primary argument: "Water managers are learning that sustainable solutions are possible as long as surface water and groundwater are recognized as linked and interdependent."

THE CALIFORNIA CENTRAL VALLEY EXAMPLE
OF GROUNDWATER PUMPING EFFECTS
ON SURFACE WATER

We are currently using a computer model of the Central Valley to better illustrate the effects of groundwater pumping on streams. Our hope is that by better illustrating and understanding these impacts we can help bring about more effective and sustainable groundwater management practices in the state. The model being used for this work is the California Central Valley Simulation Model, commonly referred to as C2VSim (Brush, Dogrul and Kadir 2013). The model is useful for describing the general behavior of the system, illustrating long-term trends, and assessing future conditions under various management options. Future studies can and should refine the findings presented here. Indeed, it is our hope that our findings will spark additional work to address these critically important issues.

The Central Valley of California is commonly divided into four hydrologic regions: Sacramento Valley, Delta, San Joaquin Basin, and Tulare Basin (figure 8.4). It is one of the most important agricultural areas in the United States and covers about 6.8 million acres. This area represents less than 1 percent of U.S. farm land but about 10 percent of U.S. crop value. About 10 maf/y of groundwater is currently being pumped from Central Valley aquifers in an average year (figure 8.5), which represents about 20 percent of total pumpage in the entire United States (Williamson, Prudic, and Swain 1989).

The vast majority of groundwater pumping in the Central Valley is for agricultural use, with about half occurring in the Tulare Basin. Our work suggests that the Tulare Basin is in a long-term condition of overdraft, on the order of 1 maf/y, making it the most heavily overdrafted basin in the state. The consequences of long-term overdraft are well known in the southern San Joaquin and Tulare Lake basins, and include loss of perennial streams and land subsidence (Galloway, Jones, and Ingebritsen 1999; Faunt 2009). However, it is in the Sacramento Valley, where groundwater conditions are generally thought to be healthy, that we may have the most to lose from the status quo, and this is the topic of our next section.

The Sacramento Valley

Groundwater levels are still relatively close to the surface in the Sacramento Valley, and numerous perennial streams and healthy ecosystems remain. Controlled reservoir releases provide robust streamflows, and widespread irrigation with surface water has supplemented natural recharge to keep groundwater levels relatively shallow over much of the valley despite large amounts of groundwater pumping. However, recent information suggests that this relative balance is being eroded, and our work suggests that it may be eroded further in the future under the status quo.

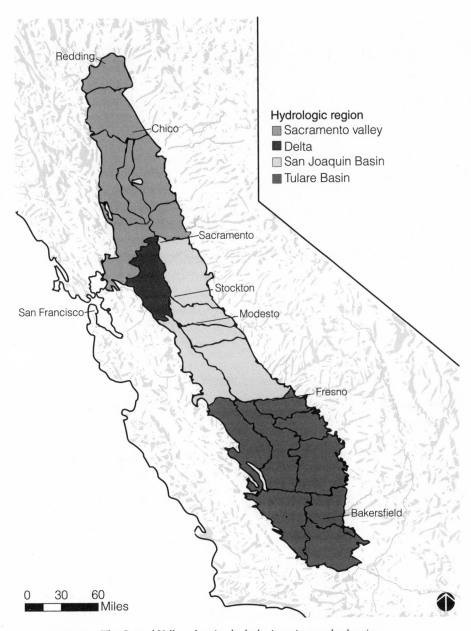

FIGURE 8.4. The Central Valley, showing hydrologic regions and subregions.

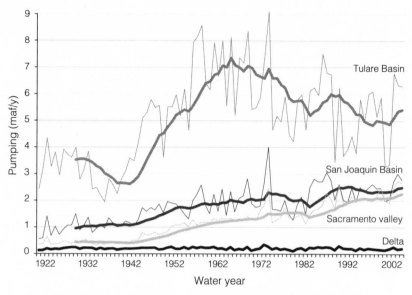

FIGURE 8.5. Historical groundwater pumping in the Central Valley by hydrologic region. Thin lines are yearly values; thick lines are 10-year moving averages.

Streamflow Impacts from Groundwater Pumping

Groundwater withdrawals in the Sacramento Valley have increased through the years, from about 0.4 maf in 1922, to 1.5 maf in 1985, to 2.2 maf in 2009 (figure 8.5). Figure 8.6 illustrates the model simulated groundwater budget of the Sacramento Valley for this historical period, as well as a simulated future period from 2010 to 2083. The vertical bars in figure 8.6 represent flow into or out of the groundwater basin in a given year. Positive numbers reflect sources of recharge to the basin such as deep percolation of rainfall, excess agricultural irrigation water, and canal seepage. Negative numbers reflect outflows from the groundwater system such as discharge to streams and groundwater pumping.

The simulation of the future period assumes that land use remains as it was in 2009, which means no growth in agricultural or urban water demand. Rainfall and surface water hydrology are assumed to be the same as they were during the 37-year period from 1973 to 2009, with this hydrology repeated twice to illustrate long-term impacts of the status quo. This period was chosen because the reservoir storage system present today was largely in operation then. As discussed below, results from both the historical and future simulations show troubling trends

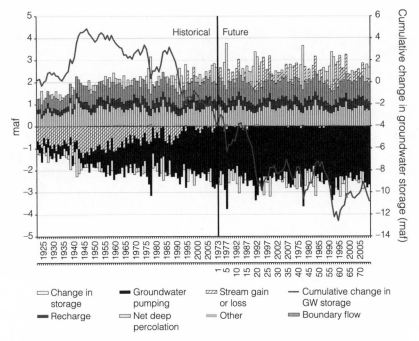

FIGURE 8.6. Historical and "status quo future" groundwater budget for the Sacramento Valley. Future years are shown as calendar years that the future hydrology is based upon, as well as number of years simulated.

with respect to effects of groundwater pumping on surface water flows in the Sacramento River and its tributaries.

Figure 8.6 indicates that most of the recharge to the Sacramento Valley groundwater basin from the 1920s to the 1940s ultimately discharged to streams. As groundwater pumping increased through time, the amount of water discharged to streams decreased, reaching a "tipping point" in recent years where overall losses of streamflow *to* groundwater on a valley-wide basis are greater than stream gains *from* groundwater. The simulation indicates that these trends will continue even under status quo conditions (figure 8.7). This is due to the delayed impacts of the increased groundwater pumping over the past several decades. This is an important point because it implies that impacts to streamflows will continue to increase into the future even if groundwater pumping is capped at current amounts.

Where figure 8.5 shows the water balance from the perspective of the groundwater system, figure 8.7 shows the situation from the perspective

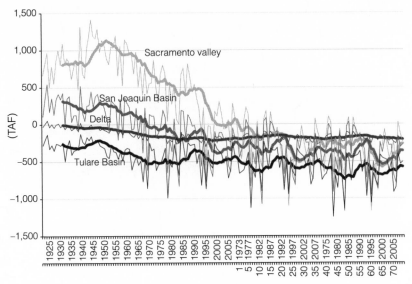

FIGURE 8.7. Historical and "status quo future" sum of annual gain and loss of streams by hydrologic region. Future years are shown as calendar years that the future hydrology is based upon, as well as number of years simulated. Thin lines are yearly values; thick lines are 10-year moving averages.

of the streams. In this figure, if the lines are above zero (positive values) then the streams in the region are net gainers of water from the groundwater system, while negative values mean streams are net losers of water in the region. This figure indicates that the San Joaquin Basin hydrologic region reached a tipping point in the mid-1960s. Although the tipping point for the Sacramento Valley appears fairly dramatic in this figure, it should not be over-interpreted. This figure does not indicate that all the streams in the valley are going dry. Rather, it indicates that the current overall valley-wide contribution of groundwater discharge to surface water is much less than historic values. An important aspect of this situation is that additional reservoir releases will be needed to meet in-stream flow requirements and water delivery calls, including export and environmental flow calls at the Delta. This is an important issue that is discussed at more length later in this section.

Figures 8.8(a) and (b) illustrate simulated results for relative change in net stream gain/loss relationships in specific stream segments for the 1920s and the 1990s. A number of streams in the Sacramento Valley change from net gainers of water in the 1920s to net losers of water in

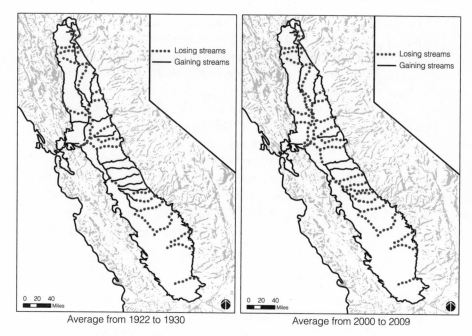

FIGURE 8.8. Maps showing net gaining and losing streams in the 1920s and 1990s. A number of streams are net losers of water in recent years. Even gaining streams are typically gaining less water than they were in the past.

the 1990s. Interestingly, the modeling suggests that some streams actually change from net losing to net gaining. These changes are related to a number of local factors, including the amount of groundwater pumping and availability of new supplies of surface water for irrigation, with resultant recharge from deep percolation of irrigation water.

Potential Impacts of Sacramento Valley Streamflow Depletion at the Delta

Returning to the simulation results provided in figure 8.7, we see that, relative to the 1940s, streams in the Sacramento Valley lost an average of about 800,000 acre-feet of water each year to the groundwater system due to groundwater pumping and may lose another 300,000 acre-feet per year in coming years under status quo conditions. This represents a loss of more than one million acre-feet of water per year of potential flow from the Sacramento Valley into the Delta.

The Delta is a hub for California water supplies, and environmental outflow requirements in the Delta are a key control for water releases from upstream reservoirs as well as water exports to the south. Any loss of flow to rivers in the Sacramento Valley is therefore an important statewide issue. The State Water Resources Control Board's Water Right Decision 1594 stated that "water is considered to be available for appropriation if the Delta is not 'in balance'" (2013, 13). Although the loss of 1 maf/y (1,400 cfs) of flow in the Sacramento River may not be significant when abundant runoff is present, during times when the Delta is "in balance" it means that additional reservoir releases are needed to meet the same downstream deliveries. Water that would have been in the stream without the groundwater pumping is no longer in the stream, so more water must now be released from upstream reservoirs to meet downstream flow requirements.

Effects of Groundwater Transfers from the Sacramento Valley

During times of drought, the transfer of water from areas of relative abundance to areas of shortage is often discussed as a means to ease impacts. One form of transfer is called a *groundwater-substitution transfer*. This occurs when water users in an area with relatively abundant surface water supplies, such as the Sacramento Valley, forgo their surface water entitlement and in its place substitute groundwater pumping to meet their needs. The surface water entitlement is then transferred to water-short areas of the state such as the San Joaquin Valley or Southern California.

We conducted model simulations using C2VSim to assess impacts in the Sacramento Valley from these types of transfers. Figure 8.9 shows water level effects near the Sacramento River from one year of groundwater-substitution transfer pumping totaling 186,000 acre-feet. Water-level declines related to this substitution pumping reach a maximum of about 0.25 feet in the shallow water table zone (OW3-L1) and about 2.5 feet in the deeper confined zone (OW3-L2). The key point illustrated by this graph is that water-level changes due to the additional pumping for the transfers persist over many years, even near recharge boundaries such as streams. As discussed below, these lower water levels translate to stream depletion.

Just one year of groundwater-substitution pumping results in stream-flow depletion over many years (figure 8.9). As we noted before, pumped groundwater is, in a sense, borrowed water that must ultimately be repaid through streamflow depletion. Figure 8.10 shows that a few years after

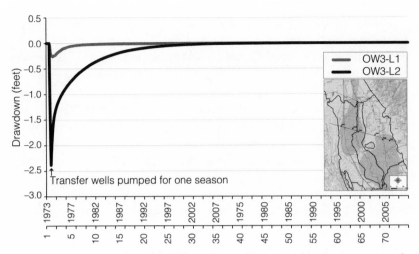

FIGURE 8.9. Long-term water-level effects of groundwater transfers. Just one season of transfer pumping lowers water levels for years.

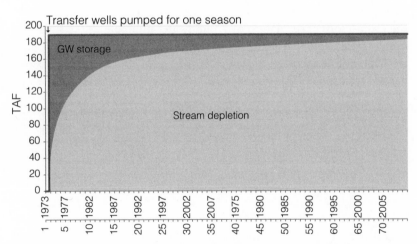

FIGURE 8.10. Long-term stream depletion effects from groundwater transfers. Stream depletion effects from just one season of transfer pumping play out over decades. Effects are distributed through time because depletion effects from deeper wells and those further from streams take longer to be expressed.

substitution pumping has ended, only about half of this pumped water has been repaid. After 10 years, about 140,000 acre-feet, about 75 percent of the amount pumped, has been repaid. Quite striking is the fact that, even 30 years after pumping ceases, streamflow is still being depleted.

The delay in stream depletion impacts from groundwater pumping depends on the distance of the well from the stream, the depth of the well, and the properties of the aquifer system. In fact, this delay in streamflow impacts during groundwater-substitution pumping can have positive aspects. For example, groundwater pumped in a year when water supplies are limited will be paid back over numerous years when surface flows are more abundant. However, because some depletion will still occur in years and seasons when flows are limited, the long-term impacts need to be explicitly considered and accounted for as part of groundwater-substitution programs.

REGULATORY AND MANAGEMENT CHALLENGES TO SUSTAINABLE GROUNDWATER MANAGEMENT

Although, as noted above, groundwater pumping interferes with surface flows, surface water rights, and groundwater-dependent ecosystems, there is no formal regulation of groundwater pumping in the state outside of adjudicated basins. Some areas, such as Orange County, have implemented cooperative groundwater management efforts, but no area has strong controls on groundwater pumping to protect surface water flows. The inherent conflict between largely unregulated groundwater pumping and a separate system of permitted surface water rights presents tremendous management challenges. These issues are explored below.

The Paramount Basin Management Objective

The concept of *basin management objectives* (BMOs) has emerged as a primary guide for groundwater management in California. BMOs are intended to be flexible guidelines describing specific actions by local stakeholders to protect and enhance the groundwater basin. To properly manage groundwater resources, we think that the *paramount BMO* for each basin needs to be a clear definition of an agreed-on sustainable yield that explicitly addresses and incorporates the trade-offs between groundwater pumping and surface water impacts.

A sustainable yield that explicitly allocates some water for the critical needs of nature is the preferred goal. In basins with severe overdraft, the simple arrest of water-level declines is of immediate concern. In these basins, BMOs should first address stabilizing water levels, then raising them above levels that might trigger subsidence, and finally, placing additional water into storage to meet drought and other water-shortage emergency needs.

The Need for Modeling Tools

The fact that groundwater pumping impacts on streams often play out over years and decades is a significant management challenge. Impacts from pumping may not be recognized until long after the pumping that caused them has been initiated and is well established, and impacts may become more severe over time—as was the case in the examples provided above for the Sacramento Valley. Intelligent use of modeling tools to support informed decision-making, coupled with robust monitoring programs, is the only remedy to this situation. This is most likely to be successful when performed in context of a proactive and adaptive approach to management activities, as discussed below.

Preference for Proactive and Adaptive Local Management

Operating basins at a sustainable yield requires strong, insightful, flexible, and adaptive management that accommodates local conditions and needs. This management is typically best performed locally, which the water management community in California overwhelming prefers. However, sustainable groundwater management ultimately requires that the amount of groundwater pumped from a basin be limited, given that, as we have noted, our ability to pump groundwater out of a basin far outweighs our ability to replenish the supply. Whether local interests can agree to and enforce such limits is open to question. Experience suggests that the "tragedy of the commons" will arise, whereby each pumper pursues their own short-term benefit to the ultimate detriment of all. If this situation can be addressed locally in a clear and firm manner, then it is all for the better. If not, then some other method of limiting pumping must be implemented.

Basin Prioritization

Many people automatically assume that overdrafted groundwater basins, especially those where land subsidence and well failures are present, have the most complex and pressing problems and therefore require the most attention and money. However, if we want to avoid problems in areas that are reasonably healthy today, it is imperative that we consider the overall value of the hydrologic system, both to people and to nature. An example of this situation exists in the Sacramento Valley, where river systems not only provide important ecosystem services but also serve as important water transmission systems. Groundwater pumping directly impacts flows in these rivers and therefore the water supplies for millions of Californians and hundreds of thousands of acres of farmland. The importance of properly managing groundwater pumping in this basin should be recognized, managed, and supported accordingly and not given lower priority simply because it is not as degraded as other more populated or heavily pumped basins.

Climate Change

Climate change adds yet another layer of complexity to sustainable groundwater management in California, and also heightens its importance. A reduction in snowpack with an attendant increase in rainfall runoff will mean that streamflows in affected areas will be "flashier," with higher peak flow but shorter duration of flow. Because recharge is a rate-limited process, a shorter period of wetting will decrease the amount of streamflow that percolates into groundwater basins. In addition, groundwater storage is likely to be called on to replace dwindling snowpack storage. Higher temperatures mean more crop evapotranspiration and therefore increased irrigation demand. Groundwater is certain to play a key role in meeting this increased demand. There is also evidence that droughts may be more common, and longer, in the future. The large amounts of water present in groundwater systems will certainly play a role in coping with these challenges.

CONCLUSIONS

If California is to maintain a vibrant economy, we must manage our groundwater supplies more effectively and recognize their intimate connection with surface waters. Only in this way can we ensure that our water resources are resilient to inevitable droughts and flexible in the face of climate change, and that we maintain the riparian and aquatic ecosystem landscape that drew so many people to California in the first place.

Proper management of groundwater requires limiting pumpage to a sustainable yield that explicitly acknowledges and addresses the inherent trade-offs between groundwater pumping and surface water flows. As part of this effort we must address the following facts in state water law and policy:

- Pumped groundwater is borrowed water that ultimately has to be paid back by streams—groundwater wells are really just another way of diverting surface water.

- Groundwater pumping is only sustainable to the degree that we accept the associated impacts to surface water systems. Stakeholders must agree to the trade-offs this entails.

- Outside of adjudicated basins, groundwater is, from a practical perspective, the de facto senior surface water right in our current regulatory system. This critical disconnect between groundwater and surface water must be addressed.

These statements are meant to be provocative and as a call to action. The technical basis for these arguments has long been known to water professionals and cannot continue to be ignored by state law and policy without the risk of periodic economic dislocation during our natural drought cycles and gradual degradation of the environment.

Time is of the essence. Groundwater pumping is increasing. Towns, businesses, and entire economies are being built and expanded on overdrafted, unsustainable supplies of groundwater. The consequences of delay will only become more severe and more costly to address in the future. We do not need perfect knowledge or perfect models. We already have enough insight into our major basins to take many actions now. Inherent uncertainties in our understanding of these systems and future conditions are best addressed by an adaptive management approach, one that engages stakeholder groups in an ongoing process, one that sets priorities and direction and incorporates a sound understanding of the way groundwater and surface water systems interact.

REFERENCES

Alley, William M., and Stanley A. Leake. 2004. "The Journey from Safe Yield to Sustainability." *Ground Water* 42(1):12–16.
Alley, William M., Thomas E. Reilly, and O. Lehn Franke. 1999. *Sustainability of Ground-Water Resources*. Circular 1186. Reston, VA: U.S. Geological Survey.

Barlow, Paul M., and Stanley A. Leake. 2012. *Streamflow Depletion by Wells: Understanding and Managing the Effects of Groundwater Pumping on Streamflow.* Circular 1376. Reston, VA: U.S. Geological Survey.

Bower, Bob, and Aristides Petrides. 2009. "Accounting for Groundwater in Watershed Management." *Southwest Hydrology* (March/April): 24–35.

Bredehoeft, John D., Stephen S. Papadopulos, and H. H. Cooper, Jr. 1982. "Groundwater: The Water-Budget Myth." In *Scientific Basis of Water-Resource Management* (51–57). Studies in Geophysics. Washington, DC: National Academy Press. http://aquadoc.typepad.com/files/bredehoeft-water-budget-myth .pdf.

Brush, Charles F., Emin C. Dogrul, and Tariq N. Kadir. 2013. *Development and Calibration of the California Central Valley Groundwater-Surface Water Simulation Model (C2VSim), Version 3.02-CG.* http://baydeltaoffice.water.ca.gov /modeling/hydrology/C2VSim/download/C2VSim_Model_Report_Final.pdf.

California Department of Water Resources. 2013. *Draft California Water Plan Update.* www.waterplan.water.ca.gov/cwpu2013/final/index.cfm.

Faunt, Claudia C. 2009. *Groundwater Availability of the Central Valley Aquifer, California.* Professional Paper 1766. Reston, VA: U.S. Geological Survey.

Galloway, Devin, David R. Jones, and S. E. Ingebritsen. 1999. *Land Subsidence in the United States.* Circular 1182. Reston, Virginia: U.S. Geological Survey.

Heath, Ralph C. 1983. *Basic Groundwater Hydrology.* Water-Supply Paper 2220. Reston, VA: U.S. Geological Survey.

Orange County Water District. 2003. *A History of Orange County Water District.* Fountain, CA. www.ocwd.com/Portals/0/About/HistoricalInformation/A%20 History%20of%20Orange%20County%20Water%20District.pdf.

Pierson, David. 2013. "In Paso Robles, Vineyards' Thirst Pits Growers against Residents." *Los Angeles Times*, September 1.

Sophocleous, Marios. 1997. "Managing Water Resources Systems: Why 'Safe Yield' is not Sustainable." *Ground Water* 35(4):561.

State Water Resources Control Board. 1983. *Decision 1594.* http://www.water- boards.ca.gov/waterrights/board_decisions/adopted_orders/decisions /d1550_d1599/wrd1594.pdf

Theis, Charles V. 1940. "The Source of Water Derived from Wells: Essential Factors Controlling the Response of an Aquifer to Development." *Civil Engineering* 10(5):277–80. http://aquadoc.typepad.com/files/theis---source- of-water.pdf.

Todd, David K. 1959. *Ground Water Hydrology.* London: Chapman & Hall.

Williamson, Alex K., David E. Prudic, and Lindsay A. Swain. 1989. *Ground-Water Flow in the Central Valley, California.* Professional Paper 1401-D. Reston, VA: U.S. Geological Survey.

Winter, Thomas C., Judson W. Harvey, O. Lehn Franke, and William M. Alley. 1998. *Ground Water and Surface Water: A Single Resource.* Circular 1139. Reston, VA: U.S. Geological Survey.

CHAPTER 9

People, Resources, and Policy in Integrated Water Resource Management

CELESTE CANTÚ

No one disagrees that water and watershed health are critical for people, the environment, and our economy. But are we using twentieth-century strategies to answer twenty-first-century challenges? Challenges such as climate change, collapsing ecosystems, and population growth and development—resulting in water scarcity, higher water and energy costs, and environmental injustice—are accumulative and synergistic. These pressures demand that we change how we think about water and, more importantly, how we manage it.

The Santa Ana Watershed Project Authority (SAWPA) in Southern California characterizes these challenges as no less than the Four Horsemen of the Apocalypse: climate change, drought on the Colorado River, the vulnerability of the Sacramento–San Joaquin River Delta, and population growth and development. The Santa Ana River watershed is among the most arid within the United States, with forecasts that overwhelmingly suggest increased aridity and more intense and frequent drought. Low rainfall is the new normal, and higher temperatures, leading to more evaporation, make the problem worse, along with changing precipitation patterns. The U.S. Bureau of Reclamation's 2012 *Colorado River Basin Water Supply and Demand Study* contains an ominous prediction about the future of the river. In 2060, it says, the average demand for water is likely to outstrip the supply by roughly eight times the current usage of Las Vegas. At the same time, the water imported from the Sacramento–San Joaquin River Delta, which serves two-thirds of California's population and irrigates millions of acres of farmland where our food is grown, has been curtailed by drought. Fish populations have declined dramatically, leading to historic restrictions in water supplied. Also, the Delta levees are fragile and vulnerable to expected earthquakes, floods, and rising sea levels. The California Department of Finance (2014) reports that California's population will cross the 50-million mark by 2055. This population gain would exceed the current population of either Illinois or Pennsylvania—enough new residents to rank on its own as the fifth-largest state in the Union. Riverside County will be the fastest-growing county in the state. These threats are not only to the Santa Ana River watershed; much of California sees the same challenges. We are facing the perfect storm as we come to understand the stark realities of the twenty-first century.

Through an integrated water resource management (IWRM) process in the Santa Ana River watershed we have learned to integrate people and functions, without artificial political or professional boundaries getting in the way, as a means of addressing these realities. But it's not easy. IWRM is challenging not only from a technical or policy perspective but for reasons you might not suspect. In this chapter we will learn about SAWPA's One Water One Watershed (OWOW) plan, a crest-to-coast, corner-to-corner integrated watershed-wide planning and management initiative considering all water resources and stakeholders in the watershed.

THE STORY OF SAWPA AND THE SANTA ANA RIVER WATERSHED

In 1968, a judge's decision settling a water-rights dispute among the largest water wholesalers in the watershed was met with, "No judge, you got it wrong!" The judge suggested that the wholesalers pool their funds and hire a professional staff to help resolve conflicts; and with that, SAWPA was created. Established as a joint powers authority in 1968, SAWPA began as a planning agency and was reformed in 1972 with a mission to develop and maintain regional plans, programs, and projects that would protect the Santa Ana River basin's water resources to maximize beneficial uses in an economically and environmentally responsible manner. Located in Riverside, California, SAWPA's headquarters is approximately in the geographic center of the watershed, and it serves an area that corresponds with the boundaries of the watershed, the appropriate primary unit for integrated management.

The watershed begins at the headwaters in the San Bernardino and San Gabriel Mountains, home to the San Bernardino National Forest. The Mediterranean climate has hot, dry summers and cooler, wetter winters. Most of the precipitation occurs between November and March in the form of rain, with variable amounts of snow in the higher mountains of the watershed. The Santa Ana River watershed drains 2,650 square miles and is home to over six million people, with major population centers including parts of Orange, Riverside, and San Bernardino Counties, as well as a sliver of Los Angeles County. The Santa Ana River flows over 100 miles, draining the largest coastal stream system in Southern California as it discharges into the Pacific Ocean.

SAWPA focuses on a broad range of water resource issues, including supply reliability, quality improvement, recycled water, wastewater treatment, groundwater, and salt management, under the umbrella of IWRM. SAWPA works with planners, water experts, engineers, and scientists in other agencies, using IWRM, collaboration, and innovation to identify issues and solutions to resolve water-related problems. Today SAWPA has grown in capability to become one of California's leading regional water agencies.

SAWPA serves at the direction of the SAWPA Commission, composed of five member agencies—Eastern Municipal Water District, Inland Empire Utilities Agency, Orange County Water District, San Bernardino Valley Municipal Water District, and Western Municipal Water District—which together represent the majority of the water management authorities and stakeholders within the watershed. They are wholesale and retail water agencies that manage groundwater production, desalination, resource management, wastewater collection and treatment, and regional water recycling. The watershed outgrew the local water supply, and imports from the Colorado River and the San Joaquin–Sacramento River Delta made up the difference. Three are members of the Metropolitan Water District of Southern California. One is a state contractor that imports water directly. Three have energy recovery/production facilities. Many have become recycled water purveyors or biosolids/fertilizer treatment providers, and all focus on water supply and salt management to protect the region's vital groundwater supplies. Several have created stormwater capture opportunities, using thousands of acres to capture flows and recharge the groundwater basin. Several coordinate with others to operate dams for water conservation and water quality improvement, along with flood control. They are considered among the most innovative water resource managers in the West.

SAWPA'S INTEGRATED WATER RESOURCE MANAGEMENT PLANS

While encompassing many approaches, IWRM is essentially an iterative, evolutionary, and adaptive process, building understanding, developing local capacity, and creating ownership. There are several fundamental principles that must be included in any description. The hydrographic watershed must be the jurisdictional boundary, rather than county, state, or city lines. A wide variety of stakeholders, who operate generally with equality, are involved. Most principal negotiations involve face-to-face interactions among the major stakeholders under agreed-on process rules designed to insure civility and engender trust. The goal of the process is to reach win-win solutions on a set of interrelated social, economic, and environmental problems so that no one goes away from the process seriously displeased. The process entails a fairly extensive fact-finding phase designed to promote understanding of the magnitude and causes of various problems. Typically, this involves a mixture of standard scientific techniques and respect for local knowledge.

It is crucial to note that IWRM practices depend on context. At the operational level, the challenge is to translate the principles into concrete action. IWRM should be viewed as an ongoing process, one that is long-term, forward-moving, and iterative rather than linear in nature. As a process of change which seeks to shift water development and management systems from their currently unsustainable forms, IWRM has no fixed beginnings or endings. Furthermore, there is not one correct administrative model. The art of IWRM lies in selecting, adjusting, and applying the right mix of these tools for a given situation. IWRM is defined by the Global Water Partnership (2000, 22) as a "process which promotes the coordinated development and the management of water, land and related resources, in order to maximize the resultant economic and social welfare in an equitable manner without compromising the sustainability of vital ecosystems." In spite of the calls for IWRM, success has been spotty, attesting to the difficulty of reconciling many conflicting agendas and challenges. UN-Water's *Status Report on the Application of Integrated Approaches to Water Resources Management* says that 64 percent of countries have developed IWRM plans and 34 percent report an advanced stage of implementation. However, the progress appears to have slowed, or even regressed. Adding to the challenges is the multiplicity of institutions, making coordination and stakeholder management even more difficult. This complex process relies on a series of champions, has very high transaction costs, and requires patience. Today IWRM is still called an emerging strategy; many believe it is the only way we can succeed with the challenges of the twenty-first century.

SAWPA has been emphasizing integration for the last 15 years. In 2007, SAWPA decided to reach for a new level of water management integration with the OWOW process. In this latest update to the IWRM plan, the consensus was that the OWOW effort would need to be bold and innovative to meet the watershed's vision.

An OWOW Steering Committee was organized by SAWPA to reflect more than the water-supply orientation and include a cross-section of many types of stakeholders, such as environmental, regulatory, municipal and county, and flood control, thus providing equal opportunity and representation throughout the watershed. Through the Steering Committee, decision-making and distribution of power and voice are provided to the stakeholders of the watershed in the planning process. The OWOW Steering Committee consists of eleven members from the watershed's three counties, serving staggered four-year terms. They are: two SAWPA Commission representatives selected by the commission, three county supervisors selected by their respective boards, three City Council members selected by a majority vote of the Council of Governments in each of the three counties, business and environmental community representatives selected by a majority vote of the eight governmental representatives on the Steering Committee (based on an application process conducted during a public meeting), and a representative selected by the Santa Ana Regional Water Quality Control Board.

The Steering Committee is responsible for the development and recommendation to the SAWPA Commission of goals and objectives for the watershed (the OWOW plan), and acts as the oversight body that performs strategic decision-making, crafts and adopts programmatic suites of project recommendations, and provides program advocacy. This includes conducting public hearings, receiving input to provide direction for the development and long-term maintenance of the plan, and developing a project prioritization process. The Steering Committee provides financial incentives for the development of multi-benefit integrated projects, identifies institutional barriers and opportunities for more efficient management that further advance the integration of water management activities, and advocates for policy changes that increase interagency effectiveness and efficiency in IWRM.

To guide the development of the OWOW plan, SAWPA staff working with the Steering Committee established a vision along with goals and objectives for the watershed that would incentivize a holistic approach to resource management. The following guiding principles were adapted from Peter Senge's *The Necessary Revolution* (2008), where he frames two concepts: *problem-solving* and *creating anew*. *Problem-solving* is about making what you do not want go away. *Creating anew* involves bringing something you care about into reality, such as sustainability, water reliability, or water quality. We are shifting the conversation from the familiar—avoiding something bad—to doing something positive and new. We are shifting the conversation from problems to possibilities.

Take a System Approach

- *See the Santa Ana River watershed as a hydrologic whole.* The planning process must be watershed-wide and bottom-up to allow for a holistic, inclusive approach to watershed management.
- *Working in concert with nature is cost-effective.* Gravity flow, percolation, wetlands, meadows, and other green infrastructure perform important functions and often have lower life-cycle costs.
- *See all problems as interrelated; seek efficiencies and synergies.* The OWOW plan and projects must pursue multiple objectives beyond the traditional objective of providing reliable water, including: ensure high-quality water for all users; preserve and enhance the environment; promote sustainable water solutions; manage rainfall as a resource; preserve open space and recreational opportunities; maintain quality of life, including the needs of disadvantaged communities; provide economically effective solutions; and improve regional integration and coordination.

Create Anew

- *OWOW is a shared vision for the watershed.*
- *Breakthrough innovations* are needed to answer emerging challenges.
- *Establish a new water ethic:* everyone knows where their water comes from, how much of it they use, what they put into it, and where it goes after it leaves them. We asked stakeholders to move from problem-solving to creating—to create a new shared vision and to realize breakthrough innovations. The change we are

seeking is not to try harder to maintain the status quo of the twenty-first century.

Collaborate across Boundaries

- *As citizens of the watershed, create solutions.* We asked everyone to check their identity at the door. All other loyalties, agendas, and priorities were to be secondary to those of the holistic view of the Santa Ana River watershed. We asked stakeholders who feel strongly to let go of cherished beliefs and views so that they could allow something bigger than themselves to develop. We asked a lot.

- *No one person can do it alone.* No one has enough understanding, credibility, or authority to connect the larger networks of people and organizations to do this work. We have to do it together. There is no reason to assume that when each agency seeks to optimize results within its jurisdiction the results will be a solution that is optimum overall. In fact, we know that just the opposite is often case. Water districts all over California answered the twentieth-century call to optimize to their district's advantage often at the expense of habitat, wildlife, and their downstream neighbors. They would "rob Peter to pay Paul" without increasing water benefits in the aggregate. We first optimize in the aggregate, and later implement at a smaller district scale.

- *Think big.* The plan must improve conditions throughout the watershed, ensuring that an improvement in the welfare of one area is not at the expense of others and that when such expenses are unavoidable, compensation is found. Quality of life must be protected, and economic impact must be understood.

With these established principles, the OWOW Steering Committee conveyed a sense of urgency, conveying direction to produce a plan that was more aggressive, for major changes in how developing, protecting, and conserving water is approached.

To manage the technical and planning work, the stakeholders organized into separate workgroups, each designated as a "pillar" within the plan's framework. The pillars identify and vet creative ideas, conduct brainstorming, and assist with regional coordination, outreach efforts, gathering and reviewing data, and developing and reviewing analysis, resulting in the OWOW chapters. The pillars include:

- *Water-Use Efficiency*, focusing on waste reduction and increased conservation
- *Water Resource Optimization*, addressing water reliability, supply, and security
- *Beneficial-Use Assurance*, focusing on water quality
- *Energy and Environmental Impact Response*, addressing the water–energy nexus and climate change
- *Natural Resources Stewardship*, addressing the environment, habitat, parks, recreation, and open space
- *Land-Use and Water Planning*, addressing the water–land nexus
- *Stormwater*, including both resource opportunities and risk management
- *Operational Efficiency and Water Transfers*, looking for opportunities to optimize water management within and among the agencies
- *Disadvantaged and Tribal Communities*, addressing issues particular to these communities
- *Government Alliance*, co-chaired by Reclamation and the Los Angeles Regional Office of the U.S. Army Corps of Engineers, and including representatives from nine federal, five state and local agencies, and two tribes, as well as the emergency support services of all three counties within the watershed.

Each pillar consists of ten to sixty volunteers, including participants from local agencies, special districts, nonprofit organizations, universities, Native American tribes, and private citizens. Expert volunteer co-chairs, responsible for facilitating the workgroup process, lead each pillar group. The pillars are asked to view watershed resources and problems from a multidisciplinary perspective that extends beyond their topic area, while considering other pillars' perspectives. For example, the Water Resource Optimization pillar considered environmental and habitat-restoration issues when developing its strategies. Through this process, synergies were developed and multi-benefit programs were identified. For example, through this approach, it was possible to incorporate the understanding that many downstream water-resource and water-quality problems could be more effectively and efficiently addressed upstream, at the source, thus requiring collaboration with other entities. Over time, this collaboration among the pillar groups provided a more unified vision, resulting in new integrated and multi-beneficial solutions to water-resource challenges, which increased collaboration among jurisdictions and geographies. To encourage collaboration between pillars, the responsibilities of each were designed to overlap.

SAWPA staff provide administrative and facilitative assistance to the pillars and the OWOW Steering Committee for overall OWOW plan development. In addition, SAWPA provides decision support tools to assist the Steering Committee and pillars in decision-making processes, provides planning documents to allow pillars to build upon existing plans, and performs significant public outreach and education about the integrated planning approach for the Santa Ana River watershed. In addition to the OWOW planning process itself, SAWPA administers ten to fifteen multi-agency task forces that support OWOW. These task forces range from surface and groundwater quality, to threatened-species preservation and restoration, to the Santa Ana River Trail, and are integrated with water resources. Taken together, these task forces constitute over 100 different agencies and organizations in the watershed. The work of these task forces—often involving retail and wholesale water agencies, groundwater management agencies, wastewater agencies, NGOs, businesses, universities, and other organizations—has been integrated into the OWOW planning process.

The resulting 2013 OWOW 2.0 Plan advances a paradigm change from a strictly water-supply to an IWRM mentality, moving from a mission of providing abundant high-quality water at the lowest possible cost to one in which water resources are managed in a sustainable manner and with regard for the needs of the environment and those downstream. Rather than investing more or working harder on the ways of the twentieth century, OWOW 2.0 seeks a new proactive approach that is lighter on the land and protects habitat and a sustainable future, for a robust economy and a healthy environment.

STAKEHOLDER INVOLVEMENT AND OUTREACH

Engaging stakeholder involvement in any large, diverse watershed is challenging, particularly when one considers every resident of the watershed a stakeholder. OWOW was designed to be a "bottom-up meets top-down" process. By encouraging the participation of different groups of people and those holding varying viewpoints from throughout the watershed, OWOW seeks to reach a larger number of stakeholders. The OWOW pillar groups represent an effective means to insure public involvement. The list of stakeholders involved is one of the most extensive ever taken by any regional water management group and includes over 4,000 representatives from 120 water agencies, including flood control, water conservation districts, and wastewater and water supply agencies. It also includes representatives from the sixty-three incorporated cities, including mayors, key department heads, city council members, and planning commissioners. Also included are representatives from county, state, and federal government, Native American tribes, the real estate community, members of the environment and environmental justice, agricultural, and development communities, consultants, trade associations, academia, nonprofit organizations, and others simply interested in water.

SAWPA has hosted multiple workshops, forums, and presentations, including a TEDx Talk, to discuss the benefits of collaboration and multi-benefit watershed projects. The annual OWOW watershed conferences attract over 400 attendees. The conferences serve as an opportunity to invite the public to become involved in OWOW, and to discuss OWOW plan development to date. They also serve to reinforce the OWOW plan goals to encourage a watershed focus, and encourage collaboration in developing multi-benefit projects.

Social media is a component of SAWPA's overall public outreach to provide leadership and information to stakeholders in reaching the goal of a sustainable Santa Ana River watershed. In its social media presence, SAWPA provides a virtual venue to invite collaboration and encourage interaction from others, to inspire and educate watershed residents, and provide Web-based information. SAWPA's website is considered "home base." By using social media tools SAWPA drives more visitor traffic to its website and expands its outreach efforts for events and services.

INDICATORS AND METRICS

Using systemic methods of data acquisition, trust emerges as an important ingredient for success. Indicators and report cards provide pragmatic perspectives for assessment and improvement of basin management performance. As valuable as the knowledge the metrics deliver is the process of joint discovery and the development of a shared understanding of common problems. GIS and graphic displays of data are key in communicating with an array of people. All must have access to data that are displayed in an accessible manner.

The water resource issues facing the Santa Ana River watershed today are complex and interlocking. Here a few IWRM examples.

Water-Use Efficiency and Water Quality

Investment in water-use efficiency inside the home resulted in an average per capita use below the statewide average. Inland, the hotter days and larger lots mean that up to 80 percent of total water consumption occurs in the yard. In coastal areas this is closer to 40 percent. About half the water used to irrigate residential landscape is not needed. It is wasted, and runs off the property, carrying pet waste, fertilizers, and salt into the stormwater collection system, eventually hitting the receiving water body with a slug of toxicity.

Cities are required to implement expensive "total daily maximum loads" meant to protect water bodies for fish and people. Taking the systems approach to solving problems, we can see that if we improve our irrigation practices immediately and ultimately improve our landscaping practices by removing turf and planting water-efficient plants, we can simultaneously improve water quality and water reliability. Waste at one end of the watershed causes water-quality problems in the river or the ocean at the other end. The disconnect is that water suppliers are not accountable for pollution resulting from water use, but cities are. Not until we get all the players from different corners of the watershed in a room can we find solutions. We could save as much as 40 percent of highly treated drinking water; and if we learned to be better landscapers, using water-efficient plants and less turf, we could save closer to 60 percent of all treated potable water! Working together we can see that adopting a budget-based water rate will improve water quality. Water-use efficiency ultimately is not only about incentives or mandates but must also be driven by a water ethic held by the consumer. We will never have the regulatory structure or the funding to fully incentivize water-use efficiency; it must also arise from the ethics of the user. This can only happen if the user has a relationship with water, understanding both its limitations and its value.

Forest First

The San Bernardino and Cleveland National Forests encompass approximately 33 percent of the Santa Ana watershed's land mass and receive

90 percent of its annual precipitation. Runoff directly affects the amount and quality of water received downstream. Yet, historically, there has been little if any relationship between forest and water management. The fire-suppression practices of the twentieth century have resulted in an unnaturally dense forest, reducing water supply and increasing fuel, resulting in more intense and catastrophic fires. Meadows, nature's sponge, retain water for groundwater recharge, but dry up due to natural and manmade channelization, blocking the potential benefit of this resource. Hard lessons were learned after the forests experienced devastating fires in the early 2000s and the aftermath of those fires directly impacted the quality of water downstream. It took years to recover. The 2013 Rim Fire, the largest ever in the Sierra Nevada and the third-largest in California history, was a direct result of these past practices. Today SAWPA and its members have partnered with the U.S. Forest Service to develop Forest First to correct this disconnect. The joint goal is to manage the forest strategically, focusing on the areas most vulnerable to catastrophic fires that could affect the downstream watershed. When trees are thinned, more water is released to the lower watershed, fire threats are lessened, and a natural fire regime, faster-moving and not as hot, prevails.

Thinning the forest to mimic a time before it was so heavily occupied by people serves to reduce fires, and to protect people and the environment, and it also serves water supply, quality, and flood missions. Restored meadows attenuate floods, allowing them to absorb water and slowly percolate it into aquifers that store this high-quality sweet water. This partnership is being pursed among downstream groundwater management agencies, flood control and water conservation districts, water supply agencies, resource agencies, and the Forest Service to find agreeable projects that can be executed in specific areas within the forest and that will have a direct effect in preserving and enhancing the quality and quantity of water resources from the source or headwaters, contributing to the overall health of the watershed. Evidence from a quantitative cost–benefit analysis is being sought to validate investment in projects that will help the Forest Service keep the forest healthy and in turn promote water quantity and quality. This investment in green infrastructure reduces carbon footprint, is less expensive than building a treatment plant, and has Mother Nature doing work for us. Integrating water and disaster management, "from prevention to cure," learning from the aftermath to establish cooperation for prevention, brings two traditionally separate communities together.

Salt Management

Almost a century of agriculture and industry has resulted in salts and other constituents infiltrating many aquifers and streams. Crops irrigated by imported Colorado River water are harvested but they leave behind a salt legacy accumulating in the groundwater. Water quality in the Santa Ana River has improved in recent years thanks to brackish-groundwater desalination and water-quality planning, but challenges remain. Drought exacerbates these challenges. Without sweet rain or Sacramento–San Joaquin Delta water to blend with native water, which is higher in salt, the watershed is closer to maximum salt limits, causing a "double whammy" of water loss due to both drought and salt.

Technologically advanced wastewater-control infrastructure has been rigorously employed, and negative impacts from agricultural runoff continue to be minimized. Nevertheless, the existing salts and contaminants present in the watershed from past practices still need to be removed, as improving water quality is inextricably linked to improving water supplies and implementing a comprehensive groundwater storage program. As part of the solution to the salt issues within the watershed, SAWPA constructed almost 100 miles of pipeline to convey high-saline brine out of the watershed. Desaltors throughout the upper watershed remove and concentrate salt from brackish groundwater, which is collected and conveyed to the coast, where it is treated before being discharged to the sea. The result is sweet water that can be used as supply, while aquifers, the river, and the people who use that water are protected from salt contamination. The lower watershed invested in the brine-line in the upper watershed (outside their jurisdiction) to protect the quality of water that would ultimately reach them.

Mill Creek Wetlands

The Mill Creek Wetlands intercept and route runoff through a series of constructed wetlands, creating habitat for creatures and recreational opportunities for people. They also improve water quality, replenish the aquifer, and protect the river from pollution. Constructed wetlands designed to treat secondary effluent will directly improve the reclaimed water supply. Water produced from the wetlands is of suitable quality to recharge aquifers. Diminishing groundwater resources can be supplemented, or, in some areas, reclaimed water can be recharged as part of a groundwater remediation program. Located along the Pacific Flyway, the critical migratory corridor connecting Canada to Latin America, Southern California wetlands provide vital habitat for migratory waterfowl. This large project is a partnership among cities, the flood control district, and water districts, and it is championed by visionary large private developers with a commitment to sustainability. The project was no single entity's responsibility; together, the partnership benefits many different sectors.

SAWPA has practiced successful IWRM for decades. Success does not lie only in the establishment of a regional agency like SAWPA but rather in a regional perspective that builds upon existing organizations and relationships working from the group up and from the inside out. SAWPA has benefited from a stable revenue stream garnered from the five member agencies and from fees for services and grants. Funding helps support the efforts, but it is not the most important factor. It can also run counter to the effort. If each agency had all the funds required to address its own problems and needs, it might not be motivated to cooperate with others. The severity of water resource problems also creates opportunities for IWRM to flourish. This may have been a strong factor in SAWPA's success, because the watershed does not produce sufficient local water and actors had to band together and cooperate to import surface water supplies and manage groundwater.

California is ripe for Integrated Water Management Planning and successful collaboration across a broad range of stakeholders with differing perspectives. This is born out in SAWPA's history, beginning with the five members, who are distinct geographically, and in their original missions (groundwater management, water importer, domestic or wastewater treatment). As time went on, each of them further diversified and worked closely with the others, and with other missions such as flood control. The latest iteration of OWOW brought in far more diverse stakeholders, including those outside the traditional water community, such as land use, energy, and recreation.

The most successful watershed organizations are not established for a single purpose and dissolved at the end of that effort. Rather, they are those which over time confronted several issues and found that their capacity increased, along with trust and legitimacy, which in turn allowed the watershed group to be more successful at solving the most challenging problems. Longevity by itself is no guarantee, but as Kemper, Dinar, and Blomquist (2007) write, there is a correlation. In fact, processes that take a long time are more likely to be perceived as successful, and have a higher level of stakeholder compliance and support.

After over 45 years of operation, SAWPA does have the benefit of longevity.

Often watershed-based groups are led by those outside of the institutionalized water industry, such as environmental groups and NGOs. They struggle to find the authority to lead or govern. Because the major water wholesalers in the watershed created SAWPA as a joint powers authority, their legitimacy and legal authorities transferred to SAWPA.

SAWPA enjoys strong support from federal and state agencies as well. While we do not enjoy perfect regulatory alignment, the agencies are aware of the need to examine this and are willing to see if alignment can be achieved. There is increasing skepticism about the ability of higher agencies at the state or federal level to render viable long-term solutions to our complex water-quality and water-resources problems on their own. Success is closer to our grasp when the hierarchy of regulatory and other resource agencies is aligned. The U.S. Council on Environmental Quality's updated *Principles and Requirements for Federal Investments in Water Resources* (2013) provides a common framework for federal agencies' funding decisions and promotes investment in integrated, multi-benefit solutions. Even before this direction, the U.S. Bureau of Reclamation's Southern California Area Office and the U.S. Army Corps of Engineers, Los Angeles District, have demonstrated strong leadership, support, and commitment to integration. These two organizations co-chair SAWPA's OWOW pillar and are working hand in glove to support the process. Reclamation invested $1 million, matched by the California Department of Water Resources and SAWPA, to conduct several studies to support OWOW, including climate change adaptation and other regional efforts. The Corps has created a pilot to support watershed-based budgeting. To avoid the piecemeal approaches of the past, the Corps has developed a strategic plan that incorporates environmental operating principles mirroring IWRM principles. The U.S. Environmental Protection Agency has also reiterated support for IWRM with their stormwater combined permit. The California legislature passed the Integrated Regional Water Management grant program, funding $1 billion to incentivize IWRM plans, processes, and multi-benefit projects for California. This follows previous substantial investments to fund IWRM projects, first at SAWPA and later elsewhere. The State Water Resources Control Board and the Regional Water Quality Control Boards (RWQCBs) are organized along watershed hydrologic boundaries and have long been committed to basin planning.

Unfortunately, watersheds are rarely the primary unit used for water governance on the local level because most political structures are organized around other units, adding another challenge to IWRM. In particular, the Santa Ana RWQCB, whose boundaries coincide with the Santa Ana River watershed and SAWPA's, has for decades worked closely with SAWPA to resolve problems in an integrated fashion, and the quality of the basin plans reflect this. The Santa Ana RWQCB participates directly in OWOW and in the various task forces. This leadership and relationship have allowed SAWPA, its member agencies, and others more flexibility and vision to solve problems in a holistic fashion, resulting in the basin plan for the Santa Ana being visionary, current, and informed by data. The California Department of Water Resources is committed to a better, more effective way of water resource management and has launched a new campaign called Water 360 to help refocus and strengthen the collective efforts of California's water management community by advancing integrated water management, and the State Water Plan promotes Integrated Water Management Planning as key.

SAWPA is no more unique than any other watershed. Every watershed has its own idiosyncratic history, challenges, and opportunities. Each enjoys past successes and suffers from old misunderstandings. Learning from both successes and failures is necessary to establish IWRM. Most have the tools for employing IWRM but frequently lack one tool that is fundamental to success: social relationship skills such as conflict management, negotiation, and collaboration. Water is very simple. We understand the natural processes of water really well. But people are complex, and that's where we run into challenges. The water industry is dominated by smart people, most of whom have had little training in conflict management or collaboration, the very skills we now find central to what we need today. These skills have been lacking in the curricula of the professions, engineering and science, most relied on to manage water. We do not have time to wait a generation for a new crop of water leaders to bring these skills. We must develop this capacity today or invite those with these skills into the field. We need to engage with those who can help us understand cooperation. What is it? In the words of the Stockholm International Water Institute (2013, 6): "What drives people, states, and organizations to 'cooperate' rather than 'defect'? How do we identify and measure the quality, aim, benefits and barriers to cooperation, and create an enabling environment for cooperation? How can more effective cooperation enable us to reach future-oriented decisions?" The surprising fact is with all our knowledge, technical expertise, and training, this key component is frequently lacking.

Other skills we need have to do with building trust and adapting to change.

Building trust is difficult and sensitive. Kemper, Dinar, and Blomquist emphasize the need to institutionalize watershed management principles and initiatives rather than relying on a charismatic champion. While a champion may be necessary to get the effort going, the process is likely to outlast the tenure of any champion. The process of management of the watershed itself must be ever responsive to change as threats and opportunities develop. Adapting to change and reform is itself very challenging; they disrupt existing institutions and practices. It is often a battle of leadership versus authority or existing practices. Time is needed for vested interests, accustomed to the status quo, to acclimate to the changes. This and political culture, established governmental structure and responsibilities, all work against rapid change; progress often has to proceed in steps or stages, so consistent commitment to the creation and implementation of management at the basin scale is vital.

The water industry has invested heavily in technology, but it may be the "soft" skills such as the ability to communicate, build trust, and resolve conflict that we find lacking.

If done correctly, IWRM challenges the management regime we have adapted to over decades. We need increased flexibility to respond in new ways to situations that are new and unfamiliar. Each situation must be tailored to respond to its unique situation. Understanding the larger context is most important.

The year 2013 was the driest, and saw the most destructive forest fire, in California's recorded history. Projected population growth, climate change, seawater rise, crashing ecosystems, energy stresses, and financial hard times are all around us. One does not need to look very carefully or be an expert to see that events are connected and that we must manage with that knowledge. California residents can see it, and will demand a more efficient, synergistic approach—and water management must respond lest it forfeit their trust. Ratepayers don't want to pay for water three times over. They are no longer willing to pay top dollar to their water supplier for water imported from far away under a huge carbon footprint, while they pay their flood manager to protect people and property by channeling what naturally falls on their community, and then also pay dearly for their excessive waste to be highly treated and dumped in the river. The management practices of the twentieth century will fall short in the twenty-first. The argument for adopting a new strategic path to water management has never been clearer. IWRM can deliver more bang for the buck, facilitate regulatory compliance, protect the environment, and manage a reliable water supply. As demand and competition increase, the argument for the integrated approach is even stronger. We need to manage the water drop through multiple stages and multiple uses. And we need to view the drop as finite but never ending.

Throughout the world, water leaders are coming to terms with a changing environment and population growth increasingly stressing water resources. Their prescription is resounding agreement that IWRM holds the most promise if we are to achieve better management. IWRM can effectively deliver a triple bottom line of economic efficiency, social equity, and environmental sustainability essential for sustainable development. In their book *Integrated River Basin Management through Decentralization*, Blomquist, Dinar, and Kemper demonstrate that the process requires time; it must be local; everyone can benefit; and partnerships at all levels are needed. In their words, "Patience is a virtue." It is hard to say which of the many challenges will drive California water policy leaders and practitioners to IWRM the fastest. California's water management leads the world in investment and in scientific and technological breakthroughs, but from the IWRM perspective there is still much left to accomplish. It is time to demonstrate leadership that will set an example for the rest of the world.

REFERENCES

California Department of Finance. 2014. *Total Population Projections for California and Counties: July 1, 2015 to 2060 in 5-year Increments*. www.dof.ca.gov/research/demographic/reports/projections/P-1/.

Global Water Partnership Technical Advisory Committee. 2000. *Integrated Regional Watershed Management*. TAC Background Papers, no 4. Stockholm, Sweden. www.gwp.org/Global/GWP-CACENA_Files/en/pdf/teco4.pdf.

Kemper, Karen E., Ariel Dinar, and William Blomquist, eds. 2007. *Integrated River Basin Management through Decentralization*. New York: Springer.

National Research Council. 1999. *New Strategies for America's Watersheds*. Washington, DC: National Academy of Sciences.

Santa Ana Watershed Project Authority. 2013. *One Water, One Watershed Plan 2.0*. www.sawpa.org/owow-2-0-plan-2/.

Senge, Peter. 2008. *The Necessary Revolution*. New York: Crown Business.

Stockholm International Water Institute. 2013. *World Water Week in Stockholm: Water Cooperation—Building Partnerships*. www.siwi.org/wp-content/uploads/2013/04/2013-Programme-and-Call-for-Registration_web.pdf.

UN-Water. 2012. *Status Report on the Application of Integrated Approaches to Water Resources Management*. United Nations Environment Programme. www.un.org/waterforlifedecade/pdf/un_water_status_report_2012.pdf.

U.S. Bureau of Reclamation. 2012. *Colorado River Basin Water Supply and Demand Study*. www.usbr.gov/lc/region/programs/crbstudy/finalreport /Study%20Report/CRBS_Study_Report_FINAL.pdf.

U.S. Council on Environmental Quality. 2013. *Principles and Requirements for Federal Investments in Water Resources*. www.whitehouse.gov/sites/default /files/final_principles_and_requirements_march_2013.pdf.

CHAPTER 10

The History of Water Reuse in California

SASHA HARRIS-LOVETT AND DAVID SEDLAK

A concrete building in one of Orange County's suburban neighborhoods has a small sign at the entrance reading Orange County Water District. Behind this sign is one of the region's vital organs: the mechanical kidneys that process and disinfect wastewater from 2.4 million residences, then pump the treated effluent back into the water supply.

This is Orange County's water recycling plant, known as the Groundwater Replenishment System. It is world-renowned for making treated sewage clean enough to add to the region's drinking-water aquifer, which the utility has been doing for over 30 years. In this semi-arid part of California, reuse is a cost-effective way to provide water for a growing population in the face of rising costs of imported water, overdrafted groundwater basins, and shrinking snowpack.

Water reuse and recycling, which are defined in this chapter as intentional reuse of treated municipal wastewater, are becoming more popular across the Golden State. Recycled water is used for irrigation of agricultural crops and urban landscaping, for industrial cooling and boiler systems, and in some cases, as in Orange County, for potable use. California utilities reused over 890 million cubic meters (724,000 acre-feet) of water in 2012 (National Research Council 2012)—enough to meet the yearly needs of approximately 3.2 million Californians (calculated from Hanak et al. 2011).

Despite its great promise, water reuse comes with its share of technical, social, and philosophical challenges. The technical challenges, like how to remove residual chemicals from wastewater, or how to assure that the technology is functioning properly at all times, may be the easiest to solve. The social challenges, like how to address public perceptions related to the reuse of wastewater, are a bit more complex. But at the core, the most difficult challenges associated with water reuse are philosophical: What do water reuse technologies reveal about humans' relationship to the environment and notions of waste?

Tracing the story of water reuse in California's history provides insight into many of these challenges. This chapter chronicles the changing technologies and attitudes toward water reuse in California, from the use of raw sewage on crops in the early 1900s to today's technologies for augmenting drinking-water supplies with treated wastewater. Historically, Californians kept wastewater "out of sight, and out of mind." This mindset continues to influence some water reuse projects across the state. During the past decade, more sensitive methods for detecting trace chemicals and pathogens have challenged the old "out of sight, out of mind" paradigm of sewage treatment. Today, as water reuse practices trend toward recycling treated wastewater for drinking-water supplies, a new guiding philosophy is needed for urban water systems. Instead of considering water "waste" after one use, it must be treated as a valuable resource. New policies and technologies to prevent toxic chemicals from entering sewage and for monitoring water quality are necessary to ensure the safety of water reuse as part of urban water supply portfolios.

DEALING WITH THE "OBNOXIOUS MATTER" OF SEWAGE: WASTEWATER BEFORE 1930

The story of water reuse in California intertwines with the story of wastewater management. Californians have long used water to get waste out of their homes and thoughts. George Davidson (1886), an engineer tasked with redesigning San Francisco's sewage system in the late nineteenth century, summed up the prevailing philosophy about waste: "We must simply but effectually get rid of the obnoxious matter in the shortest and cheapest manner." For the cities in California (and most other parts of the developed world), this meant flushing waste away with piped freshwater, instead of waiting for the rain to wash it away. Wastewater utilities built pipes to contain the odors emitted by sewage (which were believed to be toxic) and to carry the waste away from homes.

In coastal areas, pipes transported sewage to the sea, where it would be diluted enough not to offend people with its smell (Grunsky, Manson, and Tilton 1899). However, this was not possible for many of California's inland communities. An alternative to dilution was imperative for getting sewage out of sight.

In 1900, the inland city of Pasadena managed their sewage by reusing it for agricultural irrigation. They purchased a 120-hectare (300-acre) plot of land outside the city, called it the Pasadena Sewer Farm, and piped their raw sewage there to water crops. The farm produced walnuts, pumpkins, hay, and corn. Sewage farming was a profitable business. In 1903, the walnut crop alone paid for all the farm's expenses and accrued an additional $2,400 (about $63,000 in 2013 dollars) in profit for the city (Holder 1904).

Other Southern California cities turned to sewer farms as a way to make a profit on human waste while getting it away from homes. In 1909, residents of the coastal city of Redondo Beach voted down a proposed sewer outflow to the ocean and instead insisted that their city adopt the sewer farm model for reuse (Barkley 1909). Sewage, to them, was a source of water and nutrients that could make the dry landscape of Southern California produce useful crops. This wasn't a new idea; several decades earlier in Europe, Karl Marx (1906, 69) had criticized London's plan to pipe its wastes directly into the nearest large body of water: "Excretions of consumption are of the greatest importance for agriculture. So far as their utilisation is concerned, there is an enormous waste of them in the capitalist economy. In London, for instance, they find no better use for the excretion of four and a half million human beings than to contaminate the Thames with it at heavy expense."

While sewer farms in California ultimately didn't survive, due to concerns about odors and health risks associated with putting raw sewage on farm fields, the Farm Bureau continued to support the idea of sewage reuse (*Los Angeles Times* 1921b). The engineers who designed San Francisco's sewer system in the 1920s acknowledged the value of nutrients and water in the city's sewage to nearby farmers, but decided that water reuse was not feasible given the cost of pumping the sewage uphill to nearby agricultural areas (Grunsky, Manson, and Tilton 1899).

Instead of using sewage to grow commercial food crops on agricultural land, San Francisco diverted sewage from its inner-city neighborhoods to irrigate Golden Gate Park (which was then on the outskirts of the city), making it possible to grow lush, verdant meadows where before only sand dunes existed (Hyde 1937).

Around the same time, in 1921, Los Angeles voters nixed a proposal to enlarge their city's sewage system because they thought sewage should be used for fertilizer rather than squandered in the Pacific (Knowlton 1928). Even after Los Angeles eventually built an upgraded sewer outflow to the ocean in the 1920s, the *Los Angeles Times* (1921a) ran an article condemning the city government for "wasting the valuable fertilizing elements in its sewage by dumping it in the sea" instead of reusing it.

In this era, a need for fertilizer spurred part of the enthusiasm for reusing sewage in agriculture. Many farmers in the United States relied on dwindling imports of Peruvian guano and Chilean caliche to fertilize their crops (Smil 2004, 42). Sewage was also nutrient-rich, and cheaper than South American fertilizer imports. After World War II, a German company commercialized the Haber-Bosch process for converting atmospheric nitrogen to liquid ammonia for fertilizer (Erisman et al. 2008). This synthetic nitrogen quickly supplanted organic waste as fertilizer.

GETTING RID OF THE SMELL: EARLY TECHNOLOGIES FOR SEWAGE TREATMENT

As California's urban populations expanded, the practice of piping sewage away proved to be insufficient for getting rid of the smell. Coastal areas reeked. In 1922, Los Angeles responded to the stench by screening its sewage and burying the captured solids in the sand dunes before sending the remaining wastewater into the Pacific Ocean (Knowlton 1928).

In inland cities, sewer farms smelled bad and attracted flies. In Fresno in 1924, the city council decided to apply wastewater to the surface of the ground outside the city, where it would infiltrate back into the aquifer. To make the wastewater percolate quickly, they lowered the level of the groundwater by building nine additional extraction wells (City of Fresno 2013a). Because groundwater was the primary drinking-water supply for the city (City of Fresno 2013b), this project for getting sewage out of sight (and smell) essentially created the state's first planned potable water reuse system.

In San Francisco, the city grew to surround Golden Gate Park. New neighbors voiced serious complaints about the odors of sewage used for landscape irrigation. In response, in 1932 the Park Commission built a state-of-the-art activated-sludge treatment plant near the park. The new technology bubbled air through the wastewater so bacteria could break down the sewage. Chlorine killed any pathogens remaining in the effluent, so it could be used for irrigation. The treated water was also used to create an artificial brook and chain of lakes running through Golden Gate Park (Hyde 1937), reinforcing the planners' ideal of a lush environment in a naturally semi-arid area.

QUENCHING CALIFORNIANS' THIRST: EARLY WATER SUPPLY SOLUTIONS

Settlers in the semi-arid regions of Southern California originally relied on local surface water and groundwater to meet their needs for drinking, bathing, and irrigation. Farmers and cities drilled wells into the aquifers underneath what are now Los Angeles and Orange Counties, where abundant water bubbled out of the wells day and night (*San Francisco Chronicle* 1900). Towns were named Fountain Valley, Santa Fe Springs, and Artesia in recognition of the bountiful springs.

But as more people tapped southern California's groundwater, they depleted the supply. By the 1940s in Orange County, residents withdrew the coastal groundwater basin to five meters below sea level, causing the seawater to flow inland through the porous sand underground (Orange County Water District 2013). Seawater contaminated the groundwater, making coastal wells too salty to use. Southern California needed new water supplies if it wanted to encourage agricultural and suburban growth.

In response, the local, state, and federal governments funded massive infrastructure projects to satisfy the water demand of the burgeoning cities on the California coast and of the farmland reclaimed from the desert. Over the following three decades, politicians and engineers devised a network of aqueducts, reservoirs, and pumping stations to transfer water to Southern California (see Figure 0.2). These water transfers occurred, in some cases, at the expense of the ecosystems and rural communities that had previously relied upon that water for survival.

Yet, imported water wasn't enough to meet demand. During World War II, Los Angeles became the manufacturing center for wartime aircraft and other military supplies. The population soared (Kling, Olin, and Poster 1995). All these new Californians needed water.

After World War II, newspaper articles touted technology as critical for economic progress and for solving the nation's problems (*New York Times* 1956). Water shortages were no exception. When Southern California found itself short of water, it turned to technology to increase the supply.

The vast Pacific Ocean would provide an endless supply of water for California's coastal cities, if only scientists could find ways to remove

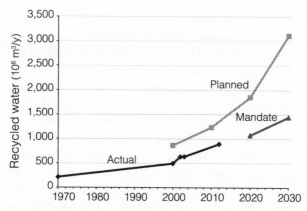

FIGURE 10.1. California water reuse, goals, and mandates. Data from California Department of Water Resources (2003), National Research Council (2012), and California State Water Resources Control Board (2009a, 2009b, 2013).

the salt. Across the country, researchers began studying desalting technologies. Electrodialysis and distillation both proved technically feasible, but extremely expensive (Fisher 1963).

Though using the ocean as a water supply was the original intention, scientists soon found that the same technologies worked far more efficiently in desalting less salty water, like the brackish groundwater in many of the state's wells. In 1959, the Central Valley town of Coalinga, where the groundwater was too salty to use, invested in the first demonstration desalting plant in the United States. The new plant in Coalinga provided a small amount of drinking water (enough for about 140 people) at a fraction of the cost of bringing it in by tanker trucks, by using electrodialysis to separate the salts out of their brackish groundwater (*Los Angeles Times* 1958a).

With technologies available for converting seawater and salty groundwater to freshwater, anything seemed possible. Even wastewater was considered a potential source of supply, given these new technologies to transform previously inaccessible sources into fresh water (Phillips 1949; *Los Angeles Times* 1958b). An engineering company's report to the California State Water Resources Control Board indicated that reusing wastewater could be the answer to the state's future water needs. The report predicted that water recycling would "save California and other States of the thirsty Southwest from economic dehydration" (*Los Angeles Times* 1955). More water would ensure growth in the driest regions of the state.

To make water reuse feasible on a larger scale, excess salt—caused in part by water softeners, detergents, and household wastes—needed to be removed more cheaply. Distillation and electrodialysis were expensive, but a promising new product—plastic—had recently come on the market. Technophiles hyped plastics as making better, cheaper products, from dolls to concrete (*New York Times* 1955; *Washington Post* 1952). If plastics could improve dolls and concrete, why not desalting technologies? In 1959, two graduate students at UCLA, Sidney Loeb and Srinivasa Sourirajan, employed a synthetic plastic membrane in a new desalting technology called reverse osmosis, making the process cheaper than ever before (Loeb 1981). Reverse osmosis worked by forcing water molecules across a membrane, thus separating them from most of the salts, nutrients, and pathogens.

The city of Coalinga, which had installed the small electrodialysis plant seven years earlier, built the nation's first reverse-osmosis treatment system in 1965 to desalt their groundwater (Loeb 1984). Though it could only produce enough freshwater for about thirty people per day (calculated from UCLA Engineering 2013), this demonstration plant proved that reverse-osmosis technology was much cheaper if used on groundwater than on seawater (Stevens and Loeb 1967; Loeb and Manjikian 1965; Loeb and Selover 1967; Rosenfeld and Loeb 1967).

The U.S. federal government took an active role in advocating a "world-wide cooperative effort" to solve global water shortages through desalination (Udall 1965). In 1961, President Kennedy gave a rousing speech to Congress about water reuse, which was reprinted in full in the *New York Times*. He said that "to meet all needs—domestic, agricultural, industrial, recreational—we shall have to use and reuse the same water, maintaining quality as well as quantity." He also allocated $75 million (1961 dollars) in federal funds to the Office of Saline Water (a program of the Department of the Interior) for increased research into technologies for reclamation of wastewater and seawater (MacGowan 1963).

Just six months later, construction began on a water reclamation plant in Los Angeles, at Whittier Narrows. The plant processed sewage and sent the treated wastewater to a sandy basin next to the facility. The treated wastewater, along with any pooled stormwater runoff, infiltrated into the groundwater. The water district then pumped the groundwater to the surface, where it became part of the local drinking-water supply. Though inland cities like Fresno infiltrated treated wastewater back into the groundwater, Whittier was the first to publicly advertise what they were doing as *water reuse*, rather than just a convenient means of waste disposal (Nelson 1961).

Wastewater reuse in California soon became a source of water for recreational purposes. In 1965, a community in inland San Diego County called Santee began using treated wastewater to fill man-made lakes used for fishing and swimming. For Santee, water reuse was cheaper than connecting their sewage pipes to San Diego's metropolitan sewage system (Hill 1965). They used activated-sludge technology to treat the sewage, then percolated it through 120 meters (400 feet) of soil for additional treatment before pumping it to the surface, chlorinating it, and releasing it into the lakes (City of San Diego 2005, section 5.5).

Before, treated wastewater was quietly reused out of the public eye for groundwater recharge or for outdoor irrigation. But in Santee, swimmers had full body contact with reclaimed water. Media accounts touted water reuse as "an inevitable fact of life as water demands increase" (White 1965). In 1968, a front-page article in the *Chicago Tribune*, titled "A Pattern for the Future: Using Water Over and Over Again," characterized water recycling as the norm for American cities, and cited Santee as a model for future development (Bukro 1967).

In arid parts of California, many communities realized that reusing wastewater was an economically feasible option for both enhancing water supply and curtailing sewage pollution. By 1970, over 123 million cubic meters (100,00 acre-feet) per year of recycled water were being used for agricultural irrigation in California—nearly a third of the capacity of Hetch Hetchy Reservoir. An additional 24 million cubic meters (20,000 acre-feet) per year were used for urban landscape irrigation in California (California Water Board 2009b), or enough water to submerge the island of Manhattan to a depth of two feet. Building on this momentum, San Diego built a reverse-osmosis facility to desalinate wastewater effluent for landscape irrigation (*Los Angeles Times* 1970a). Water recycling in Southern California garnered national attention and a positive review in a front-page article in the *Wall Street Journal* in September 1971 (Graham 1971).

The following month, Orange County announced its plan to build a new "water factory," known as Water Factory 21. The recycled wastewater produced there would keep saltwater from intruding into coastal aquifers near Newport Beach, and at a lower cost than the alternatives. For over a decade, Orange County and other water districts along the Southern California coast had bought imported freshwater to inject underground to prevent seawater from migrating inland and contaminating groundwater (West Basin Municipal Water District 2013; Orange County Water District 2013). The injected freshwater formed a barrier underground, raising the level of the aquifer at the coast and providing enough pressure to keep the seawater out of the drinking-water supply (Pryor 1971).

Water Factory 21 began operating in 1976, treating over 56,000 cubic meters (46 acre-feet) of wastewater a day (Orange County Water District 2013), enough to fill 22 Olympic-sized swimming pools. To create an effective hydraulic barrier against saltwater intrusion, engineers at Water Factory 21 realized that they needed to remove many of the salts from the treated wastewater. They treated half the waste stream with reverse osmosis, which was expensive but could remove salts; this marked the first use of reverse-osmosis technology with wastewater. They passed the other half of the wastewater through layers of anthracite coal, sand, garnet dust, and granular activated carbon (the stuff of modern-day Brita filters) to remove some of the residual chemicals in the water. Then they chlorinated the water to kill pathogens, before injecting it into the aquifer (Hammar and Elser 1980). The interior secretary, Rogers Morton, touted Water Factory 21 as an example for California and the rest of the world (Boettner 1972; *Los Angeles Times* 1972).

CALLS FOR CAUTION: RISKS OF THE UNKNOWN

Even as water recycling grew more common in California throughout the 1960s and 70s, some people called for restraint. The growth of potable water reuse coincided with a nascent awareness of the harmful impacts of some of the synthetic compounds that had been enthusiastically used after World War II. Rachel Carson's seminal work, *Silent Spring* (1962), alerted the public to the unintended health and environmental consequences of the synthetic pesticide DDT. Less than a decade later, in Southern California, the Montrose Chemical Company gained notoriety for sending DDT down the sewers into Santa Monica Bay (Dreyfuss 1971). Sewers in California had long carried waste out of sight and out of mind. As household and industrial chemicals became more ubiquitous after World War II, these potentially toxic chemicals were also thrown "away" down the drain without a second thought.

But in Santa Monica Bay, DDT did not become nontoxic when it went down the drain. Instead, it devastated the region's brown pelican population. The public worried that human health would suffer as the chemical bioaccumulated up the food chain. A 1970 *New York Times* article about the DDT in Santa Monica Bay stated, "Most humans are now believed to have DDT in their bodies. Its effects are not known, but some scientists have suggested it may cause cancer."

Given this context, it is not surprising that some Californians worried that existing wastewater treatment processes could not protect them from chemicals in sewage if the water were reused (Bengelsdorf 1965). To them, water reuse was a Pandora's Box that could wreak havoc if it allowed synthetic chemicals and viruses to make their way into water supplies by way of recycled water (Harris 1977). Citing the groundwater-recharge project at Whittier Narrows as an example, critics suggested that water reuse was harmful to the American public. The media suggested that "the nation—some say legislators and a horde of Public Health Service scientists—is rapidly poisoning its drinking water" (Mulligan 1963).

These concerns were not unfounded. Industrial, agricultural, and household chemicals passed through activated-sludge treatment plants and polluted surface waters. In many cases, these rivers supplied water for cities downstream. A 1975 study by the Environmental Protection Agency found synthetic carcinogenic chemicals in the drinking water supply of 79 of the 80 cities tested (Bukro 1975).

The California Department of Public Health voiced concerns about the safety of reused water. Henry Ongerth, then chief of the state's Sanitary Health section, said to a reporter in 1977, "Sewage is an infectious waste that has to be treated properly to protect the health of the people. . . . Health considerations—disease transmission and control—are a limiting factor [in water reuse]" (Harris 1977).

In contrast to Ongerth's perspective, engineers working for water utilities called for complete water reuse. They claimed that technologies to purify sewage to drinking-water standards already existed. While this claim was technically true, drinking-water standards assumed that sources were relatively pristine—not city sewers. At the Second National Conference on Complete Water Reuse in 1975, chairman Lawrence Cecil declared, "The technology [for complete water reuse] is here. All we have to do is do it" (Anderson 1975).

Since the advent of membrane technologies, no research had shown people getting acutely ill from reclaimed water (Pryor 1971). A 1977 study by the Los Angeles Sanitation District demonstrated that many of the common water-reuse technologies could remove 99.999 percent of the viruses from wastewater (Sanitation Districts of Los Angeles County 1977). Reverse osmosis, though originally designed to remove salts, was also found to remove the vast majority of dissolved solids, color, pesticides, nutrients, and pathogens from water (Asano et al. 2007). In Southern California, scientists found reclaimed water to be cleaner, on the basis of existing measurement techniques, than the imported water from the Colorado River on which Southern California cities had typically relied (Lee 1965).

Concerns about the health effects of reclaimed water use were pushed aside as California plunged headlong into a severe drought in the mid-1970s. Using membrane technologies to reuse wastewater continued to gain steam as California cities pursued growth in their semi-arid region.

By 1976, the State Water Resources Control Board proposed an amendment to the state water code stipulating that recycled water must be used if available. The new code, reprinted in the *Los Angeles Times,* stated that "failure to reclaim water or use reclaimed water could constitute a waste or unreasonable use of water" (Dendy 1976). By 1977, over 200 different sites in California, including golf courses, power plants, and municipal buildings, used reclaimed water (Harris 1977).

Dr. Daniel Okun, an environmental engineering professor at the University of North Carolina, Chapel Hill, continued to urge caution throughout the 1980s. He acknowledged that state-of-the-art treatment technologies for water reuse could reliably remove most pathogens and prevent acute infectious disease if the treatment systems worked properly. But he wasn't convinced that existing technologies could protect the public from chronic diseases like cancer from long-term exposure to the traces of chemicals in reclaimed water.

In Okun's estimation, the unknown health risks posed by under-studied chemicals or by newly minted synthetic chemicals were grave enough that recycled water was best reserved for non-potable purposes, like flushing toilets, watering lawns, and washing cars. In his 1980 address to the Environmental Protection Agency at their symposium on protocol development for potable reuse, he said, "It may very well be that, just as with radiation and asbestos, many decades will pass before the full impact of these organic chemicals . . . is understood" (Okun 1980).

Studies of the health effects of trace chemicals in recycled water were nearly impossible, because measurement tools were not sensitive enough to detect them. Researchers from Stanford University noted the "great difficulty of detecting analytically significant differences in the removal of trace organic materials [in Water Factory 21], which is attributed to . . . the general lack of sufficient analytical precision" (McCarty, Argo, and Reinhard 1979). What they could measure, however, met current drinking water standards.

Ongerth, then chief of the Bureau of Sanitary Engineering of the State Health Department, echoed Okun's concerns: "Studies show that the ability to control most synthetic organic compounds to current limits of detectability is good. It is recognized, however, that the majority of organic compounds in advanced wastewater treatment effluents are unidentified and of generally unknown significance" (Ongerth and Ongerth 1982).

Instead of putting recycled water back into the drinking water supply, Okun (1997) advocated for new pipes to carry recycled (nonpotable) water separate from drinking water. These dual distribution systems would allow for year-round water reuse for cooling, firefighting, and industrial boilers with minimal health risk. The downside of dual distribution is the expense of laying thousands of miles of new pipes. Costs for installing dual-distribution systems in Northern California range from $600,000 to $1.9 million (2010 dollars) per kilometer of pipe. This price tag constitutes a major barrier to increased water recycling (Bischel et al. 2012). In some places, adding new pipes to the already crowded infrastructure below the street is not physically feasible.

And even dual-distribution systems for reclaimed water are not risk-free. Studies of several such systems in the United States and Australia have documented unintentional cross-connections between the pipes for drinking water and for nonpotable reclaimed water, sometimes occurring for more than a year before they were noticed. In each of these cases, multiple households were affected, and people reported an increase in diarrheal illness and other acute infectious disease (National Research Council 2012).

WATER REUSE, EXPANDED

By the mid-1970s, water reuse projects occurred across the state, spurred by suburban expansion's competition with agriculture for water. In the agricultural Salinas Valley, extensive water withdrawals depleted groundwater supplies. Seawater intruded into coastal aquifers at a rate of nearly 150 meters (500 feet) per year (Crook and Jaques 2005), which made the groundwater suitable for irrigating only the most salt-tolerant crops, like artichokes. A water reuse program could provide the necessary low-salt water for growing fruits and vegetables. As a result, the Monterey Regional Water Pollution Control Agency built a water reclamation facility that distributed reclaimed water to farmers for irrigation (Crook and Jaques 2005).

Concerns about the safety of using recycled water on agricultural crops, many of which would be consumed raw, prompted a seven-year study to test the safety of this practice. Federal, state, and local funds provided the $7.2 million necessary to undertake a comprehensive research program, called the Monterey Wastewater Reclamation Study for Agriculture (Asano 1998).

The results of the study, which were released in 1987, indicated that reclaimed water was "safe and acceptable" for crop quality, crop growth, crop marketability, soil quality, and groundwater quality (Sheikh et al. 1990). The results of the study in the Salinas Valley gave the green light for increased reuse of water in irrigation of food crops across the state.

The media portrayed water reuse as the "green" thing to do (*Los Angeles Times* 1970b). The new term *water recycling* for the practice, which had previous been referred to as *water reuse* or *water reclamation*, helped solidify it as part of a solution to the environmental crisis. The California Water Recycling Act, signed into law in 1991, touted water reuse as "a cost-effective, reliable method of helping to meet California's water supply needs." The act also clarified the potential environmental benefits of water reuse in California, including "a reduced demand for water in the Sacramento-San Joaquin Delta which is otherwise needed to maintain water quality" (California Water Code §§ 13575–13583). It set goals of reusing 863 million cubic meters (700,000 acre-feet) per year of water in the year 2000 and 1.2 billion cubic meters (a million acre-feet) per year by 2010, though neither goal was met (see figure 10.1).

Given California's growing interest in water recycling, several professional organizations developed in the early 1990s to share information, fund research, and lobby the government for regulations amenable to water reuse. The National Water Research Institute, WateReuse Association, and WateReuse Research Foundation funded research, held professional conferences, and created materials for educational outreach (WateReuse Association 2013). As the need arose for more information pertaining to water reuse, from the chemistry of treatment processes to the marketing of new systems, these professional organizations supported water utilities as they moved forward with water reuse projects.

At West Basin Water District, just north of Orange County, water engineers pioneered the concept of "tailored water," which involves treating wastewater to different standards depending on the end use. In 1995, West Basin's facility opened and began providing water for

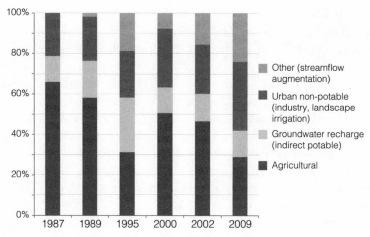

FIGURE 10.2. Uses of recycled water in California over time. Adapted from Shulte (2011, 2).

groundwater augmentation, for landscape irrigation, and for industrial cooling systems (West Basin Municipal Water District 2013). Utilities across the state looked to West Basin and Orange County as examples of successful potable water reuse projects (see Figure 10.2).

In 1998, the National Research Council, an independent body of pre-eminent research scientists, issued a new report on potable water reuse that challenged the "out of sight, out of mind" mentality of sewage management. The report noted that *unintentional* potable reuse of wastewater was common around the nation. It cited over twenty-four drinking-water facilities drawing from rivers consisting of over 50 percent wastewater at some times of year, and implied that potable water reuse would continue in the United States whether or not it was planned (National Research Council 1998). The media noticed that over 200 sewage treatment plants drained into the Colorado River, from which southern California imports much of its drinking water (Cannon 1997). Despite the generally support-ive content of the report with respect to planned water reuse, the report's executive summary stated that potable water reuse should be considered an "option of last resort," given risks from chemicals and waterborne pathogens (National Research Council 1998).

Not everyone in California was on board with the notion of water reuse, especially if recycled water was slated to become part of the drinking-water supply. In the late 90s, just after the National Research Council report came out, planned potable water reuse projects in San

Diego, Dublin-Pleasanton, and the San Gabriel Valley ground to a halt. A combination of factors likely influenced public opposition to these projects, including city politics, critical media coverage (which labeled the projects "toilet-to-tap"), lack of trust in government agencies, and lack of public outreach on the part of the utilities (Sedlak 2014). Some residents said reusing treated sewage for potable purposes was disgusting, a phenomenon social scientists dubbed the "yuck factor" (Hartley 2006). In 2002 in Redwood City, a suburb of San Francisco, a small group of residents even opposed a nonpotable water reuse plan to direct reclaimed water through a separate distribution system. They were concerned about children ingesting the water from sprinklers at parks and schools (City of Redwood City 2013).

Residents of regions that did accept increased water recycling for potable purposes, like Orange County and West Basin, had long seen water reuse as a way to curb the problem of saltwater intrusion into their groundwater. Although engineers working for the utilities knew that groundwater recharge effectively meant augmenting potable water supplies underground with treated wastewater, the utilities' public-outreach materials emphasized the saltwater-intrusion barrier over the drinking-water-augmentation aspect of the project (Boettner 1972). Furthermore, utilities in these regions spent decades building trust between citizens and water utilities through education campaigns and research programs to address health concerns (Po, Kaercher, and Nancarrow 2003).

In general, reclaiming water for nonpotable uses and distributing it through separate pipes didn't attract as much controversy as projects that slated reclaimed water for potable use. To encourage more water recycling, Irvine Ranch Water District invested in a new set of pipes to bring recycled water to their customers.

Though a dual-distribution system worked well for the city of Irvine, other cities found new pipes for reclaimed water to be too expensive. In San Jose, California, in the late 1990s, the water utility wanted to distribute nonpotable water to their customers for irrigation but they were stymied by the cost of constructing pipes to the residents in their 780-square-kilometer (300-square-mile) service area (Sedlak 2014).

CALIFORNIA AT A CROSSROADS

Orange County's indirect potable reuse project went smoothly until the year 2000, when a potent carcinogen, N-nitrosodimethylamine (NDMA), was detected in the groundwater (California Department of Public

Health 2011). To the water utility's dismay, it seemed that a significant fraction of the chemical was actually being produced during the advanced wastewater treatment process (Mitch and Sedlak 2002). The utility could detect as little as one-billionth of a gram of NDMA per liter in their water, and they knew that even this tiny amount of the substance increased their customers' cancer risk (Mitch et al. 2003). After consultation with the state health department, the water district decommissioned some of the drinking-water wells that were close to the water recycling plant (California Department of Public Health 2011).

The Orange County Water District responded to this problem by adding ultraviolet light with hydrogen peroxide to their treatment process, a technology that was previously used to treat groundwater at hazardous-waste sites (Huang, Dong, and Tang 1993). This technology would destroy the NDMA produced at the water reuse facility before the reclaimed water was introduced into the aquifer. It would also treat some other chemicals, like 1,4-dioxane, a common industrial solvent, which slips through reverse-osmosis membranes (Bellona et al. 2004).

Despite this technological mishap, the public's confidence in Orange County's drinking water remained strong, thanks in part to the water utility's proactive response to the detection of NDMA and its sophisticated media communications strategy (personal communication with Michael Wehner, assistant general manager of Orange County Water District). In 2008, Orange County Water District expanded the facility to produce 265,000 cubic meters (215 acre-feet) of reclaimed water per day, and renamed it the Groundwater Replenishment System (California Department of Public Health 2013; Orange County Water District, n. d.). Three and a half decades of operating experience had convinced Orange County residents that they would not get sick from drinking their tap water, which was part recycled water. The utility's outreach materials began to openly tout their project as "the world's largest water purification system for potable reuse" (Groundwater Replenishment System 2013).

Today, many of California's water engineers think increased water recycling will be critical to meeting the demand for water in the state. California's population is growing by over 250,000 people per year (California Department of Finance 2013), stressing existing water supplies (CDM Consulting 2010). The state's water supply is likely to diminish in the coming decades, because climate change is predicted to cause more precipitation to fall as rain rather than as snow. Scientists project a 25-percent loss in the state's average snowpack by 2050 (see Andrew, chapter 1 in this volume). Considering that over 20 million

Californians currently rely on snowmelt for part of their water supply, these changes could cause severe shortages if other sources of water are not developed (California Department of Water Resources 2013; Kiparsky and Gleick 2003).

In response to the challenges of climate change and population growth, the California Water Board's current policy is to increase recycled water use in the state in the coming decades (see Figure 10.1). The policy includes the goal to substitute "as much recycled water for potable water as possible by 2030." At a minimum, the Water Board's policy mandates that California use an additional 247 million cubic meters per year of recycled water by 2020 (over 2013 levels), and 370 million more by 2030 (California State Water Resources Control Board 2009a, 2013). The WateReuse Association and the National Water Research Institute advocate increased potable water reuse (WateReuse Association 2013), because the technology to treat sewage to drinking-water standards exists and has been tested for decades by water utilities like the Orange County Water District and the West Basin project.

In 2009, the WateReuse Association announced its Direct Potable Reuse Initiative, which aimed to identify and eliminate any barriers to direct potable water reuse in California (Smith 2010). They raised over $6 million in three years from water utilities and engineering firms to support lobbying and research efforts (Smith 2013). In 2010, the California state legislature passed a bill requiring the Department of Public Health to develop regulatory criteria for groundwater recharge with treated wastewater by December 2013, to develop rules for augmenting surface water reservoirs with treated wastewater by 2016, and to assess the feasibility of implementing a policy that sets criteria for direct potable reuse—that is, sending highly treated wastewater directly into a drinking-water treatment plant—by 2016 (California SB 918).

In parallel with these efforts, the Water Board adopted the first monitoring standards for chemical contaminants in recycled-water projects in January 2013. The monitoring standards addressed constituents of emerging concern (CECs—e.g. pharmaceutical compounds, personal care products, and hormones), a group of compounds that had raised concerns among regulators and community members when prior potable water reuse projects had been proposed. The Water Board acknowledged the need for more research on the potential presence of these substances in the drinking-water supply because many have unknown health effects. The policy read, "The state of knowledge regarding CECs is incomplete. There needs to be additional research and development of analytical methods and surrogates to determine potential environmental and public health impacts (California State Water Resources Control Board 2013)."

The new legislation will improve the state of knowledge of the presence of chemical contaminants, but it does not guarantee that chemical contaminants will never be detected in recycled water. According to the new regulation, potable water reuse utilities in California must test twice a year for a suite of regulated drinking-water contaminants and eight chemicals that are known to be present in sewage but are not included in state or federal drinking-water standards, including caffeine, DEET (a mosquito repellent), and triclosan (an antimicrobial). The eight chemicals were selected to provide an indication of the treatment plant's ability to remove chemicals commonly present in wastewater, not comprehensive information about all chemicals that could pose health risks. There are no repercussions apart from continued monitoring requirements if the concentration of the chemical detected in recycled water is less than 100 times the "monitoring trigger level," which is a health-based screening level developed by a scientific advisory panel (California State Water Resources Control Board 2013). Although no chemical contaminants have been detected in recycled water at concentrations that pose potential health risks since the Orange County Water District detected NDMA in 2000, it is possible that some future discovery could reopen the discussions about health risks associated with chemical contaminants in recycled water.

CALIFORNIA'S TWENTY-FIRST-CENTURY WATER

In a system where wastewater is treated and then returned to the water supply, sewage is no longer flushed away and forgotten. Instead, water sent down the drain is a resource that can enable Californians to meet their own needs without compromising the needs of future generations (by overdrawing groundwater supplies) or the needs of other species (that rely on having water in streams).

Going forward, potable water reuse may require California to expand its notions of water stewardship. Regulations for watershed protection, for example, may need to be extended to include city sewers. To avoid future surprises regarding chemical contaminants in recycled water, policymakers may need to focus on preventing toxic substances from going down the drain, especially those chemicals that are difficult to remove in advanced wastewater treatment plants. In an era of water reuse, "out of sight, out of mind" can no longer be a guiding philosophy for waste disposal.

A more appropriate philosophy for Californians' relationship with water and waste might be "We're all in this together." In this framework, Californians acknowledge that whatever enters the sewer will need to be removed before the water returns to the drinking-water supply.

Though water reuse is becoming increasingly important, it is not clear exactly what form it will take in California's future. Options for integrating centralized water reuse into California's cities include expanding nonpotable reuse through dual-piped distribution systems, augmenting groundwater supplies or surface reservoirs with highly treated wastewater, and piping recycled water directly into the drinking-water system. Though the state's current institutions and regulations lend themselves to the centralized solutions mentioned above, other options for recycling water exist. For example, decentralized wastewater treatment systems have strong public support in some communities (see Woelfle-Erskine, chapter 14 in this volume). In these systems, households reuse potable water on site for "cascading" uses, as in using water from the clothes washer to irrigate gardens or flush toilets. To facilitate expansion of these practices, research is needed to assess the possible health risks as well as to develop ways to reduce the current high costs of treatment.

In the future, Californians may decide to invest in dual-distribution systems for nonpotable water reuse, to turn to potable reuse of wastewater effluent, or to invest in household-scale water reuse systems. These options are not mutually exclusive. Different cities are likely to develop their own portfolios of water reuse systems that are appropriate for their topography, community values, and existing urban form. Whatever paths the state chooses, water reuse in California will continue to expand. Done correctly, with measures to prevent difficult-to-remove contaminants from entering sewers and to continually monitor water quality, water recycling will be an important part of California's toolkit for meeting the water challenges of the twenty-first century.

REFERENCES

Anderson, M. 1975. "Technology Is Here: Calls for Total Sewerage Reuse." *Chicago Tribune*, May 8.
Asano, Takashi. 1998. *Wastewater Reclamation and Reuse*. Boca Raton, FL: CRC Press.
Asano, Takashi, Franklin L. Burton, Harold L. Leverenz, Ryujiro Tsuchihashi, and George Tchobanoglous. 2007. *Water Reuse*. New York: McGraw Hill.
Barkley, S. D. 1909. "Sewers and Sanitation." *Los Angeles Herald*, October 3.

Bellona, Christopher, Jorg E. Drewes, Pei Xu, and Gary Amy. 2004. "Factors Affecting the Rejection of Organic Solutes during NF/RO Treatment: A Literature Review." *Water Research* 38(12):2795–2809.

Bengelsdorf, Irving S. 1965. "$100 Million Took Foam Out—Pollution Still Remains." *Los Angeles Times*, April 18.

Bischel, Heather N., Gregory L. Simon, Tammy M. Frisby, and Richard G. Luthy. 2012. "Management Experiences and Trends for Water Reuse Implementation in Northern California." *Environmental Science & Technology* 46(1):180–88.

Boettner, Jack. 1972. "Water Factory 21 Called Milestone by Interior Secretary." *Los Angeles Times*, January 28.

Bukro, Casey. 1967. "A Pattern for the Future: Using Water Over and Over Again." *Chicago Tribune*, August 11.

———. 1975. "More Perils to Water Are Feared." *Chicago Tribune*, May 6.

California Department of Finance. 2013. "California County Population Estimates and Components of Change by Year: July 1, 2010–2012." www.dof.ca.gov/research/demographic/reports/estimates/e-2/.

California Department of Public Health. 2011. "A Brief History of NDMA Findings in Drinking Water." www.waterboards.ca.gov/drinking_water/certlic/drinkingwater/NDMAhistory.shtml.

California Department of Water Resources. 1963. *Saline Water Conversion Activities in California.* Bulletin No. 134–62. www.water.ca.gov/waterdatalibrary/docs/historic/Bulletins/Bulletin_134/Bulletin_134-62__1963.pdf.

———. 2003. *Water Recycling 2030: Recommendations of California's Recycled Water Task Force.* www.water.ca.gov/recycling/TaskForce/.

———. 2013. "Climate Change." www.water.ca.gov/climatechange/.

California State Water Resources Control Board. 2009a. *Recycled Water Policy.* www.waterboards.ca.gov/water_issues/programs/water_recycling_policy/docs/recycledwaterpolicy_approved.pdf.

———. 2009b. *Results, Challenges, and Future Approaches to California's Municipal Wastewater Recycling Survey.* www.waterboards.ca.gov/water_issues/programs/grants_loans/water_recycling/docs/article.pdf.

———. 2013. *Policy for Water Quality Control for Recycled Water (Recycled Water Policy).* www.waterboards.ca.gov/water_issues/programs/water_recycling_policy/docs/rwp_revtoc.pdf.

Cannon, Lou. 1997. "California Acquiring a Taste for Reclaimed Water." *Washington Post*, August 31.

Carson, Rachel. 1962. *Silent Spring.* Boston, MA: Houghton Mifflin.

CDM Consulting. 2010. *Long-Term Reliable Water Supply Strategy: Phase 1 Scoping Report.* Report to the Bay Area Water Supply and Conservation Agency. http://bawsca.org/docs/BAWSCA_Strategy_Final_Report_2010_05_27.pdf.

City of Fresno. 2013a. "History of the RWRF." www.fresno.gov/Government/DepartmentDirectory/PublicUtilities/Wastewater/Fresno-Clovis+Regional+Wastewater+Reclamation+Facility/RWRFHistory.htm.

———. 2013b. "Water Division." www.fresno.gov/Government/Department Directory/PublicUtilities/Watermanagement/.

City of Redwood City. 2013. "Phase One of the Recycle Water Program." www .redwoodcity.org/publicworks/water/recycling/phase_one_program.htm.

City of San Diego. 2005. *City of San Diego Water Reuse Study 2005*.

Crook, James, and Robert S. Jaques. 2005. "Monterey County Water Recycling Projects: A Case Study." In *Water Conservation, Reuse, and Recycling: Proceedings of an Iranian-American Workshop*. Washington, DC: National Academies Press.

Davidson, G. 1886. *System of Sewerage for the City of San Francisco*. San Francisco, CA: Frank Eastman & Co.

Dendy, Bill B. 1976. "Notice of Proposed Changes in Regulations of the State Water Resources Control Board." *Los Angeles Times*, September 27.

Dreyfuss, John. 1971. "DDT Firm Agrees to End Flow of Poison Effluent into Sewers." *Los Angeles Times*, March 18, C1.

Erisman, Jan W., Mark A. Sutton, James Galloway, Zbigniew Klimont, and Wilfried Winiwarter. 2008. "How a Century of Ammonia Synthesis Changed the World." *Nature Geoscience* 1(10):636.

Graham, E. 1971. "Recycling Sewage: Advanced Treatment of Waste Seen Cutting Pollution, Saving Water: New Processes Can Produce Effluent Safe to Swim In, Sometimes Even to Drink." *Wall Street Journal*, September 10.

Groundwater Replenishment System. 2013. "About GWRS." www.gwrsystem .com/about-gwrs.html.

Grunsky, Carl E., Marsden Manson, and C. S. Tilton. 1889. *Report upon a System of Sewerage for the City and County of San Francisco*. City of San Francisco.

Hammer, Mark, and Gorden Elser. 1980. "Control of Ground-Water Salinity, Orange County, California." *Groundwater* 18(6):536–40.

Hanak, Ellen, Jay Lund, Ariel Dinar, Brian Gray, Richard Howitt, Jeffrey Mount, Peter Moyle, and Barton Thompson. 2011. *Managing California's Water: From Conflict to Reconciliation*. San Francisco: Public Policy Institute of California.

Harris, Ellen Stern. 1977. "Will Reuse Cure Water Blues?" *Los Angeles Times*, March 13.

Hartley, Troy. 2006. "Public Perception and Participation in Water Reuse." *Desalination* 187:115–26.

Hill, Gladwin. 1965. "Purified Sewage Used to Fill Swimming Pools in California." *New York Times*, July 9, 32.

Holder, Charles F. 1904. "Scientific Disposition of Sewage." Scientific American 91, no. 17 (October 22): 278.

Huang, C. P., Chengdi Dong, and Zhonghung Tang. 1993. "Advanced Chemical Oxidation: Its Present Role and Potential Future in Hazardous Waste Treatment." *Waste Management* 13(5):361–77.

Hyde, Charles Gilman. 1937. "The Beautification and Irrigation of Golden Gate Park with Activated Sludge Effluent." *Sewage Works Journal* 9(6):929–41.

Kennedy, John F. 1961. "Text of President Kennedy's Special Message to Congress on Natural Resources." *New York Times*, February 24.

Kiparsky, Michael, and Peter Gleick. 2003. *Climate Change and California Water Resources: A Survey and Summary of the Literature*. Oakland, CA: Pacific Institute.

Kling, Rob, Spencer C. Olin, and Mark Poster. 1995. *Post-Suburban California: The Transformation of Orange County since World War II.* Berkeley: University of California Press.

Knowlton, Willis T. 1928. "The Sewage Disposal Problem of Los Angeles, California." *Transactions of the American Society of Civil Engineers* 92:984–93.

Lee, Austin. 1965. "Reclaimed Water Gets a Clean Bill of Health: But Processing Must Limit Concentration of Harmful Nitrates." *Los Angeles Times,* November 21.

Loeb, Sidney. 1981. "The Loeb-Sourirajan Membrane: How It Came About." *Synthetic Membrane* 153:1–9.

———. 1984. "Circumstances Leading to the First Municipal Reverse Osmosis Desalination Plant." *Desalination* 50: 53–58.

Loeb, Sidney, and Serop Manjikian. 1965. "Six-Month Field Test of Reverse Osmosis Desalination Membrane." *Industrial & Engineering Chemistry Process Design and Development* 4(2):207–12.

Loeb, Sidney, and Edward Selover. 1967. "Sixteen Months of Field Experience on the Coalinga Pilot Plant." *Desalination* 2(1):75–80.

Los Angeles Times. 1921a. "Valuable Fertilizer Wasted in the Los Angeles Sewage." March 27.

———. 1921b. "Growers Want Fertilizer and Water in the City's Sewage." May 8.

———. 1955. "Reclaiming Waste Water Held Vital to Southwest: Improved Methods Can Save Area from Going Dry, SC Engineers Survey Finds." July 25.

———. 1958a. "Coalinga Plans First Water Desalting Plant: Equipment to Provide 28,000 Gallons for Drinking Daily, Doom Tank Cars." May 28.

———. 1958b. "Reuse of Waste Water Urged to Extend Supply." January 30.

———. 1970a. "Plant Converts Sewage Into Water." October 26.

———. 1970b. "Every Day an 'Earth Day.'" April 24.

———. 1972. "Water Reclamation Project is Prototype." August 20.

National Research Council. 2012. *Water Reuse: Potential for Expanding the Nation's Water Supply through Reuse of Municipal Wastewater.* Washington, DC: National Academies Press.

Marx, Karl. 1906. *Das Kapital,* edited by F. Engels and E. Untermann, translated by S. Moore and E. Aveling. Chicago: Kerr and Co. www.marxists.org /archive/marx/works/download/Marx_Capital_Vol_3.pdf.

MacGowan, Charles. 1963. "History, Function and Program of the Office of Saline Water." *Water Resources Review,* June 1, 24–32. United States Geological Survey. 24-32. www.usbr.gov/research/AWT/OSW/MacGowan.pdf.

Mitch, William, and David Sedlak. 2002. "Factors Controlling Nitrosamine Formation during Wastewater Chlorination." *Water Supply* 2(3):191–98.

Mitch, William A., Jonathan O. Sharp, R. Rhodes Trussell, Richard L. Valentine, Lisa Alvarez-Cohen, and David L. Sedlak. 2003. "N -Nitrosodimethylamine (NDMA) as a Drinking Water Contaminant: A Review." *Environmental Engineering Science* 20(5):389–404.

McCarty, Perry L., David Argo, and Martin Reinhard. 1979. "Operational Experiences with Activated Carbon Adsorbers at Water Factory 21." *Journal of the American Water Works Association* 71(11): 683-689.

Mulligan, Hugh A. 1963. "Pollution Poses New Peril to Water Supply." *Los Angeles Times*, January 3.

National Research Council. 1998. *Issues in Potable Reuse: The Viability of Augmenting Drinking Water Supplies with Reclaimed Water.* Washington, DC: National Academies Press.

———. 2012. *Water Reuse: Potential for Expanding the Nation's Water Supply Through Reuse of Municipal Wastewater.* Washington, DC: National Academies Press.

Nelson, Harry. 1961. "Waste Water's Re-Use Topic of Institute." *Los Angeles Times*, August 22, B3.

New York Times. 1955. "Technology Helps Make Better Dolls." March 7.

———. 1956. "Technology Called Road to U.S. Gains." November 28.

———. 1970. "Los Angeles Suit Seeks a Ban on the Dumping of DDT." October 23.

Okun, Daniel A. 1980. *Philosophy of the Safe Drinking Water Act and Potable Reuse.* Presentation at the EPA Symposium on Protocol Development: Criteria and Standards for Potable Reuse and Feasible Alternatives, Warrentown, Virginia, July 29–31. *Environmental Science and Technology* 14(11):1298–1303.

———. 1997. "Distributing Reclaimed Water through Dual Systems." *Journal of the American Water Works Association* 89(11):52–64.

Ongerth, Henry, and Jerry Ongerth. 1982. "Health Consequences of Wastewater Reuse." *Annual Review of Public Health* 3 (1):427.

Orange County Water District. n. d. "Historical Information." www.ocwd .com/About/HistoricalInformation.aspx.

Phillips, Phil C. 1949. "Salvaging of Wasted Water by Sewage Plants Urged." *Los Angeles Times*, April 12, A4.

Po, Murni, Juliane D. Kaercher, and Blair E. Nancarrow. 2003. *Literature Review of Factors Influencing Public Perception of Water Reuse.* CSIRO Land and Water Technical Report 54/03. http://citeseerx.ist.psu.edu/viewdoc /download?doi=10.1.1.197.423&rep=rep1&type=pdf.

Pryor, Larry. 1971. "Water Factory: Way to Quench Future Thirsts." *Los Angeles Times*, October 31.

Rosenfeld, Judy and Sidney Loeb. 1967. "Turbulent Region Performance of Reverse Osmosis Desalination Tubes: Experience at Coalinga Pilot Plant." *Industrial & Engineering Chemistry Process Design and Development* 6(1):122–27.

San Francisco Chronicle. 1900. "Flow of Water Will be Great: Artesian Gusher near Long Beach May Probe the Biggest of Southern California Wells: Pressure Hard to Keep under Control—Neighborhood Need Have No Fear of Drought." February 21.

Sanitation Districts of Los Angeles County. 1977. *Pomona Virus Study: Final Report.* Sacramento: California State Water Resources Control Board.

Schulte, Peter. 2011. *Using Recycled Water on Agriculture: Sea Mist Farms and Sonoma County.* Oakland, CA: Pacific Institute. www.pacinst.org/wp-content/uploads/2013/02/recycled_water_and_agriculture3.pdf.

Sedlak, David. 2014. *Water 4.0.* New Haven, CT: Yale University Press.

Sheikh, Bahman, Robin P. Cort, William R. Kirkpatrick, Robert S. Jaques, and Takashi Asano. 1990. "Monterey Wastewater Reclamation Study for Agriculture." *Research Journal of the Water Pollution Control Federation* 62(3): 216–26.

Smil, Vaclav. 2004. *Enriching the Earth: Fritz Haber, Carl Bosch, and the Transformation of World Food Production.* Cambridge, MA: MIT Press.

Smith, Dave. 2010. *California's Direct Potable Reuse Initiative.* Presentation to the WateReuse Association. www.watereuse.org/files/s/docs/Ca_Direct_Potable_Reuse.pdf.

Smith, David W. 2013. *WateReuse California Update.* Presentation to the WateReuse Association. www.watereuse.org/sites/default/files/u8/CA%20 Section%20Business%20Update.pdf

Stevens, Douglas, and Sidney Loeb. 1967. "Reverse Osmosis Desalination Costs Derived from the Coalinga Pilot Plant Operation." *Desalination* 2:56–74.

UCLA Engineering. 2013. "First Demonstration of Reverse Osmosis." www .engineer.ucla.edu/explore/history/major-research-highlights/first-demonstration-of-reverse-osmosis.

Udall, Stewart L. 1965. "Address to the First International Symposium on Water Desalination." October 4, Washington, DC. www.library.arizona .edu/exhibits/sludall/speechretrievals/addresswaterdesalx.htm.

Washington Post. 1952. "Super-Strength Concrete Developed with Plastic." March 30.

WateReuse Association. 2013. "About Us." www.watereuse.org/about-watereuse-association.

West Basin Municipal Water District. 2013. "History of West Basin." www .westbasin.org/about-west-basin/history.

White, Jean M. 1965. "Water Restoration Is Seen as Answer to Rising Demand." *Washington Post,* May 11.

CHAPTER 11

Water Justice in California's Central Valley

CAROLINA BALAZS AND ISHA RAY

BACKDROP: TOOLEVILLE AND ITS DISCONTENTS

The unincorporated community of Tooleville, California, is located at the eastern edge of Tulare County's valley floor, at the foot of the rolling Sierra Nevada foothills that are dotted with orange groves and small residential enclaves. Tooleville is a farmworker community; the roughly seventy households living here are predominantly Latino, with a median annual household income of $16,000 (about a third of the median income for California). Residents pride themselves on the beauty of their natural surroundings and their high rates of homeownership. Ms. Jimenez[1] remembers the day her father purchased a home in Tooleville—"I was so proud that we owned a house." She still lives there and is passionate about staying in her community, despite the challenges Tooleville faces.

Like most small communities in the San Joaquin Valley, Tooleville residents rely on groundwater for drinking. But since 1997, Tooleville's two wells have exceeded the Safe Drinking Water Act (SDWA) maximum contaminant level (MCL) for nitrate at least seven times. Nitrate is an acute contaminant, and, at these levels, infants are at risk of methemoglobinemia ("blue baby syndrome"), and women are at risk of adverse reproductive effects (Fan and Steinberg 1996). In some years, the drinking water has violated SDWA standards for total coliform. To deal with bacteriological contamination, residents could boil the water, but this would concentrate the nitrate.

FIGURE 11.1. Aerial map of the city of Exeter and Tooleville, California. They are less than two miles apart. The Friant-Kern Canal passes to the east of Tooleville.

Tooleville residents are frustrated that historical planning processes have limited the financial and infrastructure resources available to Tooleville. Until 2012 Tulare County's General Plan of 1973 listed Tooleville as one of fifteen communities from which public resources, including water infrastructure, should be withheld. Solutions have been hard to come by. Attempts to drill new wells have had poor results—the groundwater all around the community is high in nitrates. This has left Tooleville with a persistent compliance and exposure burden, prolonging risks from exposure as well as household coping costs. Even coping mechanisms such as purchasing bottled water are only partially protective. Most residents have drunk the contaminated well water at some point, and still use it for cooking.

Regional solutions have also been hard to achieve. For several years, residents and county officials hoped that Tooleville could physically consolidate with the nearby city of Exeter, which is less than two miles away, and has more wells and cleaner water (figure 11.1). But the city has been more interested in expanding its spheres of influence in other directions. Tooleville residents believe this to be intentional and discriminatory because theirs is a low-income neighborhood. In the 200s, Exeter cited prevailing wages as a barrier to consolidation, from which it was later exempted. In 2009, the California Department of Public Health stepped in, and has been pressuring Exeter to connect to Tooleville. In the meantime, residents continue to rely on contaminated wells, and pay twice for water—once for their utility bill, and again for bottled water they can drink.

The story of Tooleville shows that, while small size does make a system physically vulnerable, a range of political actors and social-historical factors also impact exposure and coping capacity. Tooleville's story underscores the complexity of isolating "the cause" of drinking water pollution. Finally, the composite burden we describe—of exposure and coping costs—creates place-specific environmental injustices, even in a state such as California, where safe water is regularly taken for granted as a right fulfilled for all.

INTRODUCTION

On January 1, 2013, California Assembly Bill 685, known as the Human Right to Water Bill, became effective. The new law intends to promote universal access to safe, clean, and affordable water throughout California. But what does such a bill mean in California, the richest state in the richest country in the world, where almost everyone has piped and potable water delivered to the home?

Poor drinking water quality is usually thought of as a "developing country" problem. In the main, this perception is correct. But hundreds of small communities in California and across the United States rely on unsafe drinking water sources that their modest means cannot mitigate. Research and grass-roots efforts have consistently drawn attention to high levels of contaminants in California's San Joaquin Valley (Dubrovsky et al. 2010; Harter et al. 2012); to inadequate services and infrastructure in U.S.–Mexico border *colonias* (Olmstead 2004) and rural communities in the South (Wilson et al. 2008; Heaney et al. 2011); and to bacteriological and chemical contamination in unregulated drinking water sources in the Navajo Nation (Murphy et al. 2009). Our own earlier research, conducted between 2006 and 2011, established that race/ethnicity and socioeconomic class were correlated with exposure to nitrate and arsenic contamination and with noncompliance with federal standards in community water systems (Balazs et al. 2011, 2012).

In this chapter we describe the Drinking Water Disparities Framework to explain environmental injustice in the context of drinking water in California's Central Valley.[2] The framework builds on the social epidemiology and environmental justice literatures, and is made concrete through five years of field data from California's rural San Joaquin Valley.[3] We focus on nitrate and arsenic contamination to show how race and class are correlated with contaminated drinking water in the valley. We then trace the mechanisms through which natural, built, and sociopolitical factors work through state, county, community, and household actors to constrain access to safe water supplies and to financial resources for communities.

A rich understanding of how disparities in access to safe drinking water are produced and maintained is essential for framing environmental justice concerns and developing effective public health interventions. Until recently, environmental justice research has focused predominantly on the disproportionate burdens of toxic sitings (e.g. Bullard 2005) and environmental exposures (e.g. Morello-Frosch et al. 2001; Morello-Frosch and Lopez 2006), and has been relatively silent on the topic of water.[4] It is often forgotten that the birth of the environmental justice movement in Warren County, North Carolina, included the concern that PCB-laced soil would leak into the drinking water supplies in a predominantly African American community (Cole and Foster 2001). This chapter provides a framework within which to understand environmental justice in the context of drinking water. In doing so, this chapter reflects the call by environmental justice scholars (Pulido 1996; Pulido et al. 1996) for more historically informed work on the causes and consequences of environmental injustice. We draw on a definition of environmental justice that includes both distributional and procedural elements, but more broadly defines water (in)justice as a composite burden shaped by a comprehensive set of actors, processes, and mechanisms.

Our Drinking Water Disparities Framework (Balazs and Ray 2014) shows that community constraints and regulatory failures produce social disparities in exposure to drinking water contaminants. Water system and household coping capacities lead, at best, to partial protection against exposure. This composite burden explains the origins and persistence of social disparities in exposure to drinking water contaminants.

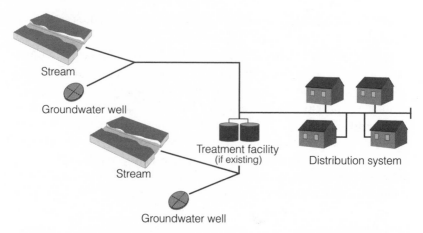

Stream

Groundwater well

Stream

Treatment facility
(if existing)

Distribution system

Groundwater well

FIGURE 11.2. Schematic of a community water system. Water from a groundwater well or stream may be treated or untreated before entering into the distribution system. Source: Balazs et al. (2011).

NITRATES, ARSENIC AND DRINKING WATER DISPARITIES IN CALIFORNIA'S CENTRAL VALLEY

We begin with the significant findings of our past research on drinking water quality in California's Central Valley. Our research explored the extent and nature of correlations between race, class, and drinking water quality, especially with respect to nitrate and arsenic contamination. For years local residents had sought to draw attention to water contamination in their communities, but had been told by local politicians that the issues were "community-specific" (personal communication). Our research sought to determine whether there was a disproportionate burden of exposure across the valley, not just in particular communities. We focused on community water systems (CWSs), which are public water systems that serve water year-round to at least 25 people or have more than 15 service connections (U.S. Environmental Protection Agency 2010). A simple schematic of a CWS is shown in figure 11.2.

Our first study (Balazs et al. 2011) asked whether CWSs predominantly serving people of color and lower-income areas were more likely to have higher levels of nitrate contamination. With its intensive irrigated agriculture, this valley has some of the highest nitrate levels in the country (Dubrovsky et al. 2010). Nearly 95 percent of the valley's residents rely on groundwater for drinking (California Department of Public Health 2008a). The valley also has some of the highest rates of poverty and minority populations—particularly Latinos—in the state (U.S. Census Bureau 2007). These communities are economically disadvantaged, making it harder for them to mitigate either the nitrates or the health consequences of nitrate contamination. With the continued use of nitrogen-based fertilizers (Dubrovsky et al. 2010), exposure may become increasingly widespread.

Our study statistically analyzed the relation between CWS demographics (percentage of the population that is Latino, and percentage of home ownership in the population) and average nitrate levels between 1999 and 2001. We classified nitrate levels for each CWS as *high* if the system-wide average exceeded the MCL of 45 mg NO_3/L; *low* if the average was less than half the MCL (<22.5 mg NO_3/L); and *medium* if it was 22.5–44.9 mg NO_3/L). Figure 11.3 shows our descriptive results. CWSs are stratified into quartiles based on percentage of Latinos living in the community. Quartiles with higher Latino percentages had a greater proportion of systems with high nitrate concentration (Latino quartiles 3 and 4 in figure 11.3). The two quartiles with the lowest rates of homeownership (a proxy for less wealth) had the largest proportions of systems in the medium and high nitrate categories (15 percent and 22 percent, respectively). In sum, we found a positive association between race and nitrate levels. A more robust statistical model then controlled for confounding variables that could mediate the relationship of interest. The model confirmed the descriptive findings. We found that CWSs that served higher fractions of Latinos and lower fractions of homeowners had higher average nitrate levels. This effect was strongest in small CWSs, indicating not only a social disparity in exposure but also a greater impact in the small communities that often have the fewest resources to cope with contamination.

Our second study (Balazs et al. 2012) explored whether CWSs that served predominantly low-income populations or people of color were more likely to have higher average levels of arsenic and a greater challenge in complying with drinking water standards. Arsenic in drinking water is linked to skin, lung, bladder, and kidney cancers (Tseng et al. 1968, 2000; Smith et al. 1992; Fereccio et al. 2000), and the most common exposure pathway is consumption of contaminated groundwater (Prüss-Ustün et al. 2011). In the valley, arsenic can reach elevated concentrations due to agricultural activities (National Research Council 2001).

In 2002, amid considerable debate, the U.S. Environmental Protection Agency (EPA) issued its revised arsenic rule reducing the allowable arsenic concentration in drinking water from 50 µg/L to 10 µg/L. It was understood from the start that systems with low economies of scale

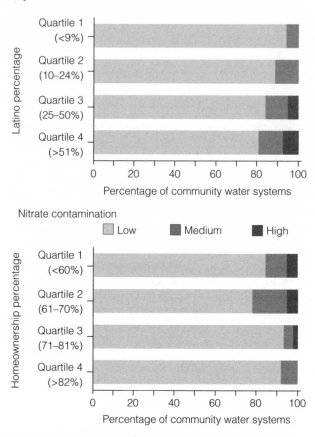

FIGURE 11.3. Percentage of community water systems with low, medium, and high levels of nitrate contamination plotted against Latino percentage and homeownership percentage in the service population. Source: Balazs et al. (2011).

would find it difficult to comply with the new rule (Stone et al. 2007; Pilley et al. 2009). But little attention was given to other inequities that could arise in exposure to arsenic or ability to comply. For this reason, we argued that a joint focus on compliance challenges and exposure to contaminants is most helpful for understanding the health, environmental justice, and policy implications of drinking water policies such as the revised arsenic rule. Our study thus employed a "joint-burden analysis" to analyze the environmental justice implications of both compliance capacity *and* exposure related to arsenic contamination. Our study period was 2005–2007, as water systems had been given until 2006 to comply with the revised rule.

We found that communities with lower rates of homeownership and greater proportions of people of color had higher odds of MCL violation. We also found a negative association between homeownership rates and arsenic concentrations in drinking water, with a stronger effect in smaller CWSs (those with fewer than 200 connections). These results indicate that communities with fewer economic resources faced a dual burden—they were not only exposed to higher arsenic levels, but were also served by systems more likely to receive an MCL violation.

What are we to make of these nitrate and arsenic findings? The association of race/ethnicity and socioeconomic status with nitrate levels could be due to several factors. Race/ethnicity could be related to the historical proximity of farm labor communities to agriculture, as well as the (in)ability of residents to participate in the governance of their CWS on account of language, citizenship status, or lack of political clout (Michelson 2000). That water quality varied by Latino percentage or homeownership matters not only on account of distributional inequities but also because elevated nitrate levels could pose a greater hazard to lower-income sub-populations that have less access to health care. Our findings also suggest a "canary in the coal mine" scenario: nitrate levels are impacting systems throughout the valley (Harter et al. 2012), not just in the small, lower-income areas that we studied. Eventually, many more towns and cities could face the mitigation and treatment costs associated with spreading nitrate contamination (Moore et al. 2011).

Arsenic in groundwater generally occurs naturally, so we should not expect a positive association between arsenic levels in CWSs and low socioeconomic status. Our results can best be understood in the broader context of the mediating role of system-level capacity. Smaller water systems often lack the economies of scale and resource base to ensure the technical, managerial, and financial (TMF) capacity to reduce contaminant levels (Committee on Small Water Systems 1997; Shanaghan and Bielanski 2003). The socioeconomic status of residents directly influences TMF capacity, because it affects the ability of a water system to leverage resources, both internal (e.g. rate increases) and external (e.g. loans; Committee on Small Water Systems 1997). Thus, in our arsenic study, CWSs with customers of lower socioeconomic status may have been less able to ensure compliance with the revised arsenic standard by 2007. Our joint-burden analysis highlights the need to consider not only exposure and current states of compliance but also the future mitigation potential of impacted water systems and the households they serve.

UNDERSTANDING WHY DRINKING WATER DISPARITIES EXIST AND PERSIST

Our work on nitrate and arsenic contamination led us to ask: Why do drinking water disparities exist, and how do they persist, despite the passage of the Safe Drinking Water Act of 1974? Designing solutions for contamination and contamination-related disparities requires a thorough historical-structural analysis (Pulido 1996; Pulido et al. 1996) of the mechanisms through which environmental injustices are produced.

Disparities in water infrastructure and "basic amenities" (Wilson et al. 2008; Wilson 2009; Vanderslice 2011) can drive adverse health effects. Historical and structural conditions shape lack of access to safe drinking water; these conditions include selective enforcement of regulations (Cory and Rahman 2009), noncompliance with federal standards (Guerrero-Preston et al. 2008; Rahman et al. 2010), inequities in access to funding (Imperial 1999), and (the absence of) a community's political power in accessing a safe water supply (Francis and Firestone 2011). Researchers have shown that the cost of service extension can drive inadequate service provision (Olmstead 2004); that municipalities provide or deny access by determining which areas to annex or exclude from their city boundaries (Wilson et al. 2008; Marsh et al. 2010); and that *de facto* segregation allows such determinations to continue (Troesken 2002).

Our Drinking Water Disparities Framework builds on this research to explain why drinking water disparities exist and persist, but draws primarily on the social epidemiology literature for its theoretical framing. Social epidemiological research uncovers how race, class, and social factors (Sexton et al. 1993; Gee and Payne-Sturges 2004) interact over multiple levels of decision-making (household, community, and region; Krieger 2001) to impact exposure to contamination (deFur et al. 2007).

Five years of primary data collection with residents, state and county drinking water regulators, water board members in unincorporated communities, participants at environmental justice meetings, and community-based organizations, in particular the Community Water Center, in the southern San Joaquin Valley provide the empirical grounding for our framework.[5] This richly nuanced dataset reveals not only the role of multi-level actors in shaping disparities but also the lived experiences of households and communities who struggle for safe water. Data on drinking water quality and Safe Drinking Water Act (SDWA) violations

(California Department of Public Health 2008a, 2008b) in CWSs across the valley complement the qualitative field data.

Our framework traces how the historical marginalization of poor communities, coupled with poor source water quality, determines the condition of their physical infrastructure and results in exposure. We thus emphasize the role of historical and structural factors, and trace the mechanisms through which these lead to exposure disparities. These structural factors are not deterministic; rather, communities and individuals exercise agency within the structures that constrain them. The extent of this agency also impacts exposure. Ultimately, our framework outlines a "composite burden" composed of exposure to contaminants plus the inability of socially vulnerable communities to mitigate contamination. We argue that this composite burden leads to persistent exposures and social disparities in exposure to poor drinking water.

THE DRINKING WATER DISPARITIES FRAMEWORK

We present the Drinking Water Disparities Framework in figure 11.4. The figure depicts the factors within the three environments (natural, built, and sociopolitical) that drive drinking water disparities across race and class. The framework shows that these factors, when mediated through the actions (or inactions) of state, county, community, and household actors, jointly impact exposure and coping capabilities. Viewed comprehensively, these multiple possible pathways, or mechanisms, at multiple levels, can result in persistent exposures to water contamination that vary by the race and class of different communities.

Three "environments" contain the factors that drive the disparities. The *natural environment* includes ecological factors such as soil types, hydrology, and climate; these cannot be altered except over a long timeframe. The *built environment* represents human-modified spaces in which "people live, work and recreate" (Roof and Oleru 2008), such as agricultural land, buildings, and water infrastructure. The *sociopolitical environment* refers to institutional and group characteristics (e.g. community or household), including historical and present-day planning policies, governance practices, and community demographics. Each environment contains factors that act across all three scales—conventionally called *levels* in the social epidemiology literature—the regional (including state and county), the community, and the household. Arrows connecting the three environments show the factors' mutual interactions. For example, citrus farming is a part of the built environment, but affects the natural environment via water quality, and farming itself is influenced by natural characteristics such as climate and soil type. The lines separating the levels indicate that specific drivers of water access can occur at, and influence, multiple levels within an environment. For example, degraded community-level water infrastructure can (but need not) interact with household infrastructure.

Factors in all three environments and across all three levels act through, and across, actors within four distinct levels relevant to the valley: the state, the county, the community, and the household.[6] The state and county levels correspond to political and geographic boundaries. State and county regulators function within their respective levels. The community is defined by the physical service area of a CWS, defined earlier. A community can be an incorporated city with its own tax base, or it can be unincorporated. Municipal employees, community organizers and community groups, water board members, and non-governmental organizations are contained within the community level. The household level is where drinking water is usually accessed, though exposure ultimately occurs at the individual level. Ordinary residents are contained within the household level.

Our Drinking Water Disparities Framework highlights the role of multi-level coping mechanisms in influencing exposure by adding *coping* to the classic exposure–disease paradigm (bottom of figure 11.4). In general, exposure to drinking water contaminants in excess of SDWA standards necessitates mitigation, and the water system is required to implement a solution. However, when a water system is incapable of doing so, or while it waits to solve the contamination problem, households must individually respond. We show that exposure and coping are mutually constitutive; the degree of exposure dictates the need for coping, and the degree to which coping mechanisms are successful directly influences exposure. This is indicated in figure 11.4 with bidirectional arrows. To the extent that the coping of one actor (e.g. a community water board) is not successful, it necessitates coping by another (e.g. the household)—this is indicated by the lines within Coping Mechanisms. Coping leads to additional costs, and these added costs also constrain future coping capacity. Jointly, these feedback cycles and the resulting exposure and coping costs define what we have called a composite drinking water burden. In the next two sections we flesh out this framework with specific examples from rural communities in the San Joaquin Valley, showing how exposure occurs and how small CWSs (try to) cope.

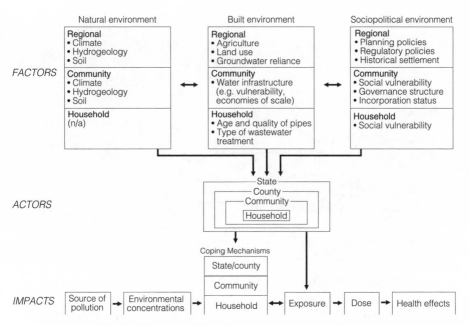

FIGURE 11.4. The Drinking Water Disparities Framework: a multi-level, multi-actor perspective. Source: Balazs and Ray (2014).

THE FRAMEWORK APPLIED: MULTI-LEVEL FACTORS AND ACTORS IMPACT EXPOSURE

To begin, the natural and built environments, such as hydrogeology and land-use practices, shape source water quality, which in turn partially defines baseline contaminant levels. For example, the climate and soil of Tulare County's eastern foothills create favorable growing conditions for citrus trees that use high amounts of nitrate fertilizer. Since the water table in this region is shallow (figure 11.5), nitrates can leach into it rapidly and travel quickly into well water (Nash 2006). As a result, communities such as Tooleville, located on the eastern side of Tulare County, have some of the highest nitrate levels in the valley (Dubrovsky et al. 1998). On the western side of the valley, in communities such as Alpaugh, the Corcoran clay layer plays a converse role. This impermeable layer requires that CWSs relying on groundwater drill deeper wells (Galloway and Riley 2006), but at these deeper levels wells are likely to draw naturally arsenic-laden water (Welch et al. 2000; Gao et al. 2007).

Built and sociopolitical factors interact with natural factors to further determine exposure levels at the community and household levels.

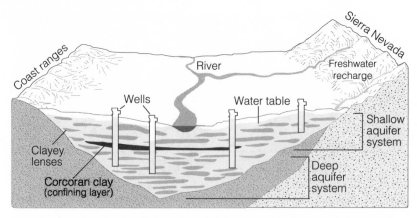

FIGURE 11.5. Cross-section of the valley, with Corcoran Clay layer on the left (west) and the shallower aquifers on the right (east). Adapted from Galloway and Riley (2006).

For example, the allocation of water rights and the development of water resources in the valley have played direct roles in determining drinking water quality. Government financing of large-scale water projects historically enabled the storage and conveyance of vast quantities of Sierra Nevada snowmelt and California Delta waters to farmlands. In this process, farmers received nearly unlimited surface water rights for agriculture (Reisner 1986), but 95 percent of the valley's residents were left to rely on groundwater for drinking (California Department of Public Health 2008a). This might not have mattered, were it not for the baseline natural conditions (e.g. arsenic in the soil) and agriculture's contamination of groundwater due to chemical runoff from pesticides and fertilizers (Dubrovsky et al. 1998; Viers et al. 2012). In 2007, 75 percent of all of California's nitrate violations occurred in the San Joaquin Valley (California Department of Public Health 2008a).

Policies at multiple levels interact to deprive communities of adequate drinking water resources. For example, the 1973 General Plan of the Tulare County Planning Department reads, "Public commitments to communities with little or no authentic future should be carefully examined beforeal action is initiated. These non-viable communities would, as a consequence of withholding major public facilities such as sewer and water systems, enter a process of long term, natural decline as residents depart for improved opportunities in nearby communities." Among the fifteen communities listed as "non-viable" are Alpaugh, Plainview, Seville, and Tooleville. Many of these communities were once labor camps, or are currently unincorporated, without their own tax base or municipal representation to draw on for infrastructure improvements. Thus this *de jure* discrimination results in the *de facto* discrimination of redlining, where the "non-viable" label is used to justify withholding of resources and to allow the perpetuation of poor infrastructure. A leader from one of the allegedly non-viable communities notes: "One of the questions a lot of people ask me is, if the water's so bad . . . why don't you move? And I'm thinking, why would you want me to move? That's my house. That's my town. I was born and raised there. . . . Do you think by moving it's going to get solved?"

As with county-level plans, selective annexation at the city level has allowed water problems to persist. The city of Exeter is less than two miles from Tooleville and has used its municipal decision-making authority to (in effect) prolong exposure in Tooleville. Exeter repeatedly cited prevailing wages as a barrier to extending pipelines to Tooleville even though two Senate bills (Senate Bill X29 and Senate Bill 110) explicitly exempted the city from paying prevailing wages. Exeter's 2020 expansion plan includes areas of undeveloped agricultural parcels and ranchette houses toward the east. The growth areas extend in some cases to at least the same distance as Tooleville, but they do not extend toward Tooleville. Residents experience this as a case of "municipal underbounding" (Marsh et al. 2010): "If we were rich, we'd raise their tax base. But we're poor, so they're not interested in us."

On balance, despite the best intentions, county drinking water regulators have been unable to ameliorate the valley's ongoing contamination problems. The SDWA promotes a system-by-system focus and provides few incentives for regulators to support regional solutions. As one county regulator noted, Tooleville residents "pay taxes in our county, they pay taxes in our stores, their children go to our school. . . . It irritates me that [Exeter] won't help those people." However, he continued, "I don't have an opinion, I'm a regulator." Clearly, this regulator can see the need for intercommunity solutions, but the solution is outside of his regulatory mandate. Residents in unincorporated places also see this problem: "Do you know how long we've been knocking on the county's door? . . . We've been doing this since my dad was a farmworker."

State and county regulatory failures add to the exposure burden produced by historically poor infrastructure and limited municipal support. In interviews, regulators agreed that because they are limited by funding and staff time, they are forced to prioritize which drinking water regulations to enforce. Violations of Tier 1 contaminants (those that can cause acute or immediate health impacts, such as total coliform or nitrate) are explicitly prioritized over a system's failure to comply with the SDWA's monitoring requirements. But prioritizing violations over monitoring leads to unforeseen exposure risks. In 2007, Fresno County returned primacy for water systems with fewer than 200 connections to state-level regulators; state officials found that many of the CWSs had failed to monitor for several years, but had not been given notice of monitoring violations by county regulators. Without water quality monitoring data county regulators had been unable to issue MCL violations, and with no notices of MCL violations residents had lacked information on whether they were being exposed to harmful levels of contaminants.

THE FRAMEWORK APPLIED: MULTI-LEVEL FACTORS AND ACTORS IMPACT COPING AND MITIGATION

If coping and mitigation strategies at the community or household level could adequately address drinking water contamination, then vulnerability to exposure could be minimized (see figure 11.4). However, our fieldwork showed that inadequate infrastructure, poor TMF capacity at the community level, failures of the regulatory system to provide information on alternatives, and inadequate funding at the state level all undermine the success of coping mechanisms.

The joint role of poor infrastructure and poor TMF in undermining mitigation is best understood through the examples of the unincorporated communities of Alpaugh and Lanare. Alpaugh, in Tulare County, had exceeded the old arsenic standard of 50 µg /L since the early 2000s (California Department of Public Health 2008a) and had experienced water outages when its backup wells broke down. In 2005, the water board obtained $4.2 million to rehabilitate its system, but it did not include plans to upgrade to the 2006 revised arsenic rule of 10 µg As/L. As one newspaper article noted, "officials were just focusing on getting water flowing. Once that was accomplished . . . they would worry about the arsenic issue" (Boyles 2005). Similarly, in the unincorporated community of Lanare, in Fresno County, the MCL for arsenic had been exceeded by 2005. In July 2006, after securing a Community Block Grant, residents celebrated the installation of a new treatment plant (Nolen 2007). Six months later the plant was closed down. A grand jury investigation found that "because of mismanagement, unacceptable arsenic levels, and the absence of any other water source, the district is in crisis" (Fresno County Grand Jury 2008).

The cases of Alpaugh and Lanare could partly be explained by poor TMF capacity, a particular problem in small water systems. One regulator explained that in small communities local residents and volunteers run the water boards: "They live there, they're residents. They don't really understand our regulatory requirements." But regulators also noted that low TMF stems from a community's low resource base. Small, low-asset communities are unable to hire full-time operators who know the ins and outs of drinking water requirements and planning.

State funding mechanisms for new water sources or treatment could offer system-level solutions, but as currently designed, they often do not promote timely solutions. Congress revised the 1996 SDWA amendments to include capacity-development programs for small systems (Shanaghan and Bielanski 2003), but, in California, TMF capacity is required before water systems can get state revolving funds (California Department of Public Health 2009). Similarly, the American Recovery and Reinvestment Act of 2009 set aside $160 million for drinking water infrastructure; it earmarked stimulus money for "high priority" projects that were "shovel ready" (California Environmental Protection Agency 2010). In both cases, the funding criteria (TMF capacity and shovel-readiness) define eligibility on the core weaknesses of resource-poor communities. Communities that lack resources lack TMF; without TMF, funding is harder to attain; and without funding, TMF cannot be developed. The funding conditions through which exposure could be mitigated are thus conditions through which disparities in exposures are prolonged.

When system-level coping fails, households assume the burden of mitigation. But a combination of disenfranchised residents, inadequate water system responses, and regulatory failures is yet another pathway toward vulnerability. Interviewees reported that local water boards sometimes discriminate against residents on the basis of language, race/ethnicity, socioeconomic status, or homeownership. In 2010, residents from the community of East Orosi testified to the United Nations Special Rapporteur on the Human Right to Water and Sanitation that, due to their Spanish accents in English, they were regularly turned away by water board administrators when seeking clarification on their water quality reports (United Nations General Assembly 2011).

Regulatory failures further undermine household-level coping mechanisms. The SDWA focuses on a contaminant-by-contaminant mode of regulation, but has no stipulations on how residents should address multiple contaminants (e.g. nitrate and total coliform). In 2007, 5 percent of the valley's 677 active CWSs received an MCL violation for both nitrate and total coliform (California Department of Public Health 2008a). A violation of the total coliform MCL triggers a boil-water order. But boiling water can increase concentrations of nitrate. Neither does the SDWA explicitly address how to cope with long-term exposures. A resident from the community of Cutler explained that for years she has received Consumer Confidence Reports indicating that dibromochloropropane levels in the water exceeded the MCL. These reports note that residents should not worry because health impacts are not based on immediate exposure but on lifetime exposure. She had lived in her community for nearly 30 years—so, she asked, should she worry or not?

In these situations, water systems simply leave residents to cope with contaminated drinking water as best they can, and SDWA regulations ultimately fail the (low-income) household. Individual coping mechanisms to reduce exposure may not be effective. Households may purchase bottled water, but individuals may not consistently drink it. Households may install water filters, but may incorrectly assume that the filter treats for the contaminant of interest (Moore et al. 2011). Yet, significant costs are incurred for these only partially protective measures. In many low-income valley communities, households pay 4 to 10 percent of their monthly income for water (Moore et al. 2011), including the utility bill and vended water, well above the EPA's affordability criterion of 2.5 percent of median household income (U.S. Environmental Protection Agency 2003). Certainly when a successful system-level mitigation strategy is developed, these costs are passed along to the household. But under those circumstances there is a higher probability of achieving water quality of adequate standards.

CONCLUDING THOUGHTS

The Drinking Water Disparities Framework traces the development of a composite drinking water burden that comprises the exposure and coping costs that many water systems and households face. The framework reveals that there is no direct causal path between race and class and disproportionate burdens; rather, race and class are imbricated in almost all the factors and actors that have historically combined, and still combine, to produce this composite burden.

The framework shows that decisions of multiple actors at every level, made intentionally or by default, prolong exposure and impede households' coping capabilities. It reveals how, alongside a baseline of contaminated source water, a series of policies have constrained access to physical and financial resources. These decisions, in conjunction with regulatory failures, a lack of community resources to mitigate contamination, and political disenfranchisement of local residents, help explain the origins of environmental injustice in the context of drinking water. These same forces also influence coping capacities, which may lead only to partial protection, which in turn exacerbates the impacts of drinking water contamination. This composite-burden analysis shows that there is no single cause or intentional action that defines environmental injustice and drinking water; rather, a comprehensive set of actors, processes, and mechanisms jointly shape it. It reminds public health practitioners and policymakers to look beyond exposure and proximate causes and include historical and structural factors in the analysis of exposure disparities. This highlights the need for a multi-pronged intervention agenda to reduce and mitigate the drinking water disparities. While this chapter focuses on the Central Valley, we believe it would be enhanced with lessons from other rural regions across the state, and with further engagement with urban settings.

From a policy perspective, the framework identifies multiple potential intervention points (Susser and Susser 1996). Numerous policies have attempted to address drinking water contamination and small-water-system challenges, including monies from the American Recovery and Reinvestment Act and the Safe Drinking Water State Revolving Fund. But unless future incarnations of these policies take seriously the disparity-producing mechanisms highlighted in this chapter, these policies are unlikely to improve drinking water conditions in the most disadvantaged communities. While new efforts to support small communities are already underway (e.g. California's Emergency Funding and Small Water System Program Plan), a concerted focus on improving TMF capacity in disadvantaged communities is critical. Funding mechanisms should not always use TMF capacity as a requirement, but should find ways to support it, or enhance other sustainable solutions. "Planning-ready" rather than "shovel-ready" funding would help small or disadvantaged systems develop their engineering and financial plans for contaminant mitigation and infrastructure needs.

Similarly, the promotion of water system consolidation—be it physical connection of a small system to a larger one or sharing of management capacities—must acknowledge the underlying political and social barriers noted here. Water policy experts often say that smaller systems block consolidation efforts for fear of losing local autonomy. But our work argues that a deeper and long-standing set of social, economic, and political processes also creates barriers. Consolidation may be more successful if it is not left to isolated cities and communities but facilitated by a regional drinking water development program. This may require abdication of some municipal authority, something many cities are loath to surrender.

Finally, future amendments to the SDWA are needed on three fronts. First, the ability of water systems to comply with monitoring and reporting violations should be given priority. Secondly, drinking water regulations should clearly address the co-occurrence of contaminants and how to adequately inform residents about long-term protective measures. And thirdly, regional or cross-system solutions will be necessary. The prevalent methods of system-by-system monitoring and contaminant-by-contaminant remediation cannot alleviate the composite burden of drinking water vulnerability in low-income communities.

POSTSCRIPT: TOWARD A WATER JUSTICE FUTURE FOR CALIFORNIA

At the time of writing this chapter, great momentum was building towards addressing the drinking water concerns we outline above. In 2012, Governor Brown created the Governor's Drinking Water Stakeholder Group, bringing together environmental justice advocates, water regulators, and agriculture and water industry experts to work on developing solutions to operation and maintenance challenges in disadvantaged communities. That same year, the California Department of Water Resources funded seven integrated regional water management (IRWM) pilot projects to address the challenges faced by disadvantaged communities and ensure more active involvement in regional planning. At the beginning of 2013, California's thriving water justice movement achieved the passage of AB 685, the first Human Right to Water bill in the country. Later that year, Governor Brown released his draft California Water Action Plan. Among the top ten priorities were the need to "invest in integrated water management and increase regional self-reliance" (priority 2), "provide safe drinking water and secure wastewater to all water systems" (priority 7), and "improve operational and regulatory efficiency" (priority 9).

When we began our research in 2005, acknowledgement and integration of these core drinking water needs was nearly nonexistent. Local community members were told that their issues were "local to their community." Agricultural interests maintained that the main source of nitrate was leaching septic tanks, not nitrate fertilizers. There was little acknowledgement of funding barriers for small systems to achieve compliance. Regional solutions were acknowledged but rarely supported by regulators or local IRWM groups.

Since then, strong environmental justice leadership and local communities have caused the ground to shift. Supported by Senate Bill X2-1 (Perata, 2008), researchers documented the primary role of agriculture in creating nitrate contamination in the valley (Harter et al. 2012). Early results from the IRWM pilot projects have highlighted that regional solutions hold much promise, given adequate funding and technical support and a willingness of traditional power-holders in IRWM planning to involve traditionally marginalized groups (Balazs and Lubell 2014). And research on cumulative impacts is developing tools to consider the multiple components of water contamination,[7] including multiple contaminants and system-level characteristics. These shifts hold much promise for ensuring that all Californians obtain access to clean and affordable water.

And yet, much work remains. Regional solutions must continue to be developed in order that systems with low economies of scale gain the efficiencies necessary to obtain clean water at affordable prices. The political participation of traditionally marginalized communities must continue to be supported so that the voices and needs of less traditionally powerful voices are adequately incorporated into water policy and shape local planning efforts. In essence, the pathways and mechanisms outlined in our Drinking Water Disparities Framework must be continually revisited, so that Californians can work toward a future that meets both the distributional and procedural goals of water justice.

ACKNOWLEDGMENTS

We thank the *American Journal of Public Health* for permission to publish substantial portions of our article (Balazs and Ray 2014). We thank *Environmental Health* for permission to cite from Balazs et al. (2012). We thank *Environmental Health Perspectives* for permission to use key figures from Balazs et al. (2011). This research was supported by the NSF Graduate Research Fellowship, the California Endowment (through a collaborative grant between the Community Water Center and UC Berkeley), the California Environmental Protection Agency (#07–020), the Switzer Environmental Fellowship, and the UC President's Post-doctoral Fellowship. We thank Rachel Morello-Frosch and Alan Hubbard, our coauthors on previous studies; Laurel Firestone, Susana de Anda, Maria Herrera (all of the CWC); Rich Haberman and Dave Spath (formerly of the CDPH); the ERG Water Group; the Morello-Frosch lab group; and Bhavna Shamasunder. The California Department of Public Health and Tulare County Environmental Health Services provided us with water-quality data.

NOTES

1. Pseudonym used in accordance with the Protection of Human Subjects protocol of the University of California, Berkeley.

2. The topic of water justice in California spans an array of issues, including traditional water uses, recreational water use, rural versus urban issues etc. (see e.g. the 2014 documentary, *Thirsty for Justice*).

3. Water-justice and drinking-water struggles are also present in urban settings, but this chapter draws on data from rural cases.

4. There is a growing international environmental justice focus on water; see e.g. Debbane and Keil (2004).

5. All interviews and focus groups were conducted in person, in English and Spanish, by Carolina Balazs.

6. National and within-state regions could potentially be included as additional levels, but for this chapter these are not central.

7. See e.g. the California Environmental Protection Agency's CalEnviro-Screen.

REFERENCES

Balazs, Carolina, and Mark Lubell. 2014. "Social Learning in an Environmental Justice Context: A Case Study of Integrated Regional Water Management in California." *Water Policy* 16:97–120.

Balazs, Carolina, and Isha Ray. 2014. "The Drinking Water Disparities Framework: On the Origins and Persistence of Inequities in Exposure." *American Journal of Public Health* 104(4): 603–11.

Balazs, Carolina, Rachel Morello-Frosch, Alan Hubbard, and Isha Ray. 2011. "Social Disparities in Nitrate Contaminated Drinking Water in the San Joaquin Valley." *Environmental Health Perspectives* 119(9):1272–78.

Balazs, Carolina, Rachel Morello-Frosch, Alan Hubbard, and Isha Ray. 2012. "Environmental Justice Implications of Arsenic Contamination in California's San Joaquin Valley: A Cross-Sectional, Cluster Design Examining Exposure and Compliance in Community Drinking Water Systems." *Environmental Health* 11(84). doi:10.1186/1476-069X-11-84.

Boyles, D. 2005. "Alpaugh Water System Work Starts." *Fresno Bee*, June 3, B5.

Bullard, Robert. 2005. "Environmental Justice in the 21st Century." In *The Quest for Environmental Justice: Human Rights and the Politics of Pollution*, edited by R. Bullard (19–42). San Francisco, CA: Sierra Club Books.

Burow, Karen R., Gregory M. Clark, J. M. Gronberg, Pixie A. Hamilton, Kerie J. Hitt, David K. Mueller, and Mark D. Munn. 2010. *The Quality of Our Nation's Waters: Nutrients in the Nation's Streams and Groundwater, 1992–2004*. Menlo Park, CA: U.S. Geological Survey.

California Department of Public Health. 2008a. *Permits Inspections Compliance Monitoring and Enforcement (PICME)*. Sacramento: California Department of Public Health.

———. 2008b. *Water Quality Monitoring*. Sacramento: Division of Drinking Water and Environmental Management, California Department of Public Health.

———. 2009. *American Recovery and Reinvestment Act (ARRA) CDPH FINAL Criteria Safe Drinking Water State Revolving Fund Projects*. Sacramento: California Department of Public Health.

———. 2009. "ARRA Funding for Public Drinking Water Systems: Funding Criteria." www.waterboards.ca.gov/drinking_water/services/funding/ARRA .shtml.

California Environmental Protection Agency. 2010. "Financial Assistance Programs—Grants and Loans: Clean Water State Revolving Fund Program." www.swrcb.ca.gov/water_issues/programs/grants_loans/srf/econ_recovery_ info.shtml.

Cole, Luke W., and Sheila R. Foster. 2001. *From the Ground Up*. New York: New York University Press.

Committee on Small Water Systems. 1997. *Safe Water from Every Tap: Improving Water Service to Small Communities*. Washington, DC: National Research Council.

Cory, Dennis C., and Tauhidur Rahman. 2009. "Environmental Justice and Enforcement of the Safe Drinking Water Act: The Arizona Arsenic Experience." *Ecological Economics* 68:1825–37.

Debbané, Anne-Marie, and Roger Keil. 2004. "Multiple Disconnections: Environmental Justice and Urban Water in Canada and South Africa." *Space and Polity* 8(2):209–25.

DeFur, Peter L., Gary W. Evans, Elaine A. Cohen Hubal, Amy D. Kyle, Rachel A. Morello-Frosch, and David R. Williams . 2007. "Vulnerability as a Function of Individual and Group Resources in Cumulative Risk Assessment." *Environmental Health Perspectives* 115(5):817–24.

Dubrovsky, N. M., C. R. Kratzer, L. R. Brown, J. M. Gronberg, and K. R. Burow. 1998. *Water quality in the San Joaquin–Tulare Basins, California, 1992–95*. Denver, CO: U.S. Geological Survey.

Fan, Anna M., and Valerie E. Steinberg. 1996. "Health Implications of Nitrate and Nitrite in Drinking Water: An Update of Methemoglobinemia Occurrence and Reproductive and Developmental Toxicity." *Regulatory Toxicology and Pharmacology* 23:35–43.

Ferreccio, Catterina, Claudia González, Vivian Milosavjlevic, Guillermo Marshall, Ana Maria Sancha, and Allan H. Smith. 2000. "Lung Cancer and Arsenic Concentrations in Drinking Water in Chile." *Epidemiology* 11(6):673–79.

Francis, Rose, and Laurel Firestone. 2011. "Implementing the Human Right to Water in California's Central Valley: Building a Democratic Voice through Community Engagement in Water Policy Decision Making." *Willamette Law Review* 47:495–537.

Fresno County Grand Jury. 2008. *Fresno County Grand Jury 2007–2008 Final Report*. Fresno, CA: Fresno Superior Court.

Galloway, Devin, and Francis S. Riley. 2006. "San Joaquin Valley, California: Largest Human Alteration of the Earth's Surface." In *Land Subsidence in the United States*, edited by D. L. Galloway, D. R. Jones, and S. E. Ingebritsen (23–34). Circular 1182. Menlo Park, CA: U.S. Geological Survey.

Gao, Suduan, K. K. Tanji, and G. S. Bañuelos. 2007. "Processes and Conditions Affecting Elevated Arsenic Concentrations in Groundwaters of Tulare Basin, California, USA." In *Trace Metals and Other Contaminants in the Environment*, edited by P. Bhattacharya, A. B. Mukherjee, J. Bundschuh, R. Zevenhoven, and R. H. Loeppert (383–410). Amsterdam: Elsevier.

Gee, Gilbert C., and Devon C. Payne-Sturges. 2004. "Environmental Health Disparities: A Framework Integrating Psychosocial and Environmental Concepts." *Environmental Health Perspectives* 112(17):1645–53.

Guerrero-Preston, Rafael, José Norat, Mario Rodríguez, Lydia Santiago, and Erick Suárez. 2008. "Determinants of Compliance with Drinking Water Standards in Rural Puerto Rico between 1996 and 2000: A Multilevel Approach." *Puerto Rico Health Sciences Journal* 27(3): 229–35.

Harter, Thomas J., et al. 2012. *Addressing Nitrate in California's Drinking Water with a Focus on Tulare Lake Basin and Salinas Valley Groundwater: Report for the State Water Resources Control Board*. Report to the legislature, University of California, Davis.

Heaney, Christopher D., et al. 2011. "Use of Community-Owned and -Managed Research to Assess the Vulnerability of Water and Sewer Services in Marginalized and Underserved Environmental Justice Communities." *Journal of Environmental Health* 74(1):8–17.

Imperial, Mark T. 1999. "Environmental Justice and Water Pollution Control: The Clean Water Act Construction Grants Program." *Public Works Management and Policy* 4(2):100–18.

Krieger, N. 2001. "Theories for Social Epidemiology in the 21st century: An Ecosocial Perspective." *International Journal of Epidemiology* 30: 668–77.

Marsh, Ben, Allan M. Parnell, and Ann Moss Joyner. 2010. "Institutionalization of Racial Inequality in Local Political Geographies." *Urban Geography* 31(5):691–709.

Michelson, M. R. 2000. "Political Efficacy and Electoral Participation of Chicago Latinos." *Social Science Quarterly* 81(1):136–50.

Moore, Eli, Eyal Matalon, Carolina Balazs, Jennifer Clary, Laurel Firestone, Susana De Anda, Martha Guzman, Nancy Ross, and Paula Luu. 2011. *The Human Costs of Nitrate-Contaminated Drinking Water in the San Joaquin Valley*. Oakland, CA: Pacific Institute.

Morello-Frosch, R., and R. Lopez. 2006. "The Riskscape and the Color Line: Examining the Role of Segregation in Environmental Health Disparities." *Environmental Research* 102(2):181–96.

Morello-Frosch, Rachel, Manuel Pastor, and James Sadd. 2001. "Environmental Justice and Southern California's 'Riskscape': The Distribution of Air Toxics Exposures and Health Risks among Diverse Communities." *Urban Affairs Review* 36(4):551–78.

Murphy, M., L. Lewis, R. I. Sabogal, and C. Bell. 2009. *Survey of Unregulated Drinking Water Sources on Navajo Nation.* American Public Health Association 137th Annual Meeting and Exposition, Philadelphia, PA.

Nash, L.L. 2006. *Inescapable Ecologies: A History of Environment, Disease, and Knowledge.* Berkeley: University of California Press.

National Research Council. 2001. *Arsenic in Drinking Water 2001 Update.* Washington, DC: National Academy Press.

Nolen, E. 2007. "Lanare Treatment Plant Dedicated." *Twin City Times*, April 18, News:1.

Olmstead, S.M. 2004. "Thirsty Colonias: Rate Regulation and the Provision of Water Service." *Land Economics* 80(1):136–50.

Pillley, A. K., S. Jacquez, R. W. Buckingham, R. P. Satya, K. Sapkota, S. Kumar, A. Graboski-Bauer, and T. Reddy. 2009. *Prevalence of Arsenic Contaminated Drinking Water in Southern New Mexico Border Colonias.* American Public Health Association 137th Annual Meeting and Exposition on Water and Public Health, 7–11 November, 2009, Philadelphia, PA. http://apha .confex.com/apha/137am/webprogram/Paper204703.html.

Prüss-Ustün, Annette, Carolyn Vickers, Pascal Haefliger, and Roberto Bertollini. 2011. "Knowns and Unknowns on Burden of Disease Due to Chemicals: A Systematic Review." *Environmental Health* 10(9), published online January 21. doi:10.1186/1476-069X-10-9.

Pulido, Laura. 1996. "A Critical Review of the Methodology of Environmental Racism Research." *Antipode* 28(2):142–59.

Pulido, Laura, Steve Sidawi, and Robert O. Vos. 1996. "An Archaeology of Environmental Racism in Los Angeles." *Urban Geography* 17(5):419–39.

Rahman, Tauhidur, Mini Kohli, Sharon Megdal, Satheesh Aradhyula, and Jackie Moxley. 2010. "Determinants of Environmental Noncompliance by Public Water Systems." *Contemporary Economic Policy* 28: 264-74.

Reisner, M. 1986. *Cadillac Desert: The American West and Its Disappearing Water.* New York: Viking.

Roof, K., and N. Oleru. 2008. "Public Health: Seattle and King County's Push for the Built Environment." *Journal of Environmental Health* 71:24-27.

Sexton, Ken, Kenneth Olden, and Barry L. Johnson. 1993. "'Environmental Justice': The Central Role of Research Is Establishing a Credible Scientific Foundation for Informed Decision Making." *Toxicology and Industrial Health* 9(5):685–95.

Shanaghan, P., and J. Bielanski. 2003. "Achieving the Capacity to Comply." In *Drinking Water Regulation and Health*, edited by F. Pontius (449–62). New York: John Wiley and Sons.

Smith, Allan H., Claudia Hopenhayn-Rich, Michael N. Bates, Helen M. Goeden, Irva Hertz-Picciotto, Heather M. Duggan, Rose Wood, Michael J. Kosnett, and Martyn T. Smith. 1992. "Cancer Risks from Arsenic in Drinking Water." *Environmental Health Perspectives* 97:259–67.

Stone, D., J. Sherman, and E. Hofeld. 2007. "Arsenic in Oregon Community Water Systems: Demography Matters." *Science of the Total Environment* 382:52–58.

Susser, M., and E. Susser. 1996. "Choosing a Future for Epidemiology: From Black Box to Chinese Boxes and Eco-Epidemiology." *American Journal of Public Health* 86(5):674–77.

Troesken, W. 2002. "The Limits of Jim Crow: Race and the Provision of Water and Sewerage Services in American Cities, 1880–1925." *Journal of Economic History* 62(3):734–72.

Tseng, Chin-Hsiao, Tong-Yuan Tai, Choon-Khim Chong, Ching-Ping Tseng, Mei-Shu Lai, Boniface J. Lin, Hung-Yi Chiou, Yu-Mei Hsueh, Kuang-Hung Hsu, and Chien-Jen Chen. 2000. "Long-Term Arsenic Exposure and Incidence of Non-Insulin-Dependent Diabetes Mellitus: A Cohort Study in Arseniasis-Hyperendemic Villages in Taiwan." *Environmental Health Perspectives* 108(9):847–51.

Tseng, W., et al. (1968). "Prevalence of Skin Cancer in an Endemic Area of Chronic Arsenicalism in Taiwan." *Journal of the National Cancer Institute* 40:453–63.

Tulare County Planning Department. 1973. *County of Tulare General Plan: Water and Liquid Waste Management, Policies, Programs.* County of Tulare, CA.

U.S. Census Bureau. 2007. "USA Counties." http://censtats.census.gov/usa/usa .shtml.

U.S. Environmental Protection Agency. 2003. *Recommendations of the National Drinking Water Advisory Council to U.S. EPA on its National Small Systems Affordability Criteria.* www.epa.gov/safewater/ndwac/pdfs /report_ndwac_affordabilitywg_final_08-0803.pdf.

———. 2010. "Public Drinking Water Systems: Facts and Figures." http:// water.epa.gov/infrastructure/drinkingwater/pws/factoids.cfm.

United Nations General Assembly. 2011. *Report of the Special Rapporteur on the Human Right to Safe Drinking Water and Sanitation.* C.d.A.A/ HRC/18/33/Add.4. UN Human Rights Council.

Vanderslice, J. 2011. "Drinking Water Infrastructure and Environmental Disparities: Evidence and Methodological Considerations." *American Journal of Public Health* 101(S1):S109–14.

Viers, J. H., D. Liptzin, T. S. Rosenstock, V. B. Jensen, A. D. Hollander, A. McNally, A. M. King, G. Kourakos, E. M. Lopez, and N. DeLaMora. 2012. "Technical Report 2." In *Addressing Nitrate in California's Drinking Water with a Focus on Tulare Lake Basin and Salinas Valley Groundwater.* Report for the State Water Resources Control Board to the Legislature. Center for Watershed Sciences, University of California, Davis.

Welch, Alan H., D. B. Westjohn, Dennis R. Helsel, and Richard B. Wanty. 2000. "Arsenic in Ground Water of the United States: Occurrence and Geochemistry." *Groundwater* 38(4):589–604.

Wilson, Sacoby M. 2009. "An Ecological Framework to Study and Address Environmental Justice and Community Health Issues." *Environmental Justice* 2(1):15–23.

Wilson, Sacoby M., Christopher D. Heaney, John Cooper, and Omega Wilson. 2008. "Built Environment Issues in Unserved and Underserved African-American Neighborhoods in North Carolina." *Environmental Justice* 1(2):63–72.

CHAPTER 12

The Incendiary Mix of Salmon and Water in Mediterranean-Climate California

MATTHEW J. DEITCH AND G. MATHIAS KONDOLF

It is generally appreciated that California enjoys a Mediterranean climate, with warm dry summers and cool wet winters, and that the high seasonality and interannual variability of the rainfall (and thus streamflow) create challenges to water management. These water supply challenges (and responses thereto) have direct parallels in other Mediterranean-climate regions, such as Spain, which rivals California in density of dams and degree of hydrologic alteration (Kondolf and Batalla 2005; Batalla, Gomez, and Kondolf 2004). What may not be so widely appreciated are the implications of the presence in California rivers of native anadromous salmon. Arguably the most charismatic of California's remaining megafauna, salmon serve as key indicators of environmental health in aquatic environs (including streams and estuaries) as well as in entire watersheds. They also impose constraints on water supply exploitation that are arguably unique among Mediterranean-climate regions. Water that remains in the stream and is not claimed for human uses is not wasted—it is necessary to sustain the salmon, which require water in streams for various life-cycle stages throughout the year.

Anadromous salmonids in California can be viewed as non-Mediterranean species, evolved in the cold waters of the North Pacific, and established in Mediterranean-climate California in periods of colder climate, during the Pleistocene. As cold-water organisms, salmon have managed to persist in what is not a fundamentally favorable climate through remarkable adaptation to local hydrologic and climate conditions (Beechie et al. 2006) and through phenotypic plasticity (Thorpe 1989; Williams 2006). Cold water is critical to their survival. California's geography is propitious, with high mountains supplying cool snowmelt to coastal rivers. While Atlantic salmon and sea-run brown trout persist in northwestern Iberia (with an Atlantic climate) and formerly ascended the Rhone River system to cool Alpine waters, there is currently no comparable Mediterranean-climate region in which native salmonids figure so prominently.

Most Mediterranean-climate rivers have very low native fish diversity (Almaça 1995), so the occurrence of native anadromous salmonids in the rivers of California creates a unique overlapping of Mediterranean hydrology and channel dynamics with anadromous salmonid life histories. The degree of human alteration to the riverine habitats of California has been extensive and severe, yet most of the runs have managed to persist by evolving "extraordinary life history diversity to persist in the face of stressful conditions that often approach physiological limits" (Katz et al. 2012).

CALIFORNIA'S MEDITERRANEAN CLIMATE

Coastal California is blessed with a Mediterranean climate, with cool winters (average low temperatures above freezing) and warm summers (average high temperatures below 30 °C), similar to the Mediterranean Basin itself, as well as portions of coastal Chile, South Africa, and Australia, which are also characterized as having a Mediterranean climate. More so than temperature, however, the defining feature of the Mediterranean climate is the seasonality of precipitation. While Mediterranean-climate regions are often considered dry, their annual rainfall can be comparable to those of Atlantic-climate regions. The key ecological difference is the predictable summer drought. The overwhelming majority of the year's rainfall occurs during the winter (with very little snowfall because of moderate winter temperatures). In California, from as far north as Fort Bragg all the way south to San Diego, more than 90 percent of the annual rainfall occurs from November to April (figure 12.1). Mediterranean-climate regions are also characterized by high year-to-year variability. Healdsburg, in Sonoma County, has about the same *average* annual precipitation (40 inches) as Lafayette, Indiana, but its annual rainfalls vary from 84 inches in a wet year to 12 inches in a dry year; Lafayette's variability is only half that (figure 12.2).

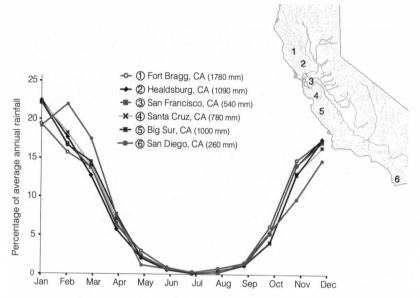

FIGURE 12.1. Average monthly rainfall as percentage of total annual rainfall across coastal California (with total annual rainfall in parentheses).

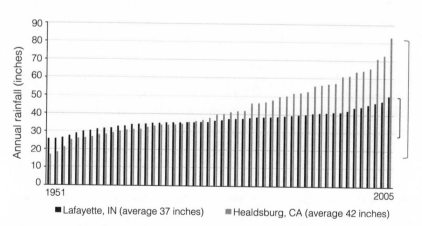

FIGURE 12.2. Distribution of annual rainfall, 1951–2005, in Healdsburg, CA, and Lafayette, IN (with range in brackets, right). Both average 40 inches precipitation annually, but annual rainfall varies more widely in Healdsburg—from 12 to 84 inches.

Streamflow follows rainfall almost identically, with more than 90 percent of flow during winter and less than 3 percent during the summer, and similar large interannual variability as well. The fluctuation between regular flooding in winter and drought in summer typical of coastal California streams is believed to represent among the harshest conditions for stream ecosystems in the world (Gasith and Resh 1999), creating an array of challenges to the organization and resilience of aquatic biota (Beche, McElravy, and Resh 2006; Resh et al. 2012).

SALMON IN CALIFORNIA

If one were to randomly interview residents of California's coastal cities, probably few would associate their state with big salmon runs. Yet, prior to the over-harvesting and habitat loss that accompanied the nineteenth-century influx of European settlers to California, between 2 and 3 million adult salmon migrated upstream through the Golden Gate annually, swimming upstream to their natal spawning grounds. Many rivers were literally alive with salmon, an experience that now requires a trip to Alaska.

There are three primary salmon species in California. Chinook salmon (*Oncorhynchus tshawytscha*), also called king salmon, occur to some degree in coastal rivers, but predominantly spawn in the inland waters of the Sacramento–San Joaquin system. Coho salmon (*Oncorhynchus kisutch*), also called silver salmon, tend mostly to occur in coastal streams, and their populations have been severely depleted by habitat loss and harvesting. The most adaptable species is the steelhead trout (*Oncorhynchus mykiss*). It is genetically identical to the rainbow trout, but like the salmon's, the steelhead's "life form" is anadromous, meaning that it spends its adult life at sea and reproduces in fresh water.

There are three principal "runs" of Chinook salmon in California rivers, named after the seasons in which they typically migrated upstream as adults: spring, fall, and winter. The spring run was by far the largest historically, spawning mostly in higher-elevation tributaries of the Sacramento and San Joaquin Rivers (Moyle 2002). The life history of these fish was well adapted to the seasonal hydrology of the river system. During spring snowmelt, abundant cold water flowed from the upper reaches of the basin to the ocean, providing longitudinally connected paths for salmon migration to a myriad of natal spawning streams. The adults would hold over the hot summer months in pools with consistently cold water temperatures, maintained by a combination of groundwater and snowmelt. As air temperatures dropped in the early fall and the threat of high water temperatures passed, the adults would spawn. The extensive construction of dams has blocked access to nearly all these natal spring-run salmon streams in the system (Kondolf and Batalla 2005; Lindley et al. 2006), leaving only remnant populations in the smaller tributaries that were spared dam construction in the twentieth century, such as Mill, Deer, and Butte Creeks.

In contrast, the winter run formerly spawned in the upper reaches of the Sacramento River system. These reaches are now completely cut off by Shasta Dam, but ironically the run persists by spawning in the artificially cold waters released by Shasta Dam, directly downstream of the dam (this is referred to as a tailwater fishery). The principal run that persists today is the fall run, whose adults wait to ascend the lowland rivers until late October or early November, after water temperatures have dropped, spawning in gravel beds in valley or foothills reaches of the rivers. Unlike the winter and spring runs, much of the historical fall-run habitat remains accessible, but it has been degraded by sediment starvation and loss of gravel substrate, channel simplification, flow alteration, and degraded water quality. Moreover, the fall run is now largely supported by hatcheries, whose progeny have essentially overwhelmed wild fish (Williams 2006).

Remnant runs of spring- and winter-run salmon figure prominently in management decisions for the large water projects in the Central Valley, and are often viewed as the "tails wagging the dogs" of large public infrastructure projects, occasionally inspiring calculations of the large sums spent per individual fish to maintain the species. Nevertheless, they serve as a key feature in determining how river systems can be managed to maintain environmental benefits.

The gradual and persistent decline of anadromous salmonids in California has led many to be listed under the federal and state Endangered Species Acts, which has led, in many parts of the state, to substantial restrictions on how water and land in salmon-bearing watersheds can be used. For example, the listing of coho salmon in its native coastal waters has created conflicts and pressures throughout Central and Northern California. Juvenile coho salmon (and federally threatened steelhead trout) emerge from redds in winter and spring, and spend at least one full summer in freshwater streams before migrating to the ocean; steelhead may spend as many as three summers in freshwater streams (Moyle 2002). Despite the persistent low flow during the summer, steelhead and coho were documented in coastal California streams throughout the nineteenth and twentieth centuries (Steiner 1996; Leidy, Becker, and Harvey 2005), presumably taking advantage of areas of cool upwelling water and complex habitats associated with large wood.

Steelhead trout are particularly impressive in their adaptability and have been observed in coastal streams as far south as Baja California (Needham and Gard 1959). They have adapted to the variable runoff in these systems, migrating upstream right after a storm peak, during the falling limb when the water is clearing but there is still sufficient flow for them to pass. After promptly spawning, many (though not all) redescend and return to the ocean before flows drop too far for them to pass. Their progeny require less water to pass downstream, and so can migrate seaward over a wider range of flows than were required by their parents to migrate upstream.

As is clear from the discussion above, salmon of California have an indispensable relationship with the flow regime to which they have adapted. The flow regime of a river, which characterizes the patterns of streamflow in terms of magnitude, frequency, duration, timing, and rate of change of various flow features (Poff et al. 1997), provides a useful framework for understanding the hydrologic conditions necessary for the salmon life cycle. Peak-flow events, termed *channel maintenance flows,* maintain adequate sediment characteristics, prevent vegetation encroachment, and maintain in-channel morphological features (Kondolf and Wilcock 1996). In coastal streams, peak flows occur a few times a year in winter and early spring and recede quickly once the rainfall event has ended. In the streams of the Sierra Nevada and the Cascade Range, they typically occur during spring snowmelt periods and recede gradually through the spring and early summer, with occasional occurrences as rain-on-snow events during winter. These peak-flow conditions are considered a type of *environmental flow,* a flow regime feature necessary for the preservation of particular ecological processes (Tharme 2003).

Streamflow conditions following peak flows also provide important environmental flows for salmon. In coastal California, peak flows act as signals to begin upstream migration; most migration to suitable spawning grounds and spawning occurs during the subsequent streamflow recession and winter base-flow conditions. In Sierra Nevada and Cascade streams, the snowmelt period can last for four or five months, providing consistently cool water for Chinook salmon to migrate and oversummer in accordance with their migration periods. Summer base flows also are important environmental flows for sustaining salmon populations. In coastal California, coho and steelhead typically spend at least one full summer in freshwater streams before migrating to the ocean, so maintaining consistent flow through summer is important for the survival of salmon in this region. Modification of any of these environmental flows can have deleterious effects on salmonids.

RESTORATION EFFORTS

Restoration efforts along the West Coast of North America are dominated by efforts to restore habitat for anadromous salmonids, especially federally protected runs of Pacific salmon and steelhead trout. The threat of extinction is especially serious in the southern end of the Pacific salmonid distribution, in California, where water temperatures are often marginal at best, and where habitat alteration and flow diversions have been extensive. The listing of these salmon runs has created profound conflicts between efforts to preserve the species and human uses of rivers and surrounding landscapes. River restoration investments in California have been largely directed at improving habitat for these species, and many projects are undertaken to mitigate the effects of new development (Kondolf, Podolak, and Grantham 2012).

In most rivers, habitat has been degraded by multiple human alterations, but restoration projects tend to address only some of the stressors: those that can be changed most easily, which entail the least financial cost, or encounter the least political resistance. Thus, "the trajectories of ecological restoration are rarely parallel with degradation trajectories" (Kondolf et al. 2006). This has certainly been the case in restoration efforts for salmon in California, which have been dominated by attempts to physically improve habitat, either without restoring natural flow regimes, or ignoring the highly variable nature of flows in some streams. On Selby Creek, a tributary to the Napa River, a 2006 project involved installation of multiple structures, including 102 "boulder wing deflectors," 11 boulder weirs across the channel bed, 18 "boulder streambank protection sites," and 3 "boulder armor sites" over 1.5 miles of channel (Bioengineering Institute 2009). The boulder deflectors, which extended from the banks into the channel, were intended to create scour pools (figure 12.3). However, this alluvial fan reach naturally dries up each summer from rapid infiltration of flow into its coarse sediment, so the structures do not create the intended habitat in the summer season, when deep pools are most needed by the fish.

FIGURE 12.3. A habitat "restoration" project on Selby Creek, Napa Valley, California. Over 100 boulder structures were installed in the channel of the creek with the intention of inducing bed scour and creating pools for fish. However, this alluvial fan reach naturally dries up each summer, so the fish could not benefit from the intended habitat during the season when it was needed. (a) View after construction in 2006. Photo courtesy of Laurel Marcus, California Land Stewardship Institute. (b) The same view in 2012, showing little scour has occurred, some boulder structures partly buried in finer sediment, and most importantly, the channel is dry. Photo courtesy of Mia Docto, University of California, Berkeley.

WATER MANAGEMENT IMPACTS

The seasonal pattern of precipitation in Mediterranean-climate regions means that water availability and demand are out of phase. Precipitation and runoff occur almost exclusively in the fall and winter, when plants are dormant and the demand for irrigation (and hydroelectric generation for air conditioning) is lowest. Thus, seasonal and interannual water storage is often needed to meet the basic needs of human populations, to support industrial-scale agriculture, and in some cases for flood control. As a result, Mediterranean-climate rivers tend to be more highly regulated than humid-climate rivers of comparable size. For example, Spain (whose climate is Mediterranean except in the northwestern provinces) has 1,200 large dams, more than any other country in Europe, 2.5 percent of the population of dams of the world, collectively impounding 40 percent of the country's average annual runoff. This is a much higher rate of impoundment than typically encountered in more humid regions. In contrast, reservoir capacity in other river systems is 5–18 percent of the annual runoff on the Elbe, Rhine, and Wesser Rivers in Germany, and less than 20 percent of annual runoff on the Potomac River (Kondolf and Batalla 2005).

The Sacramento–San Joaquin River system is the state's largest, receiving most of its runoff from the Sierra Nevada range, with elevations exceeding 4 km and substantial orographic precipitation falling as snow at high elevations. Snowmelt runoff from the Sierra constitutes the principal developed water source in California, with all major tributaries to the Sacramento and San Joaquin Rivers impounded since the late nineteenth and early twentieth century, to divert water for gold mining in the foothills mining districts and for agriculture on the valley floor. Through the twentieth century new reservoirs were constructed, and small ones replaced with larger ones, to provide irrigation and municipal water, such that by the end of the century most tributaries of the Sacramento and San Joaquin Rivers had large reservoirs, many impounding more than 100 percent of the annual runoff (figure 12.4). In addition, large water transfers move water southward through the canals and aqueducts of the Central Valley Project and the State Water Project (Kondolf and Batalla 2005).

Streamflow data collected from these streams by the U.S. Geological Survey demonstrate that large reservoirs have substantially altered the flow regime from its natural state. For example, the survey has operated a streamflow gauge on the American River near Fair Oaks since 1904,

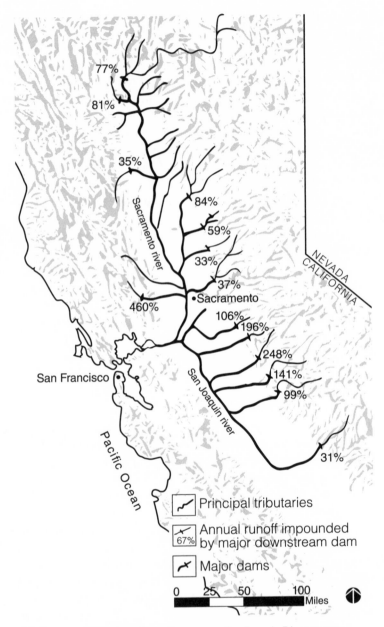

FIGURE 12.4. Major dams in the Sacramento–San Joaquin River system, showing degree of impoundment, expressed as percentage of average annual flow stored in the reservoir.

before and then after the completion of Friant Dam and Reservoir in 1955. Two years with similar overall discharge (2.7 million acre-feet), one before and one after dam completion, demonstrate the changes to the flow regime. In 1950, peak flows occurred in winter, indicating rain-on-snow events; and snowmelt elevated streamflow from March through July. Afterward, streamflow receded by orders of magnitude, to as low as 200 cfs (cubic feet per second) in summer. In contrast, streamflow in 1995 was more consistent through the year. The loss of peak flows has the potential to alter sediment dynamics and increase vegetation encroachment, reducing the quality and quantity of spawning habitat. The consistent discharge through spring and summer also alters the environmental flows associated with adult salmon oversummering and juvenile rearing until appropriate conditions for upstream migration.

In the Coast Ranges, water management has tended to be more dispersed, with multiple small diversions and dams operated by private landowners and local water districts (Deitch, Kondolf, and Merenlender 2009a). Cumulatively, these can be significant in their impact, especially in times of base flow, when there is naturally very little water to maintain flowing water or even perennial pools (Grantham, Merenlender, and Resh 2010). The influence of water management on flow regime in coastal streams is discussed further below.

Through the examples above we can form a conceptual model to characterize the challenges of water management in coastal California. In cases where snowmelt runoff is stored in dams for delivery elsewhere in the state, such as the large dams of the Sierra Nevada that form the Central Valley Project and the State Water Project, water must be moved from the river system to the concentrations of agricultural and municipal need. This can be described as *incongruence of space*: the snowmelt that is stored in dams is shipped through canals, pipes, and other conveyances to where it is used. In cases where water is obtained locally, such as in the many small ponds that meet agricultural needs in coastal California, sufficient water is naturally present only in winter, so it must be stored until summer for uses such as irrigation. This can be described as *incongruence of time*: rainfall and ample runoff is stored in small ponds across seasons and is available for use when needed later in the year. This conceptual framework is useful for exploring the challenges that water managers face in dealing with the complexities of providing water in Mediterranean-climate regions.

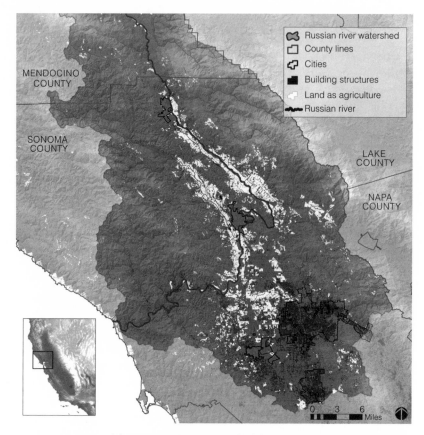

FIGURE 12.5. Map of the Sonoma County portion of the Russian River watershed, showing building structures, city outlines, and land as agriculture.

CASE STUDY: THE RUSSIAN RIVER

Human development in Sonoma County provides another example of the challenges of managing water given interest in protecting anadromous salmonids. Sonoma County supports a profitable agricultural sector, mainly in wine-grape cultivation, as well as a growing residential component; this residential growth has occurred in cities and towns (including Santa Rosa, Petaluma, Rohnert Park, and Sonoma) as well as in rural development dispersed across the rugged mountains and agricultural valleys (figure 12.5).

Sonoma County is unique among counties in the San Francisco Bay Area because it is regionally self-sufficient relative to water use. Its communities do not rely on water from the Central Valley Project or the State Water Project to meet municipal or agricultural needs, instead relying on infrastructure from within and adjacent to its basin to obtain water. This means that the interactions between the mechanisms used to obtain water and the impacts that they have on salmonids are local to coastal California, rather than exported elsewhere to the mountain tributaries of the Central Valley. The approximately 500,000 residents of Sonoma County are thus effectively stewards of their own salmonid populations (though, as described below, they affect salmon in the nearby Eel River watershed and are thus stewards of those salmon as well).

Urban Water Supply

The municipalities of Sonoma County generally obtain water through different mechanisms than rural residents and agriculture. Cities including Rohnert Park, Santa Rosa, Sonoma, and Petaluma contract with the Sonoma County Water Agency (SCWA) to provide water to their residents. The SCWA provides water for an estimated 600,000 people in Sonoma and parts of Marin County, principally through infrastructure on the Russian River (figure 12.6). The SCWA co-manages two reservoirs in the Russian River watershed: Lake Mendocino, formed by Coyote Valley Dam on the East Fork Russian River north of Ukiah in Mendocino County (capacity approximately 120,000 acre-feet); and Lake Sonoma, formed by Warm Springs Dam on Dry Creek in northern Sonoma County (capacity approximately 350,000 acre-feet). Each of these reservoirs utilizes the stream channel (the Russian River and Dry Creek, respectively) as a conduit for delivering water from the dam to a location on the mainstem Russian River where it can be extracted via large wells in gravel bars beside the river and delivered to municipalities. (The SCWA does not directly contract with agricultural users to provide water.)

In providing water to municipalities, the SCWA effectively faces incongruities both of space and of time. Reservoirs store water in winter, and demands for water are greatest in summer. Thus, reservoirs must have sufficient water storage to meet spring and summer needs, because reservoir inflow is generally very low during the dry season. The SCWA also faces the issue of needing to deliver water from its location of storage to its distribution center, and then to its locations of use. Lake Sonoma and Lake Mendocino are 22 miles and 60 miles, respectively, from treatment and distribution hubs, and water must be conveyed from each reservoir for it to be useful.

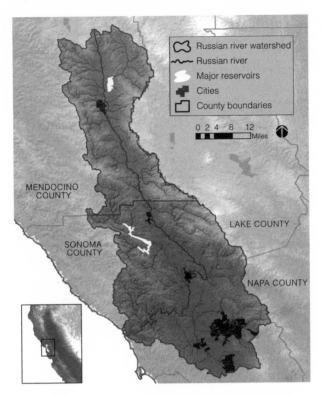

FIGURE 12.6. Map of the Russian River basin, showing key hydrologic features.

Use of the mainstream channels as conduits for water delivery creates some challenges for salmonids in these streams. The reservoirs store winter rainfall and thus reduce winter flood peaks (e.g. Dry Creek flood flows have been reduced from 40,000 cfs to 10,000 cfs), but otherwise they maintain winter base-flow conditions as they store flow from upstream (Moran 2012). More significantly for the salmonids, by releasing stored water from the winter, the reservoirs affect the flow regime, augmenting flow during the dry season. The Russian River and Dry Creek both historically had such low summer flows that in dry years their flow might not be continuous, more likely alternating between reaches with continuous flowing water and reaches with isolated pools (Langridge 2002, Moran 2012). However, thanks to higher water tables throughout the valley, these isolated pools were fed by upwelling groundwater, and so maintained cool water temperatures beneficial to fish. Under current conditions, groundwater levels have been lowered by pumping, and reservoirs typically release a combined total of more than 100 cfs through the summer. Most of the released water was runoff stored in the reservoirs, but releases from Lake Mendocino also include water from the Eel River, transferred via the Potter Valley Diversion (Langridge 2002).

The altered summer flows were the main factor considered in a "biological opinion" on the operation of Warm Springs Dam and Coyote Valley Dam issued by the National Marine Fisheries Service (2008) under the Endangered Species Act. The Russian River and Dry Creek hold the distinction among streams in coastal California of having *too much* water to support salmonids during the dry season. According to the opinion, the high summer streamflow in the main-stem Russian River and Dry Creek create conditions that are unsuitable for oversummering juvenile salmonids because the velocity in the channel (necessary to provide sufficient flow to the central diversion point) is too great. For example, to support a total flow of 100 cfs, the water velocity in Dry Creek is typically 2–3 feet per second; this is approximately five times the ideal water velocity for juvenile coho salmon. As part of its compliance with the decision, the SCWA is coordinating a $20,000,000 project to physically alter the channel form in Dry Creek (from Warm Springs Dam to the Russian River) to create low-velocity areas suitable for juvenile salmonids during the summer.

Large seasonal water storage and transfer to diversion points (such as SCWA's use of Dry Creek as its conduit for moving water from Warm Springs Dam to its diversion site on the main-stem Russian River) is standard fare in Mediterranean-climate rivers worldwide. However, the presence of federally protected salmon in the Russian River basin makes such "business as usual" impossible, and requires extraordinary measures to make water diversions compatible with juvenile salmonid habitat. The requirements outlined in the Russian River biological opinion may prove difficult or rigid for water management, but the rules and protocols followed by the SCWA may provide some lessons for more ecologically compatible water management elsewhere in California.

Rural Water Supply

No large water providers meet agricultural and rural residential water needs in Sonoma County. Instead, those water needs are mostly met individually through groundwater pumping, extraction from springs, or diversion from streams or adjacent shallow aquifers through individual, small-scale methods. This decentralized water management regime has advantages and disadvantages relative to maintaining streamflow for salmonids. Extraction methods (e.g. groundwater wells or surface diversions) tend to be dispersed, so their effects are spread out over the watershed instead of concentrated in one location. They also tend to be small, compared to the large water diversions that meet municipal needs: residential surface water diversions or groundwater pumping for small vineyards may be on the order of 10 gallons per minute (0.02 cfs). Development in rural Sonoma County also tends to be low, with agricultural coverage in larger Russian River tributaries generally less than 10 percent of total catchment area and rural residences in low numbers as well.

Agriculture and residences in rural Sonoma County are challenged by incongruences in time, but not in space. The region typically receives more than 40 inches of rain (approximately half of this in a very dry year), which is sufficient to meet agricultural needs on an annual scale; but, as described above, rainfall is opposite the time of water need. Direct diversions from streams are often employed to meet water needs in summer, but because streamflow is typically low during spring and summer, diversions from streams have the potential to drastically reduce streamflow and eliminate portions of salmonid habitat. Relative to the impacts they may have on the flow regime, these diversions tend not to be large enough to affect streamflow during winter; but they may substantially alter streamflow during the dry season, when flow naturally recedes toward intermittence (Deitch, Kondolf, and Merenlender 2009a, 2009b). Reduction of streamflow in summer can substantially reduce oversummering habitat and eliminate downstream drift for salmon.

Temporal Patterns of Diversion

Ironically, in the Russian River basin and throughout the wine country north of San Francisco, the most problematic and controversial use of water by vineyards has been not for irrigation but for frost control. The grapes produced in the Russian River Valley are only minimally irrigated, to maintain their intensity and high quality, and multiple growers rarely irrigate simultaneously. While even small diversions can have large effects during the low flows of summer, the effects are somewhat mitigated by the fact that this water demand is infrequent and not simultaneous: vines are typically irrigated once or twice per week for a few hours, so effects on water sources are often attenuated. Deitch, Kondolf, and Merenlender (2009b) document that diversions from

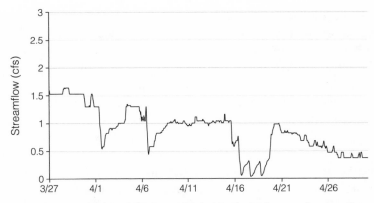

FIGURE 12.7. Hydrograph for Franz Creek, March–April 2004, showing effect of diversions for frost protection on streamflow. The drainage area above this point is about 5 square miles, of which about 5 percent is in vineyard.

streams to meet these needs can cause streams to become intermittent when minimum air temperature is especially low (figure 12.7).

Recent efforts to develop regional policies to regulate water use illustrate the technical and social complexities of sustainable water management. In 2008, severe frosts occurred during a dry year, when streamflow was already low, and dead salmon observed in dried-out streams were attributed to diversions for frost protection by upstream vineyards (which caused the streamflow in the Russian River to be reduced by 30 percent, or 60 cfs). This prompted a recommendation by the National Marine Fisheries Service (which oversees the protection of salmon under the Endangered Species Act) for the State Water Board (which oversees water rights in California) to recommend that frost protection no longer be a reasonable use of water because it has the potential to cause widespread streamflow reduction, loss of salmon habitat, and fish kills. Outrage over water-use restrictions, and evidence provided by grower groups and other stakeholders that it is possible for water to be used for frost protection in ways that do not affect streamflow, caused the state and federal agencies to reconsider the initial recommendation. A policy was then proposed that would require grape growers using water for frost protection to demonstrate that their methods for obtaining water do not adversely affect salmonid habitat. If growers could demonstrate over multiple years that their frost-protection water use did not affect nearby streamflow, they could continue to use water for frost protection. This policy was invalidated by a court in Mendocino County under the argument that the state had exceeded its authority and evidence of its need was insufficient; but it was then upheld by a state appellate court (Anderson 2014).

Despite the initial court decision, growers throughout the region have reduced their reliance on streamflow on cold spring mornings to meet frost-protection water needs. Where appropriate, frost-protection fans have been installed to mix warmer air at higher elevations with cool air that falls to the bottom of the valley. Though cost-effective, the fans are not suitable for all locations. Many grape growers have constructed some kind of water storage for frost protection that can be filled by a combination of rainwater, groundwater, and streams when conditions are appropriate. Reservoirs may cost more than ten times as much, per area served, as frost-protection fans, but their widespread application and certainty of supply is of substantial value. The growing presence of reservoirs in the landscape has raised concerns about the cumulative effects of storing water on discharge, especially if the reservoirs are located on ephemeral (seasonally flowing) headwater streams and store early-winter discharge until they fill. If large and widespread enough, they could reduce early-season winter "attraction flows" (those conditions that draw steelhead and coho into tributaries to spawn early in winter). Cumulative effects analyses of reservoirs indicate that these impacts are in general not significant because reservoirs still tend to be distributed across the watershed and are far enough into the headwaters that their effects are offset by the substantial unimpaired drainage network (Deitch, Merenlender, and Feirer 2013). When we compare other important Mediterranean-climate grape-growing regions of the world, none face these kinds of constraints for the protection of native salmon.

Spatial Patterns of Diversion

Extraction methods for meeting rural water needs tend not to be concentrated in one location; they also are not evenly distributed in space. For example, in Mill Creek (a tributary to Dry Creek in Sonoma County), vineyards are concentrated in the headwaters and the lower portion of the watershed, where physical conditions are appropriate, and residences tend to be concentrated along the creek in the middle and lower reaches (figure 12.8). This concentration of residential development can impair streamflow as well, as water is needed for household uses (figure 12.9). While dispersed development has the potential to substantially alter streamflow during the dry season, winter rains, if

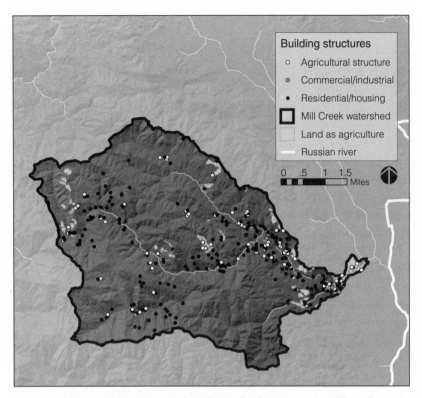

FIGURE 12.8. Map of agricultural and residential development in the Mill Creek watershed, Sonoma County, from Trout Unlimited and CEMAR (2014).

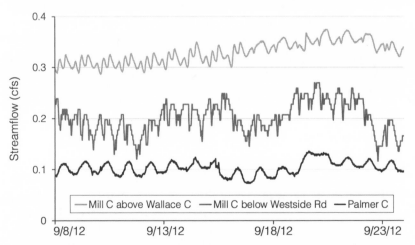

FIGURE 12.9. Streamflow at three locations in the Mill Creek drainage network, one showing the typical day–night fluctuations that typically occur as a result of evapotranspiration, and the other two showing intra-daily fluctuations that could be a result of residential in-stream diversions upstream of each streamflow gauge.

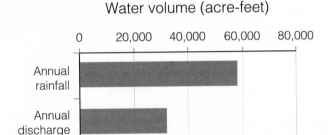

FIGURE 12.10. Comparison of rainfall, streamflow, and human water demand in the Mill Creek watershed.

stored, may provide enough water to meet human needs and reduce the impacts on streams.

To evaluate this possibility, it is useful to compare rainfall and discharge (the input to and output from the watershed) to human water needs. If annual human water needs represent a small fraction of the annual rainfall that reaches a watershed (or the discharge that leaves the watershed), then the natural availability of water may be sufficient to meet human water needs throughout the year. In the case of Mill Creek, the available water on an annual basis far exceeds current human demands (figure 12.10), implying that with sufficient storage, demands for residential and agricultural water could be met without serious harm to the native salmon. The challenge for individual water managers is to figure out how to overcome the incongruence of time and tap into the abundance of winter water.

SALMON IN A CHANGING CLIMATE

Despite adaptation to the warmer temperatures of California, many salmon runs are temperature-limited, and persist only barely, at the limit of their thermal tolerance. As the climate changes and average temperatures increase, some of these currently marginal habitats will become unsuitable. Moreover, the anticipated future trends of warmer winter temperatures will result in increased winter rainfall and reduced snowpack in the Sierra Nevada (which provides most of California's agricultural and residential water supply). As a result, reservoirs collecting

Sierra Nevada runoff are expected to receive less water from snowmelt in spring and summer, and more inflow via rainfall in winter. If water managers are charged with providing for agricultural uses and releasing water from reservoirs for salmonid needs, the reduction in spring and summer snowmelt will mean less reservoir inflow in summer months and less water available for these competing uses.

The shift to increased reservoir accrual in winter and less in spring and summer means that these water managers will face an incongruence of time in addition to the existing incongruence of space. Yet, despite this new challenge, there may be opportunities to learn from other places in California where this condition already occurs. The shift to increased winter storage more closely resembles the flow regime in the Coast Ranges of California, and the conditions already experienced by groups like the Sonoma County Water Agency. Thus, water managers may benefit from looking to the Russian River, to the Warm Springs and Coyote Valley Dams, for lessons on successful and unsuccessful means for managing water to maintain downstream salmonid populations in a future with less snow.

CONCLUSION

California represents the juxtaposition of Mediterranean-climate hydrology and water development with anadromous salmon (and we might add, the Endangered Species Act). Its water management challenges are arguably exceptional, if not unique. As salmon cling to survival in an environment that is at best only marginally suitable for them, the legal protection for these listed species creates constraints on water supply, both for large reservoirs and their releases, and for small diversions for farm and rural residences dispersed across the landscape. While by now we have perhaps grown accustomed to taking these factors into account, if we compare the California situation to other Mediterranean climates, we find many strong parallels, but a big difference: native salmon.

REFERENCES

Almaça, Carlos 1995. "Freshwater Fish and their Conservation in Portugal." *Biological Conservation* 72:125–27.

Anderson, Glenda. 2014. "Appeals Court Upholds Russian River Frost-Protection Rules." *Santa Rosa Press-Democrat*, June 18.

Batalla, Ramon J., Carlos M. Gomez, and G. Mathias Kondolf. 2004. "River Impoundment and Changes in Flow Regime, Ebro River Basin, Northeastern Spain." *Journal of Hydrology* 290:117–36.

Beche, Leah A., Eric P. McElravy, and Vincent H. Resh. 2006. "Long-term seasonal variation of benthic-macroinvertebrate biological traits in two mediterranean-climate streams in California, USA." *Freshwater Biology* 51: 56–75.

Beechie, Timothy, Eric Buhle, Mary Ruckelshaus, Aimee Fullerton, and Lisa Holsinger. 2006. "Hydrologic Regime and the Conservation of Salmon Life History Diversity." *Biological Conservation* 130:560–72.

Bioengineering Institute. 2009. *Final Project Report, Selby Creek Stream Restoration and Riparian Revegetation Project.* Laytonville, CA.

Deitch, Matthew J., G. Mathias Kondolf, and Adina M. Merenlender. 2009a. "Surface Water Balance to Evaluate the Hydrological Impacts of Small Instream Diversions and Application to the Russian River Basin, California, USA." *Aquatic Conservation: Marine and Freshwater Ecosystems* 19:274–84.

———. 2009b. "Hydrologic Impacts of Small-Scale Instream Diversions for Frost and Heat Protection in the California Wine Country." *River Research and Applications* 25:1118–34.

Deitch, Matthew J., Adina M. Merenlender, and Shane T. Feirer. 2013. "Cumulative Effects of Small Reservoirs on Streamflow in Northern Coastal California Catchments." *Water Resources Management* 27:5101–18. doi: 10.1007/S11269-013-0455-4.

Gasith, Avital, and Vincent H. Resh. 1999. "Streams in Mediterranean Climate Regions: Abiotic Influences and Biotic Responses to Predictable Seasonal Events." *Annual Review of Ecology and Systematics* 30:51–81.

Grantham, Theodore E., Adina M. Merenlender, and Vincent H. Resh. 2010. "Climatic Influences and Anthropogenic Stressors: An Integrated Framework for Streamflow Management in Mediterranean-Climate California, U.S.A." *Freshwater Biology* 55:188–204.

Katz, Jacob, Peter Moyle, Rebecca Quiñones, Joshua Israel, and Sabra Purdy. 2012. "Impending Extinction of Salmon, Steelhead, and Trout (Salmonidae) in California." *Environmental Biology of Fishes* 96: 1169-1186.

Kondolf, G. Mathias, and Ramon J. Batalla. 2005. "Hydrological Effects of Dams and Water Diversions on Rivers of Mediterranean-Climate Regions: Examples from California." In *Catchment Dynamics and River Processes: Mediterranean and Other Climate Regions*, edited by C. Garcia and R.J. Batalla (197–211). London: Elsevier.

Kondolf, G. Mathias, Andrew Boulton, Scott O'Daniel, Geoffrey Poole, Frank Rahel, Emily Stanley, Ellen Wohl, Asa Bang, Julia Carlstrom, Chiara Cristoni, Harald Huber, Saija Koljonen, Paulina Louhi, and Keigo Nakamura. 2006. *Process-Based Ecological River Restoration: Visualising Three-Dimensional Connectivity and Dynamic Vectors to Recover Lost Linkages. Ecology and Society* 11. www.ecologyandsociety.org/vol11/iss2/art5/.

Kondolf, G. Mathias, Kristin Podolak, and Theodore E. Grantham. 2012. "Restoring Mediterranean-Climate Rivers." *Hydrobiologia* 719:527–45. doi:10.1007/s10750-012-1363-y.

Kondolf, G. Mathias, and Wilcock, Peter R. 1996. "The Flushing Flow Problem: Defining and Evaluating Objectives." *Water Resources Research* 32:2589–99. doi:10.1029/96WR00898.

Langridge, Ruth. 2002. "Changing Legal Regimes and the Allocation of Water between Two California Rivers." *Natural Resources Journal* 42: 283–331.

Leidy, Robert, Gordon Becker, and Brett Harvey. 2005. "Report on the status and distribution of salmonids in the San Francisco Bay." Center for Ecosystem Management and Restoration, California Coastal Conservancy, Oakland, CA USA.

Lindley, Steven T., Robert S. Schick, Aditya Agrawal, Matthew Goslin, Thomas E. Pearson, Ethan Mora, James J. Anderson, et al. 2006. "Historical Population Structure of Central Valley Steelhead and its Alteration by Dams." *San Francisco Estuary & Watershed Science* 4:1–19.

Moran, Thomas. 2012. "Water into Wine." *Boom: A Journal of California* 2(1). www.boomcalifornia.com/2012/07/water-into-wine/.

Moyle, Peter B. 2002. *Inland Fishes of California*. Berkeley: University of California Press.

Needham, Paul R., and Richard Gard. 1959. *Rainbow Trout in Mexico and California with Notes on the Cutthroat Series*. Berkeley: University of California Press.

National Marine Fisheries Service. 2008. *Biological Opinion for Water Supply, Flood Control Operations, and Channel Maintenance Conducted by the U.S. Army Corps of Engineers, the Sonoma County Water Agency, and the Mendocino County Russian River Flood Control and Water Conservation Improvement District in the Russian River Watershed*. Consultation under the U.S. Endangered Species Act, Section 7. Santa Rosa, CA: National Marine Fisheries Service, Southwest Region.

Poff, N. LeRoy, J. David Allan, Mark B. Bain, James R. Karr, Karen L. Prestegaard, Brian D. Richter, Richard E. Sparks, and Julie C. Stromberg. 1997. "The Natural Flow Regime: A Paradigm for River Conservation and Restoration." *Bioscience* 47:769–84.

Resh, Vincent H., Leah A. Beche, Justin E. Lawrence, Rafael D. Mazor, Eric P. McElravy, Alison P. O'Dowd, Deborah Rudnick, and Stephanie M. Carlson. 2012. "Long-Term Population and Community Patterns of Benthic Macroinvertebrates and Fishes in Northern California Mediterranean-Climate Streams." *Hydrobiologia* 719:93–118.

Steiner Environmental Consulting. 1996. *A History of the Salmonid Decline in the Russian River*. Potter Valley, CA: Steiner Environmental Consulting.

Tharme, Rebecca E. 2003. "A Global Perspective on Environmental Flow Assessment: Emerging Trends in the Development and Application of Environmental Flow Methodologies for Rivers." *River Research and Applications* 19:397–441.

Thorpe, John E., 1989. "Developmental Variation within Salmon Populations." *Journal of Fish Biology* 35 (Supp. A): 295–303.

Williams, John G. 2006. "Central Valley Salmon: A Perspective on Chinook and Steelhead in the Central Valley of California." *San Francisco Estuary and Watershed Science* 4(3). http://escholarship.org/uc/item/21v9x1t7.

CHAPTER 13

Adaptive Management in Federal Energy Regulatory Commission Relicensing

KRISTEN PODOLAK AND SARAH YARNELL

Between 2015 and 2025, twelve hydroelectric projects in California's Sierra Nevada will submit applications for new operating licenses to the Federal Energy Regulatory Commission (FERC). These new licenses will establish the flow release schedules that affect power generation as well as aquatic and riparian ecosystem health and recreational resources downstream. The process of relicensing provides an opportunity to incorporate environmental considerations and recent scientific understanding to reduce the impacts of hydropower operations on aquatic and riparian species. Once a license is set, there are few ways to change it during the 30-year license period. One way is a reopener clause, which FERC includes in most licenses to allow changes to license conditions (Mount et al. 2007). Other agencies with mandatory conditioning authority also routinely include specific reopeners in a license. In practice, however, FERC typically does not use the reopener clause, and there is little push to reopen licenses from the courts. An alternative is to incorporate an adaptive management approach into the license whereby hydropower owners and stakeholders may modify flows during the term of the license within defined parameters.

The relicensing of the Mokelumne River Project (FERC No. 137) on the North Fork Mokelumne River and of the Rock Creek–Cresta Project (FERC No. 1962) on the North Fork Feather River in the Sierra Nevada were the first two settlement agreements to incorporate an adaptive management approach. The Mokelumne River Project was the longest relicensing in FERC history. In view of outdated study data, the settlement agreement included adaptive management with a mechanism to monitor flow impacts, adjust the monitoring, and implement flow changes in the future if needed. The Mokelumne license was a "pioneer" and an "early experimental effort" in allowing for change over time (Hydropower Reform Coalition 2009a; Rheinheimer et al. 2007), and it set a precedent for relicensing projects that followed. Months later, the Rock Creek–Cresta Project relicensing incorporated a similar adaptive management format.

For hydropower projects, a primary benefit of adaptive management is that scientific findings on ecosystem responses to changes in streamflow can drive the flow management. Additionally, scientific uncertainty or gaps in baseline data are not necessarily detrimental, as they do not preclude changes to future flow conditions. On the other hand, there may be concern that an adaptive management approach does not provide enough certainty to the licensee, inhibiting long-term planning and potentially exposing the utility to unexpected costly changes. Thus, the critical balance for an adaptive management approach is to provide agencies confidence that license conditions will be changed if needed during the term of the license, while simultaneously assuring them that license conditions will change only within acceptable bounds and only when specified circumstances are met.

In this chapter, we briefly describe the history of the Mokelumne River Project relicensing (1974–2001) and the Rock Creek–Cresta Project relicensing (1979–2000) and review how adaptive management is working under the new licenses (2001–2013). The history leading up to these settlement agreements and the collaborative planning involved set the stage for adaptive management in the FERC relicensing process. An adaptive management approach allowed competing ecological and recreational interests to make streamflow decisions, despite an initial lack of data, and to agree to flow modifications based on future studies. Reviewing the progress of the Mokelumne River Project and the Rock Creek–Cresta Project almost halfway through the new 30-year licenses can yield insights for other relicensing efforts and highlight some of the challenges and successes. This new adaptive management model in FERC relicensing holds promise to increase the monitoring of biological responses to streamflow management, increase collaboration between stakeholders and hydropower utilities, and improve flow management for people and nature. We begin with a discussion of the natural flow regime and the impacts of flow management on aquatic organisms.

THE NATURAL FLOW REGIME

The natural flow regime (flow magnitude, duration, timing, frequency, and rate of change—Poff et al. 1997) in the North Fork Mokelumne (NF Mokelumne) and North Fork Feather (NF Feather) Rivers is typical of a Mediterranean-climate river (figure 13.1). Cool, wet winters and warm, dry summers result in water availability that is mismatched with the human demand for water in the summer and early fall. As a result, all but one of the Sierra Nevada watersheds are extensively dammed for water storage to meet the summer–fall irrigation and urban water needs of the Central Valley as well as for generation of electricity via hydropower. Some variability exists in the hydrology of Mediterranean-climate rivers due to prevailing hydro-climate drivers like the El Niño Southern Oscillation and the Pacific Decadal Oscillation, which change runoff characteristics on both annual and decadal time scales. As a result, while the seasonal flow pattern has remained consistent from year to year, the flow regime can vary in magnitude, timing, and storm-event frequency in wet and dry years, with periodic winter and spring flood flows rearranging the channels and riparian vegetation (Kondolf, Podolak, and Grantham 2012). Spring snowmelt from the high elevations of the Sierra Nevada lends some predictability to the flow regime, however, with consistent receding flows in late spring and early summer creating a steady transition from the flood flows of winter to the low minimum streamflows of summer and fall (Yarnell, Viers, and Mount 2010).

Native aquatic species necessarily have evolved adaptations to survive the harsh extremes of winter flood and summer drought, with many relying on the singular annual event of the spring snowmelt recession as a predictable reprieve. With gradually declining flows and a low frequency of pulses, the spring recession provides a stable transition from high abiotic pressures (e.g. scour, turbidity) during winter high flows to high biotic pressures (e.g. competition, predation) during late-summer and fall low flows (Yarnell, Viers, and Mount 2010). During the spring recession, predictable flow conditions coincide with high resource availability, resulting in high reproductive success, growth rates, and survivorship for species adapted to this seasonal flow regime. In regulated river systems where both winter flood flows and the spring snowmelt recession are often captured behind dams or diverted for

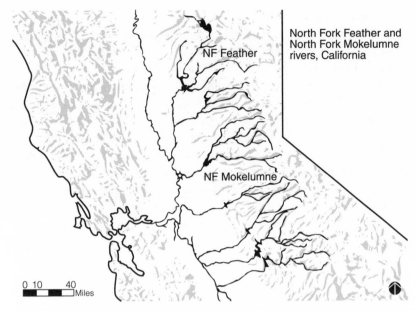

FIGURE 13.1. The North Fork Mokelumne River and North Fork Feather River watersheds, in California's Sierra Nevada.

hydropower, streamflows below many dams have been altered drastically, creating an acute impact on freshwater ecosystems, particularly in the Sierra Nevada (Moyle 1996).

One example of a species that is directly affected by hydropower development is the Foothill yellow-legged frog (*Rana boylii*, FYLF), a river-breeding amphibian native to mid-elevation streams in California and southern Oregon that is designated a California Species of Special Concern due to declining populations. Individuals breed annually in early spring following the start of the spring snowmelt recession, timing their reproduction to minimize the risk of egg scour caused by unpredictable late-spring storms and to maximize growth during summer low flows. Most FYLFs lay egg masses on open, newly scoured cobble bars or in moderately deep locations attached to large boulders in areas protected from fast current. The eggs must remain submerged for more than five days (up to two weeks, depending on the temperature of the water and egg mass) until the tadpoles hatch. The tadpoles then graze in shallow, warm, near-shore environments throughout the summer until metamorphosis occurs in fall. Although they are well adapted to the natural seasonal cycle of flow in Mediterranean climates, the egg masses are still vulnerable to scour from late-season storms and to desiccation from rapid decreases in spring flow; and the tadpoles are susceptible to scour from rapid changes in flow during the summer (Kupferberg et al. 2009; Yarnell, Lind, and Mount 2012). As a result, the frogs are extremely vulnerable to altered flow regimes and have been extirpated from most highly regulated river systems (Lind 2005; Kupferberg et al. 2012).

In some regulated rivers, including the NF Mokelumne and NF Feather, where winter flows are diverted or captured behind reservoirs and minimum streamflows dominate the remainder of the year, small frog populations have persisted in the main-stem river downstream of dams despite alterations to the hydrograph. In dry years, when flows remain close to minimum streamflow year-round and temperatures are warmer, frog populations can successfully reproduce. However, in wet years, when high flows spill over dams, spring flow conditions can be unpredictable and devastating. Rapid changes in flow not only alter in-stream habitat conditions but can miscue or eliminate the environmental flow cues utilized by frogs to initiate breeding (Wheeler and Welsh 2008). If flows are stable and temperatures warm in early spring, the frogs may initiate breeding, only to have their egg masses scoured by a subsequent late-spring spill flow over a dam. If egg masses are laid during high spill flows, the abrupt curtailment of flow when spill ceases can result in sudden decreases in flow depth downstream, which can result in desiccation of egg masses and tadpoles. Due to the longevity of the frogs, small populations can persist in these moderately regulated main-stem rivers if unpredictable spring spill flows are infrequent or rare and summer minimum streamflows are stable.

THE NORTH FORK MOKELUMNE AND NORTH FORK FEATHER RIVERS

The NF Mokelumne River watershed upstream of Pardee Reservoir is a primary water supply for the East Bay portion of the San Francisco metropolitan area, while the lower watershed begins at Camanche Reservoir at the edge of the Central Valley. The NF Mokelumne was extensively developed for hydropower by Pacific Gas and Electric (PG&E) starting in the early twentieth century and includes seven storage reservoirs with a capacity of about 217,000 acre-feet, four powerhouses that supply 205 megawatts, 20 miles of flumes, and 16 miles of tunnels to transport the water. Built in 1902 by the Standard Electric Company, the Electra Powerhouse was one of the earliest hydropower developments in California. In 1931, the hydropower

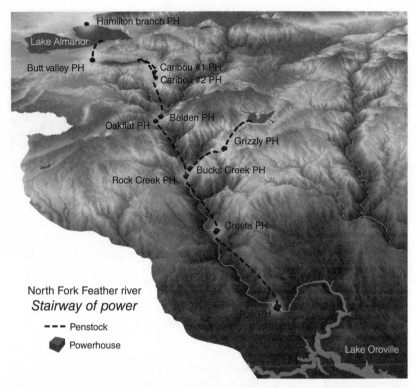

North Fork Feather river
Stairway of power

– – – Penstock

◆ Powerhouse

FIGURE 13.2. Schematic of the North Fork Feather River watershed showing all current hydropower facilities.

capacity was increased with the Mokelumne River Project under a 50-year license that expired in 1975.

In one of the largest watersheds in the Sierra Nevada, the NF Feather River drains the northern end of the mountain range from near Lassen Peak into the Central Valley. With a large runoff volume (mean annual flow of 3,000 cubic feet per second, or cfs) and an elevation drop of 35 feet per mile, the watershed was developed for hydropower beginning in the early twentieth century and became known as the Stairway of Power. A drop of water landing near the crest of the Sierra in the watershed can pass through up to seven powerhouses on its journey to the valley floor (figure 13.2). With a total generation capacity of 756 MW, the dams and powerhouses in the watershed have significant economic value and make an important contribution of energy to the electric grid. As such, the watershed was operated to maximize hydropower, with predominately flatline minimum streamflows year-round in each of the river reaches. In

FIGURE 13.3. Mike Latendresse navigates a class IV rapid on the most popular section of whitewater, Tiger Creek, on the North Fork of the Mokelumne River downstream of the Tiger Creek powerhouse. Photo by Paul Martzen.

1979, PG&E filed an application to renew the first of the modern project licenses in the watershed, and negotiations regarding the relicensing of the Rock Creek–Cresta Project began. Both the NF Mokelumne and NF Feather Rivers provide a variety of recreational opportunities during spring and summer. Sport fishermen value the trout populations that are widespread in both river systems, and whitewater enthusiasts value the class II–V whitewater runs that occur in both project reaches (figure 13.3). There are four commonly run whitewater sections on the NF Mokelumne River upstream of Pardee Reservoir, and both the Rock Creek and Cresta runs on the NF Feather are well regarded in the kayaking and rafting communities. These whitewater runs are within driving distance of other popular whitewater such as the South Fork and North Fork of the American River, and often attract regional use by paddlers on extended trips.

GRIDLOCK: DECADES OF RELICENSING
Mokelumne River Project

Relicensing of the Mokelumne River Project was atypical. It was the longest-running relicensing in history, going for more than 29 years, with annual renewal of the license since 1975. The reasons for the extended relicensing were threefold. First, there was a long-standing disagreement between the utility (PG&E), federal and state entities, and nongovernmental organizations over the amount of information needed to establish new streamflow guidelines for the license. Second, a competing license filed in the 1970s by the city of Santa Clara complicated the relicensing for 12 years until it was resolved in court. Third, in the 1980s, Amador County proposed a new hydroelectric project within the Devil's Nose reach of the project area, which FERC decided to wrap into the Mokelumne River Project relicensing, further complicating the process.

PG&E applied to renew its federal operating license with FERC in 1972. In 1974, before FERC could issue a new license, the city of Santa Clara tried to take over the project, leading to ownership negotiations that reached the Supreme Court and took until 1990 to resolve. The city of Santa Clara owns and operates a municipal electric utility, and after having to purchase more power from PG&E at a higher rate than the Western Area Power Administration (WAPA), it decided to pursue development of its own generation properties, such as the Mokelumne Project, to gain independence from PG&E. At the same time, they refused to pay the higher PG&E rates and set aside more than $80 million from 1971–1979 (Moore and Winters 1982). A settlement between the city of Santa Clara, WAPA, and PG&E resolved the rate difference, and Santa Clara (now Silicon Valley Power) went on to develop hydropower and other alternative energy sources in various locations throughout the state, not including the Mokelumne River Project.

In 1987, Amador County, in the Mokelumne River watershed, applied to construct a new hydropower project in the Devil's Nose reach of the North Fork. The construction would be in the same area as the existing Mokelumne River Project. FERC decided to include the proposed project within the same relicensing, extending the scope of the relicensing. The logic was that Devil's Nose would catch up with the rest of the relicensing and be more efficient as a single project. FERC looked at the economics of the Devil's Nose proposal ahead of the environmental studies that might have blocked the development and decided to reject Amador County's proposal in 1995. Yet, 20 years into the relicensing, the various stakeholders had still not reached agreement on the new license conditions.

In June 1999, 24 years past the 50-year expiration date, the Mokelumne River Project relicensing reached a turning point when FERC called a meeting of all the interveners in the relicensing. There were forty entities at the initial meeting, ranging from whitewater recreational and fishing interests to conservation organizations focused on ecological impacts, hydropower interests, and resource agencies. FERC stated that the group had one year to reach a settlement or they would issue a license based on an environmental assessment completed ten years earlier. Many of the entities felt that the existing biological information on the hydropower project's impact on the river ecosystem was inadequate to support flow decisions for the new license. However, FERC needed to take action on the relicensing, so the group prioritized which studies could be conducted during the year they had in which to reach agreement on streamflows. The entities formed a new settlement group called the Mokelumne Relicensing Collaborative to work through the streamflow issues and present a plan to FERC.

Rock Creek–Cresta Project

Discussions regarding the Rock Creek–Cresta Project relicensing on the NF Feather began in 1979, and due to competing interests faced similar challenges and problems as the Mokelumne River Project. When the original license expired in 1982, PG&E had competitors for the Rock Creek–Cresta Project. A group of municipalities filed a competing application to operate the project, creating a legal conundrum that was not resolved until 1993 (Hydropower Reform Coalition 2009b). In 1991, PG&E partially resolved disagreements with the California Department of Fish and Game over minimum flow releases and entered into a side agreement that established minimum streamflows and hourly ramping rates to minimize stranding of fish. Yet, disagreements over ecological resources and recreational uses continued.

After extensive disagreement about operation of the Rock Creek–Cresta Project under a future license, PG&E convened settlement negotiations in 1998. Stakeholders from various state and federal resource agencies, PG&E, local interests, whitewater enthusiasts (American Whitewater), and fishermen (California Sportfishing Protection Alliance) formed the Rock Creek–Cresta Relicensing Collaborative with the sole intent of negotiating and coming to agreement on terms for the new project license.

BREAKING THE GRIDLOCK: RELICENSING COLLABORATIVES

Mokelumne River Project

The Mokelumne Relicensing Collaborative held 97 meetings over a 12-month period to determine the new license flow conditions. In both collaboratives, the time commitment for the stakeholders was immense. Member Pete Bell, working with the Foothill Conservancy, lived in the upper Mokelumne Watershed, and this meant driving about 100 miles for each meeting; the collaborative was "all he did, seven days a week for a year." But the outcomes were substantial. The collaborative requested that FERC grant a 30-day extension in June 2000, and met the extension deadline with a final settlement in July 2000.

The Instream Flow Incremental Methodology (Stalnaker et al. 1995) had been used in numerous relicensing projects to define streamflow levels to support habitat for aquatic species. However, the methodology did not meet all of the concerns of the Mokelumne ecological subgroup, so they developed their own method to determine streamflow patterns that would likely benefit aquatic habitat and whitewater recreation. The goal was to establish flows that would match as closely as possible the natural flow regime. The group used streamflow records from the preceding 27 years and separated them into five water-year types: wet, above normal, below normal, dry, and critically dry. They then used daily precipitation records to identify 15-day periods without rainfall and determined a minimum of 20 cfs as the lowest streamflow boundary. Cross-sectional surveys and streambed sediment sizes helped to determine the pulse flows in spring to mobilize the sediment and maintain dynamic channel geomorphic processes. These were set at 500, 1,100, and 1,800 cfs for the dry, below-normal, and above-normal water years, respectively. On the NF Mokelumne, in wetter years, when there would be spills at Salt Springs Reservoir, recreational releases would occur between May 15 and June 15 on two non-consecutive weekend days on the Devil's Nose Run, three weekends and two consecutive weekend days on the Tiger Creek run, and one weekend day on the Ponderosa Run (Mokelumne River Project 2000, § 11). In dry years, when the reservoir would not spill, the group recommended a reduced schedule of weekend whitewater boating flow releases on each run.

FERC issued a 30-year license for the Mokelumne River Project on October 11, 2001, 76 years after the original 1925 license.

Rock Creek–Cresta Project

Over two years (1998–2000), the Rock Creek–Cresta Relicensing Collaborative met monthly to review project data and negotiate new license conditions. On the NF Feather, in all water-year types, monthly recreational releases of 1 to 4 weekend days were recommended from June through September, with the magnitude of flow varying depending on water-year type (lower releases in drier years—Rock Creek–Cresta Project 2000, § 5). The Rock Creek–Cresta Relicensing Collaborative successfully reached agreement in September 2000, leading to a new license.

Unlike previous negotiations in each of the relicensing projects that had been unsuccessful, the collaboratives took a new approach to decision-making. The goal of the collaboratives was "to develop streamflow and non-flow measures for protection of ecological resources, while providing for other beneficial uses including whitewater recreation and hydroelectric power generation" (Mokelumne River Project 2000). To do so, the groups agreed to use a consensus-based decision-making process, meaning that all parties could "live with" the decisions. They followed a three-step process to collaboration: (1) identify interests, (2) develop objectives, and (3) develop protection, mitigation, and enhancement measures. In this manner, the groups took a holistic approach to watershed management instead of focusing on hydroelectric generation only. They considered streamflow regimes in light of fisheries, geomorphology, water quality, macroinvertebrates, amphibians, riparian habitat, recreational uses, and hydropower.

Two subgroups formed in each of the relicensing collaboratives to better define specific flow measures: an ecological resources subgroup developing a flow regime to address aquatic and geomorphic factors, and a recreation subgroup developing a flow regime to meet recreational needs (McGurk and Paulson 2001). For ecological streamflows, the subgroups established flow measures that moved beyond simple year-round flatline minimum in-stream flows to include varying minimum streamflow depending on water-year type, spring pulse flows to mobilize sediment and maintain geomorphic processes, and hourly ramping rates to limit stranding of fish. The proposed flow regimes of the two subgroups did not align, and reconciliation required considerable negotiation. The recreational subgroups determined flow measures by assessing the number of weekend days that streamflow would be suitable for boating using the unimpaired flow record. Based on the water-year type and the associated likelihood of reservoirs spilling, the subgroups developed guidelines for flow releases of varying magnitude and duration.

In both collaborative agreements, spring flow recommendations from the two subgroups could be made complementary. For example, if a June pulse flow release was designated, the release could be scheduled to be coincident with a June recreation flow day (Rock Creek–Cresta Project 2000, § 5). However, concerns were raised regarding the impacts of the timing and duration of the recreational releases on the breeding and rearing habitat of the FYLF. As flows declined in spring following a pulse or spill flow, frogs might begin breeding only to have eggs or tadpoles scoured away several weeks later during a weekend recreational release. However, at the time of the collaborative negotiations, there were few data available on the effects of pulsed or managed flows on the frogs as well as the extent of the populations in the two watersheds, particularly the NF Mokelumne, where no formal surveys had been completed. In the NF Mokelumne, recreational flows were not allowed until it was determined that there would be no adverse effects on the FYLF. In contrast, in the NF Feather, recreational flows were allowed to occur simultaneously with the monitoring of the FYLF.

Further negotiation between the two subgroups in each collaborative finally resulted in settlement agreements and provided for future monitoring studies. Monitoring studies funded by PG&E would be completed, not only to determine potential impacts of the new flow schedules on the frog population but also to assess potential impacts on other aquatic biota such as fish and macroinvertebrates and whether the recommended recreational flows were adequate for the various user groups. The collaboratives agreed to monitor and make decisions about the recreational flows within the first 3 years of the newly issued license and to adjust the flow schedules according to the new study data if necessary. The adaptive management approach allowed agreement between the two subgroups within each collaborative, primarily because streamflows could be adjusted in the future based on monitoring. Several of the people involved in the Mokelumne Relicensing Collaborative mentioned that the trust within the group and good working relationships were also key to reaching an agreement (personal communications, Pete Bell and David Steindorf, November 2013; McGurk and Paulson 2001).

The Settlement Agreements

The settlement agreements for the two hydropower projects established ecological streamflows designed not only to follow the pattern of the unimpaired hydrology but also to meet the interests of certain stakeholders, such as American Whitewater. The agreements and subsequent licenses provided funding for studies of the impacts of the new flow schedules on downstream aquatic biota and resources, as well as the formation of an Ecological Resources Committee (ERC) comprised of interested stakeholders who would evaluate the monitoring results and suggest alterations to the flow schedules if needed. The establishment of an ERC for each project as a means of facilitating ongoing adaptive management was a novel and progressive approach within hydropower licensing.

In the process of the settlement agreements, PG&E agreed to flow changes on all affected reaches on the NF Mokelumne and both the Rock Creek and Cresta reaches on the NF Feather. The settlement agreements established new minimum streamflows for each month of the year depending on water-year type; pulse flows in the winter or spring in all water-year types for the NF Feather and in the spring in all except critically dry years in the NF Mokelumne; hourly ramping rates to minimize stranding of fish; and on the NF Mokelumne, specific operation requirements for the upper lakes (Mokelumne River Project 2000). Additionally, there were detailed water-quality requirements for each reach. Below Salt Springs Dam on the NF Mokelumne and below the Rock Creek and Cresta Dams on the NF Feather, a water temperature boundary condition was set such that mean daily water temperature would be maintained at less than 20 °C when possible. Dissolved oxygen would be greater than 7.0 ppm in cold-water reaches and greater than 5.0 ppm in warm-water reaches within six months after the license. Additional mitigation measures were included on the NF Mokelumne, including removal of three dams on tributaries: the West Panther Creek, East Panther Creek and Beaver Creek dams (§ 8), and ceasing to use the river as a conduit when flushing the PG&E canal system during maintenance operations (§ 9). Finally, PG&E created new or improved access points for whitewater recreation on the four NF Mokelumne recreational runs. Gage data for each of the recreational runs were made available for paddlers to know when the water would be at ideal flows, and public notifications were to be made when recreational releases were determined at the start of each spring season.

To facilitate adaptive management, each settlement agreement required the formation of an ERC (§ 6 and 8 of the Mokelumne and Rock Creek–Cresta agreements, respectively) to monitor the impact of the streamflow changes on the aquatic ecosystems and adjust management based on the findings as needed. Each committee consisted of the

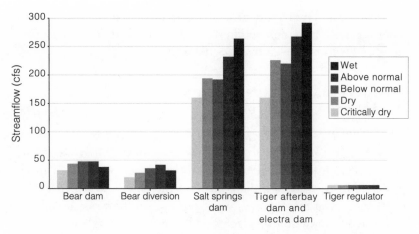

FIGURE 13.4. Minimum streamflow amounts set for adaptive management adjustments for specific hydropower dams and diversions in the North Fork Mokelumne River, by water-year type.

signatories to the settlement agreement, and the Forest Service participated as a liaison. The settlement agreements specified a Protection, Mitigation, and Enhancement Fund established by PG&E to support the monitoring studies for the first 15 years of the NF Mokelumne license. For the NF Feather, monitoring occurred on a different schedule depending on the focal resource. PG&E had to submit a report of the activities of the ERC by May 31 each year and provide FERC with notice of any decisions by the ERC or Forest Service regarding project operations.

There were additional specific adaptive management sideboards for each of the projects. For example, on the NF Mokelumne, the upper-lakes operation had a limit of 1,500 acre-feet in Upper and Lower Blue Lakes for a single above-normal or wet year, with no carry-over to subsequent years. For minimum flows, specific flow allocations were made with blocks of water for each facility for each water-year type (figure 13.4). These blocks could be used over a minimum of three months with no more than 50 percent of the total amount released in any one month. Pulse flows and minimum pool depth for Salt Springs Reservoir adaptive management blocks of water were also included. These sideboards on the flow changes set some certainty for FERC and PG&E in the licensing process and limited what the ERC could recommend and the Forest Service decide for flow changes in the future of the license (personal communication, Alvin Thoma, PG&E, 2014). Later, providing such boundaries on flow requirements or other factors in adaptive management became a provision within the FERC process (FERC Policy Statement on Hydropower Licensing Settlements 2006).

ADAPTIVE MANAGEMENT IN ACTION: ECOLOGICAL RESOURCES COMMITTEES

Mokelumne River Project

After the new license was established in 2000 for the Mokelumne River Project, PG&E had 4 years to make the modifications to their hydropower infrastructure and dams to meet the new flow schedules. The expected cost of the upgrades was $5–6 million; however, it ended up being significantly more. PG&E lost 3.7 percent of its annual power generation following the relicense (California Energy Commission 2003). Although the Stream Ecology Monitoring Program would not take effect until 2005, the licensee conducted water temperature and amphibian monitoring studies in 2000–2004 (table 13.1). FYLF surveys were conducted in one dry year (2001) and two below-normal years (2002 and 2003; Ibis Environmental and PG&E 2004). In 2004, the combination of snow and the Power wildfire resulted in monitoring at only the four river sites with the highest number of frogs.

Study results showed that frogs were present at about half of the surveyed river sites each year, but that the number of juvenile frogs observed in fall surveys had decreased by 43 percent from 2001 to 2003. This was unexpected, as these years had been drier and thus historically important for successful reproduction of frogs due to a lower likelihood of high spring flows that might scour egg masses or early tadpoles. However, the flow schedules implemented under the new license had created higher flows later into spring during each of these years than would have occurred naturally. Therefore, the ERC and the Forest Service determined that recreation and pulse flows should not occur on the Devil's Nose reach between April and June when frogs are breeding (April–May) and tadpoles are newly emerged (May–June; Ibis Environmental and PG&E 2004).

The impacts of pulsed flows on fully developed tadpoles and newly metamorphosed juveniles were still unknown, so researchers initiated a study in September 2003 to evaluate the effects of a summer pulse flow on the frogs. The study found that many tadpoles were stranded in isolated pools following the pulse and the abundance of tadpoles decreased. The effect of the pulse on juveniles was more ambiguous, with abun-

TABLE 13.1 THE MONITORING SCHEDULE FOR THE ECOLOGICAL RESOURCES COMMITTEE, SET TO BEGIN IN 2005 FOLLOWING PG&E UPGRADES TO INFRASTRUCTURE TO MEET FLOW REQUIREMENTS IN THE NORTH FORK MOKELUMNE RIVER.

Monitoring metric	Year																				
	99	00	01	02	03	04	05	06	07	08	09	10	11	12	13	14	15	16	17	18	19
Water temperature	B	M	M	M	M	M	M	M	M	M	M	M	M	M	M	M	M	M	M	M	M
Water quality							M	M	M	M	M	M	M	M	M	M	M	M	M	M	M
Fish	B								M	M	M			M	M	M			M	M	M
Macroinvertebrates		B							M	M	M			M	M	M			M	M	M
Mountain yellow-legged frog			M	M	M	o															
Foothill yellow-legged frog		B	M	M	M	o	M*	M*	M	M*	M	M									
Yosemite toad			M	M	M	o															
Riparian vegetation					B		B	M	M	M	M	M				M					
Geomorph: Cole Ck sediment							B	M	M			M				M					
Geomorph: sensitive channels							M*				M					M					
Recreation and pulse flows							M	M	M	M*											
Short-term power releases							M														
Upper lakes recreation survey																					

NOTE: B = baseline, M = monitoring, * = ERC-added, o = missed due to Power fire and snow dance decreasing after the first pulse but increasing after the second pulse (Jones and Stokes 2004). In late May 2004, stranding of tadpoles was again observed, resulting in an additional recommendation that recreational releases be curtailed in dry and critically dry years to limit impacts on the frogs. The main entity representing recreational releases, American Whitewater, supported the decision and recommended that it be "reconsidered as new information comes in from one year to the next—we recognize that all of us are on a steep learning curve with specific biological needs of FYLF" (Gangemi 2004).

The 2012 survey of frogs showed the greatest increase in population since the 2001 survey between Salt Springs and Tiger Creek Powerhouse. Although 2012 was a dry year and an incredibly productive and successful year for frogs across most Sierra Nevada rivers, with some of the highest numbers of individuals observed in a decade (Yarnell et al. 2013), the adaptive management approach on the NF Mokelumne appeared to be working. Most recently, the ERC decided to increase summer flows downstream of the Tiger Creek Afterbay, but established a stage-discharge study to assess the impacts of flow management on the frogs. Following spring spill, the down-ramping of flows is typically faster than would occur under unimpaired conditions, so the ERC and the Forest Service are currently reviewing whether any changes are necessary to address potential impacts. Similar ramping concerns exist in other hydro-projects with frogs, and resource managers have been working to implement spill-cessation schedules that more accurately mimic a natural spring snowmelt recession and provide a range of ecological benefits (Yarnell et al. 2013).

Rock Creek–Cresta Project

In a parallel path and timeline to the Mokelumne ERC, monitoring studies to determine appropriate levels for recreational releases and potential ensuing impacts on aquatic biota began following the Rock Creek–Cresta Relicensing Settlement agreement in 2001. Of particular concern to resource agencies was the presence of a population of FYLF in the Cresta reach. The impact of changing the spring and summer flow regimes on the frog populations was unknown, as in the Mokelumne ERC, so studies were designed to compare conditions in the Cresta reach with the adjacent regulated reach just downstream of the Poe Dam, where there was a large stable frog population.

Per the license and settlement agreement, three 5-year test flow periods were established that included minimum flows, pulse flows, and recreational flows. Recreational flow releases occurring over one weekend per month began in the summer of 2002 in the Cresta reach. Field biologists tracked the frog populations over the next 3 years in both the Cresta and Poe reaches and by 2005 began to see a marked decline in the Cresta frog population (Garcia and Associates 2005). In 2006, the ERC and the Forest Service suspended all recreational releases for a year as they reviewed study data and waited to see whether the population might rebound. No single study completed during the first several years of recreational releases could prove that the summer recreational pulse flows were causing the Cresta population decline. However, data from the multiple studies, combined with new academic research initiated in 2005 by the University of California, Davis, presented a weight of evidence that strongly suggested that altered summer flow regimes were preventing successful frog reproduction (Kupferberg et al. 2009). As a result, in 2007, the ERC again cancelled summer flow releases on the Cresta reach; but it did create a 3-year recreational flow plan for the upstream Rock Creek reach, which did not contain frogs.

In 2008, the ERC and the Forest Service faced a difficult decision about how to deal with the negotiated recreational flows that had been provided in the new license for the Cresta reach. While whitewater enthusiasts did not want to harm sensitive species like the frog, they did want to boat the river and would not lightly give up their hard-won allocation of water. Yet, fishing enthusiasts and resource agency personnel wanted to see a more natural flow regime that would benefit the native aquatic ecology. After much discussion, the ERC and the Forest Service landed on a compromise that would have far-reaching effects. The allocation of water for recreational streamflows would be "stacked" in the springtime, when flows would naturally be higher, leaving the later summer months at constant low flows, similar to the natural flow regime. Boaters would lose their monthly flow releases during the warm late-summer months, but would gain month-long releases throughout May, June, and July in wetter years. Flows would then gradually step down to summer minimum streamflows so that frog egg masses and tadpoles would not be stranded by a sudden drop in water level. The more natural flow regime also met the interests of many fishermen, who sought better in-stream habitat conditions not only for fish but also for the benthic community on which fish rely. The new flow schedule was implemented by PG&E in the Cresta reach in 2009 as part of the second 5-year test flow period.

As of 2013, the frog population in the Cresta reach remains critically low. It might not recover from the previous impacts. In February 2014, PG&E, along with the ERC and the parties to the settlement agreement, recommended streamflow changes for the third 5-year flow period. These include changes to ramping rates for the rising and falling limbs of the pulse flows, baseflow changes in relation to water temperature, new riparian flows, and specifications for hydropower operations during power outages and for cleanup of debris. These flow changes "should provide for improved protection of the FYLF, while allowing for an extended spring season of whitewater recreation" (US Department of Agriculture 2014). PG&E can only modify ramping rates to remove debris collected during the previous winter from flow gates. In critically dry years, from June to October, the baseflow will be higher by a set amount when the water temperature rises above 20 °C. Riparian test flows are a new addition; from March 15 to April 30 the flows will be about half of the pulse flows, to improve riparian habitat, with monitoring to determine the impact. The receding flows will now be coordinated on the NF Feather and the East Branch of the NF Feather, with the goal of matching the recession rate on the tributary and the main stem. These flow changes continue to address the balance between FYLF, whitewater recreation, and hydropower, with the addition of riparian habitat in this third 5-year period.

Support for this new approach to designing more natural flow regimes has been widespread. American Whitewater, alongside resource agency personnel, continues to advocate restoring spring flows in regulated rivers by allocating water for higher flows in spring to meet ecological and geomorphic needs as well as providing recreational boating opportunities. As a result, new spring flow regimes that better mimic a natural flow regime have been negotiated in regulated reaches of the Pit and McCloud Rivers in the Cascade range, and the Middle Fork American and Upper Yuba rivers in the Sierra Nevada (Yarnell et al. 2013).

CHALLENGES AND SUCCESSES IN ADAPTIVE MANAGEMENT

The terms set in a new hydropower license limit adaptive management in some cases. The determination of water-year types may change midyear, adding complications for monitoring and flow management. From January to March 2003, the forecast for the NF Mokelumne watershed called for a dry year, so the ERC made the decision not to monitor the impacts of flows on aquatic ecology. There was more than 250 percent of normal precipitation in April, and the water-year type changed to below normal. The estimates for water-year type come from snowpack information and estimates of accumulated unimpaired runoff, but there is uncertainty in both. As a result, there was a scramble to schedule the monitoring. The reservoir operations changed from a no-spill to a spill condition for Salt Springs Reservoir, but the release had to be delayed to avoid flooding downstream campgrounds along the river and potentially endangering campers. Additionally, the frog surveys needed to be completed before the spill and after, adding a delay beyond the weekend. The result was a faster ramping rate for the spill below the reservoir. To help solve these types of water-year issues it would be helpful to add additional water-year types, more equipment to monitor snowpack, and a range of forecasts made by different forecasters (McGurk 2003).

Some challenges arise as the state of knowledge changes, both with surveyors and with time. On the NF Mokelumne, the consultant who completed the initial frog surveys did not detect egg masses in spring (a more consistent year-to-year metric) and so focused on newly metamorphosed frogs in fall. A different consultant was used for subsequent surveys, and more than 80 egg masses were found, using a different survey method and timing. The early survey results influenced decisions within the license, for better or worse. On the NF Feather, increasing knowledge through time as additional studies were completed on pulsed flow effects on frogs required continued adjustment of the flow schedules. A significant advantage of the adaptive management approach is that as new information emerges through studies and monitoring efforts, changes to flow schedules can be made without the expensive and time-consuming legal costs of reopening a license.

For an ERC to work effectively, all of the members must be active within the group. Since the Mokelumne and NF Feather ERCs formed, they have met most months for the past 13 years. Not all members attend all meetings, but if they are not present, the meeting notes and any relevant studies are shared digitally. The ERC would have difficulty making decisions and adapting flows if members of the group did not work well together. In both of the project ERCs, good working relationships allowed each committee to work through unanticipated impacts to frogs that were not foreseen in the settlement agreements due to the lack of scientific study at the time of relicensing.

As time continues, additional challenges will arise for each project. The historically dry conditions during the 2013–14 water year presented conditions more extreme than those considered during relicensing. Licensees, resource agencies, and other stakeholders throughout the state worked together to develop alternative management strategies for numerous stream reaches. For example, reduced minimum in-stream flow requirements were developed to prevent draining threatened reservoirs to minimum pool and zero flow release, protecting the species and habitat. Collaborative problem-solving and constructive relationships were critical to dealing with these unexpected conditions.

On the NF Feather, issues of trying to control large spill gates to a fine level of flow adjustment, combined with incomplete communication and understanding of flow needs, led to a drastic decrease in flow downstream. In May 2012, spill flows in the Poe reach were reduced from 1,600 cfs to 275 cfs in a single day, resulting in desiccation of 40 percent of the frog egg masses laid that spring (personal communication, Andie Herman, PG&E, 2012). The quick response of the NF Feather ERC and the Forest Service to the flow drop in 2012 led to improved communication protocols and additional training across the NF Feather projects, as well as upgrades to some of the hydropower equipment.

CONCLUSION

The NF Mokelumne and NF Feather watersheds are typical of western-slope Sierra Nevada watersheds, with extensive hydropower infrastructure and similar aquatic species of concern. There are also popular whitewater stretches in these mountainous rivers with their granite bedrock. Relicensing of the Mokelumne River Project on the NF Mokelumne took the longest of any relicensing effort in the United States, in large part because of the uncertainty in determining streamflow guidelines and the lack of data. Additional legal complications from the city of Santa Clara's trying to take over ownership of the hydropower and a new hydroelectric proposal by Amador County further delayed the relicensing. This long history lead FERC to push for the 1-year settlement agreement, and a group of people worked collaboratively to determine the suggested flow amounts and timing, along with the terms for adaptive management due to the lack of current study data at the time of relicensing. The settlement agreement included formation of an ERC responsible for managing the adaptive monitoring and adjustments to future flows. A similar history and set of actions occurred in the relicensing of the Rock Creek–Cresta Project on the NF Feather, with the relicensing collaborative negotiating a settlement agreement that established an adaptive management process as part of the implementation of the new license.

The focus of the collaboratives in the NF Mokelumne and NF Feather relicensing settlement agreements was on developing streamflow guidelines and other measures to protect ecological resources while meeting the needs of both hydropower generation and whitewater recreation. The ERCs attempted to determine the unimpaired streamflows and mimic them in hydropower operations and flow scheduling. The collaboratives split into ecological and recreational subgroups to determine the best flows for their respective interests. Although aspects of the flows preferred by the subgroups conflicted, the collaboratives reached a settlement that created flexibility in the future to adjust blocks of monthly water, up to a specified amount, based on the water-year type and ecological demands. While the collaboratives set a wide array of objectives, they did not know at that time what a key role the FYLF would play in monitoring and adapting flows.

Adaptive management in the Mokelumne River and Rock Creek–Cresta Projects has improved knowledge of both flow impacts on frogs and flow management of the river. The best indicators of success of adaptive management are the increase in understanding of frog ecology stemming from both projects and the increase in the frog population in the NF Mokelumne alongside continued hydropower generation. Although adaptive management is not a panacea, and these licenses clearly have room to improve, for example in incorporating climate change projections (Viers 2011), it is a promising example of effective relicensing. As changes in flow management and subsequent impacts were monitored, uncertainty declined over time. Adaptive management required long-term commitment of both human and financial resources. Ongoing monitoring required stable and sufficient funding, and stakeholders committed to the process of balancing short-term actions in the context of long-term learning and benefits. As more time passes during the current Mokelumne River and Rock Creek–Cresta Project licenses, there is an opportunity to evaluate the adaptations and lessons learned over a longer timeframe and to compare this approach with other similar adaptive management approaches to flow management for people and nature.

ACKNOWLEDGMENTS

Four people directly involved in the relicensing of the Mokelumne River Project and the Rock Creek–Cresta Project made this chapter possible. Thank you to Pete Bell, who works with the Foothill Conservancy, and Dave Steindorf, from American Rivers, for informative interviews that provided an inside view of relicensing and the challenges of planning with multiple stakeholders. Alvin Thoma, director of the Power Generation Department for PG&E, reviewed an early draft of the chapter and helped make the story more accurate from the hydro-generation perspective. Beth Livingston, the hydroelectric coordinator for the Eldorado National Forest, improved the section on recent flow modifications and negotiations regarding foothill yellow-legged frogs. Finally, thanks to Sarah Kupferberg, Vince Resh, Mary Power, and Middy Tilghman for raising our awareness of the intertwining stories of hydropower relicensing, frogs, and whitewater paddling.

REFERENCES

California Energy Commission. 2003. "California Hydropower System: Energy and Environment." Appendix D in *2003 Environmental Performance Report*. Prepared in support of the *Electricity and Natural Gas Report* under the Integrated Energy Policy Report Proceeding (02-IEP-01). www.energy.ca.gov /reports/2003-10-30_100-03-018.PDF.

Federal Energy Regulatory Commission. 2006. "Policy Statement on Hydropower Licensing Settlements." *Settlements in Hydropower Licensing Proceedings under Part I of the Federal Power Act*. Docket No. PL06-5-000.

Gangemi, John. 2004. "AW Response to Mokelumne 9/16 ERC Meeting. Rec Flow Decision Documents. Letter to Ecological Resources Committee." In Mokelumne River Project (FERC Project No. 137), *Ecological Resources Committee Annual Report 2009*, Appendix E, September 15.

Garcia and Associates. 2005. *Results of 2004 Surveys and Monitoring for Foothill Yellow-Legged Frogs (Rana Boylii) within the Rock Creek-Cresta Project Area, North Fork Feather River and 2002-2004 Recreation and Pulse Flow Biological Evaluation Summary*. Prepared for Pacific Gas and Electric Company, San Ramon, California, Job 332/80.

Hydropower Reform Coalition. 2009a. *Mokelumne River Project: North Fork of the Mokelumne River, California. Hydropower Reform Coalition Success Story*. www.hydroreform.org/sites/default/files/Mokelumne_FINAL _0_0.pdf

———. 2009b. *Rock Creek-Cresta Project: North Fork of the Feather River, California. Hydropower Reform Coalition Success Story*. www.hydrore-form.org/sites/default/files/NFFeather_FINAL_0.pdf

Ibis Environmental and PG&E. 2004. *Brief Summary of Amphibian Surveys: Mokelumne River Project 2001 to 2004. Preliminary Findings and Implica-*

tions for Future Operations Associated with Recreation and Pulse Flow Releases. Prepared for Pacific Gas and Electric Company, San Ramon, CA.

Jones and Stokes. 2004. *Evaluation of the Effects of Short-Term Power Generation Water Releases on Foothill Yellow-Legged Frogs (Rana boylii) and Their Habitat, Mokelumne River Project, FERC No. 137.* Prepared for Technical and Ecological Services, Pacific Gas and Electric Company, San Ramon, CA.

Kondolf, George M., Kristen Podolak, and Theodore E. Grantham. 2012. "Restoring Mediterranean-Climate Rivers." *Hydrobiologia* 719:527–45. doi:10.1007/s10750-012-1363-y.

Kupferberg, Sarah J., Amy J. Lind, Sarah M. Yarnell, and Jeffrey F. Mount. 2009. *Pulsed Flow Effects on the Foothill Yellow-Legged Frog (Rana boylii): Integration of Empirical, Experimental and Hydrodynamic Modeling Approaches.* Report no. CEC 500-2009-002. Public Interest Energy Research Program, California Energy Commission. Davis, CA.

Kupferberg, Sarah J., Wendy J. Palen, Amy J. Lind, Steve Bobzien, Alessandro Catenazzi, Joe Drennan, and Mary E. Power. 2012. "Effects of Flow Regimes Altered by Dams on Survival, Population Declines, and Range-Wide Losses of California River-Breeding Frogs." *Conservation Biology* 26:513–24.

Lind, Amy J. 2005. *Reintroduction of a Declining Amphibian: Determining an Ecologically Feasible Approach for the Foothill Yellow-Legged Frog (Rana boylii) through Analysis of Decline Factors, Genetic Structure, and Habitat Associations.* PhD thesis, University of California, Davis.

McGurk, Bruce J. 2003. *Pitfalls of Forecasting Water-Year Type Classification in New Hydropower Licenses.* Paper presented at the 71st Annual Western Snow Conference, April 2003, Scottsdale, AZ.

McGurk, Bruce J., and Beth Paulson. 2001. "Mokelumne River Hydroelectric Project Relicensing: Collaborative Process Integrates Multiple Resource Concerns." In *Proceedings of the Eighth Biennial Watershed Management Conference*, edited by R. Coats. Report No. 101. University of California Water Resources Center, Berkeley.

Mokelumne River Project. 2000. *Mokelumne Relicensing Settlement Agreement.* FERC Project No. 137. Washington, D.C.

———. 2009. *Ecological Resources Committee Annual Report.* FERC Project No. 137. San Francisco, CA.

Moore, Edwin J., and Lingel H. Winters. 1982. American Microsystems (Plaintiffs) v. City of Santa Clara (Defendants), Docket No. 52481. 137 Cal.App.3d 1037, 187 Cal Rptr. 550. Court of Appeals of California, First District, Division One. www.leagle.com/decision/19821174137CalApp3d1037_11087.

Mount, Jeffrey F., Peter B. Moyle, Jay R. Lund, and Holly Doremus. 2007. *Regional Agreements, Adaptation, and Climate Change: New Approaches to FERC Licensing in the Sierra Nevada, California.* Project report. Center for Watershed Studies, University of California, Davis.

Moyle, Peter B. 1996. "Status of Aquatic Habitat Types." In *Sierra Nevada Ecosystem Project: Final Report to Congress. Volume II: Assessments, Commissioned Reports, and Background Information* (945–49). Centers for Water and Wildland Resources, University of California, Davis.

Poff, N. Leroy, J. David Allan, Mark B. Bain, James R. Karr, Karen L. Prestegaard, Brian D. Richter, Richard E. Sparks, and Julie C. Stromberg. 1997.

"The Natural Flow Regime: A Paradigm for River Conservation and Restoration." *Bioscience* 47:769–84.

Rheinheimer, David E., Jay R. Lund, Marion W. Jenkins, and Kaveh Madani. 2007. *Concepts and Options for Mitigation of Hydropower Dams in the Sierra Nevada: A Preliminary Review.* Draft project report prepared for the California Hydropower Reform Coalition. Center for Watershed Sciences, University of California, Davis.

Rock Creek-Cresta Project. 2000. *Rock Creek-Cresta Relicensing Settlement Agreement.* FERC Project No. 1962.

Stalnaker, Chris, Berton L. Lamb, Jim Henriksen, Ken Bovee, and John Bartholow. 1995. *The Instream Flow Incremental Methodology: A Primer for IFIM.* Biological Report 29. National Biological Service, U.S. Department of the Interior. www.dtic.mil/dtic/tr/fulltext/u2/a322762.pdf.

U.S. Department of Agriculture. 2014, May 30. "Request for Change to Section 4(e) Condition 5, Part A in the Rock Creek—Cresta License (FERC No. 1962)." Letter to Kimberley D. Bose, Secretary, Federal Energy Regulatory Commission.

Viers, Joshua H. 2011. "Hydropower Relicensing and Climate Change." *Journal of the American Water Resources Association* 47:655–61.

Wheeler, Clara A., and Hartwell H. Welsh, Jr. 2008. "Mating Strategy and Breeding Patterns of the Foothill Yellow-Legged Frog (*Rana boylii*)." *Herpetological Conservation and Biology* 3:128–42.

Yarnell, Sarah M., Amy J. Lind, and Jeffrey F. Mount. 2012. "Dynamic Flow Modelling of Riverine Amphibian Habitat with Application to Regulated Flow Management." *River Research and Applications* 28:177–91.

Yarnell, Sarah M., Ryan A. Peek, David E. Rheinheimer, Amy J. Lind, and Joshua H. Viers. 2013. *Management of the Spring Snowmelt Recession: An Integrated Analysis of Empirical, Hydrodynamic, and Hydropower Modeling Applications.* Final Report. Publication no. CEC-500-2013-TBD. California Energy Commission.

Yarnell, Sarah M., Joshua H. Viers, and Jeffrey F. Mount. 2010. "Ecology and Management of the Spring Snowmelt Recession." *BioScience* 60: 114–27.

CHAPTER 14

Emerging Cultural Waterscapes in California Cities Connect Rain to Taps and Drains to Gardens

CLEO WOELFLE-ERSKINE

During the critically dry summer of 2009, an unlikely coalition converged around an uncharismatic fluid called graywater. Fish lovers and kayak guides joined forces with agricultural water districts to promote the reuse of laundry and bathwater to irrigate residential gardens. Rogue "graywater guerrillas"—do-it-yourself gardeners who had rerouted their washing machines and bathtubs to drain into their tomato patches, in violation of state plumbing codes—surfaced, and they packed meetings convened by governor Arnold Schwarzenegger's special task force to rewrite the arcane code. California's Department of Public Health, which had previously viewed graywater as a source of waterborne illness, testified that graywater reuse would benefit public health by averting water scarcity. Graywater and rainwater had hit the mainstream.

In the lobby outside that 2009 hearing, policymakers, plumbing inspectors, water managers, and graywater advocates debated how the new code would affect municipal water consumption statewide. How would graywater's assimilation into mainstream water governance affect the spread of alternative water practices among California residents? Since cofounding the Graywater Guerrillas in 1999, my colleagues and I had taught hundreds of residents how to reroute their bathtubs and washing machines to water their gardens and how to turn plastic barrels into rainwater tanks. In these workshops, we presented graywater and rainwater systems as part of a strategy to reduce reliance on large dams and groundwater extraction. Workshop participants took this idea to heart and organized others in their social networks to attend meetings and write letters in support of a more permissive graywater code. Now, with the new code legalizing simple graywater systems, people could hire a contractor to install and maintain a graywater system for them. Would they make the same connection between individual water use and the systemic effects of large water works, and share the desire to engage politically in transforming California's waterscape?

In this chapter, I explore how small, decentralized water systems are spreading through California cities and may eventually disrupt the hegemony of large, centralized water systems. I approach this from two directions: first, discussing the perhaps surprisingly far-reaching consequences of household water-system experiments; and second, examining the interplay of infrastructural constraints with new cultural potentials for California cities in the twenty-first century.

In the first part of the chapter, I show that residential rainwater and graywater systems reduce household water use, though by how much depends crucially on how people use their systems. Graywater is a gateway to other changes in water practices, and often inspires political participation, for example in the 2009 code-change hearings and debates over whether to expand Northern California's Pardee Dam. Municipal water utilities, once skeptical of water savings and worried that relaxing plumbing regulations would lead to disease outbreaks, have changed their tune and now offer free classes on and rebates for installing rain barrels and graywater systems. Many utilities see residential graywater and rainwater catchment as part of a new, twenty-first-century approach to water management that maximizes conservation and diversifies supply, reducing dependence on the state's interconnected water system.

These new plans challenge orthodoxies embedded in the twentieth century development of California's waterworks:

- that large waterworks and centralized water management schemes can expand without limit, and with little regard for aquatic ecosystems;
- that centralized waterworks can best protect California residents from droughts, floods, disease, and economic stagnation;
- that decentralized water supply and wastewater treatment have no place in modern urban water provision.

Increasingly, these orthodoxies are no longer orthodox to water managers. Yet, official responses to impending "mega-droughts" show how far California is from crafting a water infrastructure—and indeed, a water culture—that can cope with the unpredictability of the state's variable rainfall and protect its riverine and estuarine ecosystems. Governor Jerry Brown's 2014 drought emergency declaration waived Endangered Species Act mandates to preserve streamflow to protect salmon, smelt, and dozens of other aquatic organisms that teeter on the brink of extinction. Nor were the treaty rights of tribes or the livelihoods of fishers considered. It focused mainly on agricultural diversions and massive new Delta tunnels and advocated modest, voluntary cutbacks in profligate residential practices like hosing down sidewalks.

Against this backdrop, the second part of the chapter looks at infrastructural constraints and the revolutionary opportunities that will shape water use in future cities. In exploring this emerging twenty-first-century California waterscape, I do not mobilize engineering logic or economic arguments, nor do I explore the supposed ecosystem services or other ecological benefits these systems might provide. Instead, I begin from the standpoint of everyday household water practice, tapping a "social turn" in water governance scholarship that focuses on social strategies for coping with water scarcity. In this vein, Australian feminist scholars Zoe Sofoulis, Yolande Strengers, and Cecily Maller argue that what Sofoulis (2005) calls Big Water imposes a psychological distance between urban water users and the rivers and aquifers that supply their water. At this remove, people have little material reason to conserve. By living with "small water"—by monitoring rainfall and the daily flows of their bath and rainwater to their gardens—many people say they come to recognize water's life-sustaining role in their local environment *and* in the riverine ecosystems that yield their household water.

That recognition is fostering new water cultures and pushing policy debates, but so far on a small scale. Small water will only challenge Big Water when decentralization becomes the new normal, and when conserved water stays in the river, rather than fueling new development or agriculture elsewhere. As I describe in the conclusion, the political ripple effects of small water changes can propagate savings beyond the municipal water sector. When graywater users came together to protest expensive expansions to the Pardee Dam in 2009, they argued that utilities and agricultural water works should follow their example and invest in water conservation and restoring natural water storage in wetlands and floodplains. It is in these political challenges that parallels—of wastefulness, of over-engineering, of disconnection from life-sustaining resources, of privatization of commons—between municipal water and the agricultural and industrial water sectors become clearer.

I explore changes afoot in household water practices by drawing insights from 15 years of engagement with California urban water issues.

This engagement began when I and housemate Laura Allen designed a constructed wetland to treat graywater in our small Oakland backyard in 1999. A repurposed bathtub filled with gravel and cattails provided clear irrigation water, attracted birds and salamanders, and sparked our interest in the infrastructure that brought water to our taps. A trip to its source in Sierra Nevada—the Mokolumne River backed up behind the Pardee Dam—inspired us to self-publish a 'zine called the *Graywater Guerrillas' Guide to Water*. This treatise was part tongue-in-cheek call to disobey plumbing codes, part do-it-yourself plumbing guide, and part dead-serious grappling with the politics and inequities cemented into California's water works. It eventually grew into the 2007 anthology *Dam Nation: Dispatches from the Water Underground*. My standpoint within the strange world of graywater aficionados—as a participant, as an educator, and as a theorist and researcher—gives me an unconventional view into the every-day water practices of people who reconfigure their home water systems, the concerns of engineers and regulators in local and state water agencies, and the aspirations of NGOs involved in river and climate adaptation advocacy. It is from this standpoint at that I "study up" to the scale of western water works and state and regional water governance processes.

SMALL, DECENTRALIZED WATER SYSTEMS IN CALIFORNIA CITIES

Behind most California taps lies a centralized municipal water system, which looks pretty much the same anywhere in the world. A large pipeline brings water from a river, reservoir, or large well to a treatment plant, where chemicals added to the water cause particles to clump together; the clumps settle; and the remaining water is treated with chlorine, chloramine, and/or ultraviolet light. Water then either flows by gravity or is pumped through networks of branching pipes, which decrease in size until they reach a tap. In the United States, piped water reaches individual households continuously. In many cities in the global South, water arrives only intermittently, and may need to be fetched at a distance from the house; sewage treatment is often lacking or insufficient.

Similarly, most California toilets feed a centralized sewer system. Water that flows down drains or is flushed down toilets mixes before it leaves the household—the combination is called "wastewater"—and flows through larger and larger sewer pipes, which converge at treatment plants. There, sewage undergoes a clumping-and-settling process similar to the supply treatment process; the liquid that remains is aerated in clarifier tanks, then discharged into a stream, estuary, or ocean. As Harris-Lovett and Sedlak note in chapter 10 of this volume, typical secondary sewage treatment does not remove many anthropogenic chemicals and pharmaceutical compounds.

Where is a concerned water user to start subverting centralized water and sewage systems? The answer depends on where a person lives, and what they use the most water for. Decentralized water systems can take many forms, and range in scale from a single household fixture to systems serving a complex of buildings or even an entire neighborhood.

Decentralized water supply systems draw from small wells or springs, or collect rainwater from rooftops or other hard surfaces in barrels or large tanks. Decentralized wastewater systems can handle graywater (bath, sink, and laundry water), blackwater (from the toilet), or a combination of the two, as septic systems do. Indoor graywater systems collect bath and laundry water for use in flushing toilets. Graywater irrigation systems divert bath, laundry, or sink water outside to irrigate plants; there are gravity-fed "direct" systems, and more complicated setups involving pumps, filters, and automatic irrigation systems. Blackwater systems treat sewage (a mix of laundry, shower, sink, and toilet water) in biological "machines" modeled on wetland ecosystems. Waterless toilets and urinals dispense with water altogether, instead collecting urine for use as a liquid fertilizer, and composting or drying out the brown stuff for use as a soil amendment. The Bronx Zoo's Eco-Restroom reclaims the composted poop and pee of 500,000 visitors annually for use as a fertilizer (Praeger 2007). Fear of excrement and regulatory obstacles still limit installation of blackwater-treatment wetlands and waterless toilets in U.S. urban areas, hence this chapter's focus on graywater and rainwater systems.

Decentralized graywater and rainwater harvesting systems vary greatly in complexity, capital cost, maintenance requirements, energy use, and the amount of interaction they require from users. Their designers make use of these differences to adapt systems to local particulars of climate, soil, and the built environment. A high-tech graywater system featuring storage tanks and automated pumps, sand filters, and drip irrigation typically costs upwards of $10,000 and requires constant energy input and professional maintenance (Greywater Action 2013). The simplest graywater systems are a dishpan in the sink or a bucket placed under the sink drain; these require no changes in plumbing and cost less than $20. Someone who lives with the high-tech system interacts with it only when she programs the drip-irrigation timer; the low-tech system's user will hand-carry buckets of water outside daily to hand-water plants. Between

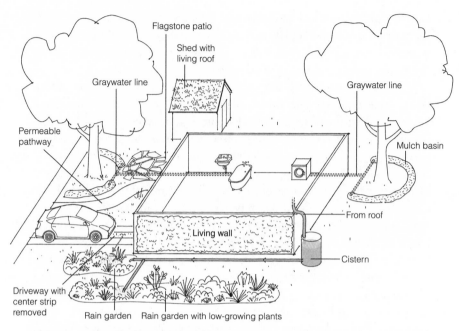

Flagstone patio

Shed with
living roof

Graywater line

Graywater line

Permeable
pathway

Mulch basin

Living wall

From roof

Cistern

Driveway with
center strip
removed

Rain garden   Rain garden with low-growing plants

FIGURE 14.1. An integrated home waterscape, for a water-efficient single-family house. Illustration by Arthur Mount, taken from *Creating Rain Gardens: Capturing the Rain for Your Own Water-Efficient Garden* by Cleo Woelfle-Erskine and Apryl Uncapher. Copyright 2012, published by Timber Press, Portland, OR. Used by permission of the publisher. All rights reserved.

these extremes, simple "laundry-to-landscape" and "branched-drain" direct graywater systems (see figure 14.1) require some modifications to plumbing, require no energy inputs, and cost between $250 and $1,000 (Allen, Bryan, and Woelfle-Erskine 2012). Someone who lives with these systems will periodically inspect the graywater outlets in the garden, replace wood chips in mulched basins that receive and infiltrate graywater, and perhaps repair flexible tubing punctured by a stray garden fork.

Home rainwater catchment systems range in size from 55-gallon barrels to 100,000-gallon tanks that irrigate playing fields. Use of untreated rainwater for irrigation tends to be uncontroversial, and the Rainwater Capture Act of 2012 allows storage of up to 5,000 gallons of rainwater for irrigation purposes. Currently, most municipal building codes nationwide prohibit the potable use of rainwater, but Texas, Oregon, and Ohio permit potable use after treatment, and Washington allows toilet flushing and laundry use. In contrast to restrictive mainland codes, the U.S. Virgin Islands requires most buildings to feature a self-sustaining potable water system such as a well or rainwater-collection system.

Living roof

Graywater irrigation

Rain garden

FIGURE 14.2. A home in Portland, Oregon, featuring several alternative water systems, including a rain garden. Photograph by Peter Erskine.

Rain tank costs vary depending on size and construction material (plastic, concrete, or steel); they are typically between fifty cents and a dollar per gallon, plus plumbing and pumps.

Another rainwater harvesting strategy forgoes tanks by diverting gutters to vegetated depressions in the soil that capture and infiltrate water. Called rain gardens, swales, bioswales, or dry wells, these structures create climate-adapted landscape features that often incorporate native wetland and streamside plants (figure 14.2). Gallon for gallon, rain gardens are much less expensive than rain tanks, so they can typically handle much larger volumes of water. Dozens of municipalities across the United States encourage or subsidize installation of rain gardens because they trap storm runoff and pollutants that would otherwise harm local waterways.

How Much Water Can Graywater and Rainwater Systems Save?

Die-hard graywater enthusiasts Laura Allen and Leigh Jarrard report that their households use only 40 gallons per person per day, while raising kids in diapers and maintaining lush landscapes full of edible fruit trees and vegetables (Laura Allen, personal communication; Jarrard 2013). They realize such substantial savings by combining graywater and rainwater irrigation systems with water-saving practices like turning off the shower while soaping up. In marshaling evidence for the 2009 graywater code rewrite, Allen and other proponents argued that simple graywater irrigation systems alone could meet all residential irrigation needs and reduce per capita daily water use to 81 gallons per day, a reduction of 40 percent (Department of Housing and Community Development 2009a; California Department of Water Resources 2005).

In California, stored rainwater is less useful than graywater for irrigation, since almost no rain falls during the summer, when plants need the most irrigation. Assuming that gardens need some irrigation between winter storms, a 300-gallon rain tank might fill and be emptied two or three times during the rainy season, saving perhaps 1,000 gallons of municipal water. Rainwater's real savings potential comes from nonpotable indoor use. A typical single-family home with a 2,000-gallon rain tank could replace all the municipal water used for flushing toilets, laundry, and bathing between October and May, saving 20,000 gallons of municipal water, even without graywater reuse. (This may seem like a large tank, but creative landscape architects have designed a 2,000-gallon tank made from a steel culvert that is 3 feet in diameter by 10 feet tall and fits in even a small urban lot.)

What impact can these potential savings have at the regional and state levels? According to a UCLA study, if 10 percent of households in the South Coast Hydrologic Region reused all their graywater, the volume of water saved would equal or exceed the capacity of a large modern seawater desalination plant (Cohen 2013). California-wide, if all single-family homes installed graywater systems, annual savings would reach 2.5 billion gallons. Adding a 2,000-gallon rain tank at each single-family house brings the potential savings to 3 billion gallons (U.S. Census Bureau 2013). (This back-of-the envelope calculation uses 2010 census population data for the number people living in single-family homes, and the California Department of Water Resources estimate of 183 gallons per person for daily use.) Many multifamily buildings in California cities have shared yard space and are suitable for simple graywater systems, so the potential statewide savings are even higher.

However, actual water savings are complex, and they change depending on how people use infrastructure. Few U.S. studies have investigated residential graywater systems *in situ*, and those that have only evaluated a handful of systems (Whitney, Bennett, and Prillwitz 1999; Little 2000; City of Los Angeles 1992). Two recent studies from Spain and Australia investigate infrastructural and behavioral aspects of decentralized household water systems, and report water savings between 40 and 90 percent, with the highest savings at a new development designed with rainwater harvesting, graywater reuse, and waterless toilets (Domènech and Saurí 2010; Mitchell 2006). Following the 2009 code change, many California water districts looked to graywater as a conservation measure. They wanted better estimates of the savings such systems were likely to realize, and of any long-term soil problems that might result.

Graywater in Single-Family Homes: Northern California Cases

Greywater Action's 2012 study investigated how home graywater systems perform in a range of residential settings, and how much water, if any, they save (Allen, Bryan, and Woelfle-Erksine 2012). The study sites featured single-family or duplex homes with yards, and had graywater systems that irrigated garden plants with untreated bath, lavatory sink, or laundry water. I co-authored the study and designed the investigation of water conservation and social effects. We examined water bills to compare water use before and after graywater system installation. We used structured interviews to discover what other factors (e.g. changes in landscaping, changes in household membership) might account for changes in water use. The study evaluated water use, soil and plant health, system performance, and cost. Here, I present a new analysis of data on how people's ways of using water and sense of hydro-ecological citizenship changed after they built a graywater or rainwater catchment system.

Does Decentralizing Infrastructure Affect Water-Use Practices?

After installing a graywater system, per capita water use decreased, on average, from 65 to 48 gallons per day, or by 26 percent (Allen, Bryan, and Woelfle-Erskine 2012). This average hides wide variation between households, however. In the most water-conserving household, per capita use *decreased* by 122 gallons per day. In the most profligate household, per capita use *increased* by 32 gallons per day. These results raise several questions about how graywater system infrastructure (which includes both new landscaping and new plumbing) affects water-use practices. In the few households where water use increased, we suspect that people installed new plantings that required supplemental irrigation from the municipal supply. Where water use decreased, we can attribute about half of the decrease to installation of a graywater system, based on users' reports of their laundry, showering, and irrigation practices. The rest of the savings comes from changes people made in how they used water, such as installing rain barrels, purchasing water-saving

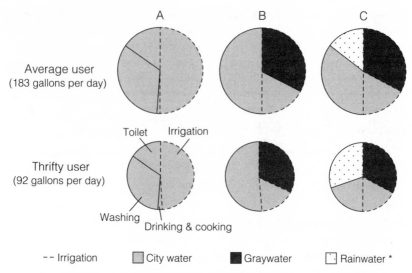

Average user
(183 gallons per day)

Thrifty user
(92 gallons per day)

A   B   C

Toilet   Irrigation

Washing
Drinking & cooking

-- Irrigation   ☐ City water   ■ Graywater   ⬚ Rainwater *

* Rainwater scenario shows winter water use, with rainwater used to flush toilets.

FIGURE 14.3. Per capita daily water use for two households, one that uses the California average of 183 gallons per day, and one that conserves water through low-flow appliances, drought-tolerant landscaping, and thoughtful use. a. A typical urban home uses city-supplied water for all indoor and outdoor use. Half of water is used outside, for irrigation, one-third for bathing and laundry, and one-sixth for toilets. Potable use for drinking and cooking is a very small fraction of the total water used. b. Diverting graywater outside reduces city water use by one-third, to 123 gallons of total use (61 gallons for irrigation) per day. c. Storing rainwater in a 2,000-gallon tank and using it to flush toilets during the winter months reduces winter water use to 95 gallons and 34 gallons, respectively, per day. If the "thrifty" house installed a dry toilet and used rainwater for washing clothes and bathing, it would require no city water during the winter months.

appliances, and turning off the water while washing dishes (figure 14.3). Several people relandscaped their yards when they installed their graywater systems, removing thirsty lawn and replacing it with fruit trees and perennial ornamentals. These changes in landscaping reduced overall water need, and also changed how people used water outdoors: some people stopped watering altogether, while others turned off part of their drip-irrigation system. These findings suggest that utilities wanting to maximize water savings should promote synergistic landscaping changes—including graywater, rain gardens, xeriscaping, and native-plant landscaping—that make irrigation with municipal water unnecessary. Other researchers have documented increased water savings from combinations of infrastructure changes and water-conserving practices (Li, Boyle, and Reynolds 2010; Burkhard, Deletic, and Craig 2000; Mitchell 2006; Ferguson 2009; Ghisi and Ferreira 2007), and we suspect that in our study, these combinations result in the highest water savings.

The study revealed that in the real world, simple graywater systems require some tweaking and adaptation from the user. About half of the households surveyed reported no drawbacks to their graywater systems, while the other half reported small hassles including clogging, runoff, and uneven irrigation. Only two of sixty-six respondents reported that graywater systems took "a lot of work," suggesting that most systems need little maintenance. Despite the problems, no respondent thought that anyone could get sick from their graywater system, and all were satisfied with system performance.

For most households, their first experiments in decentralized infrastructure were a gateway to further projects (figure 14.4). Twenty-nine households (44 percent) had both rainwater and graywater systems. Rather than rainwater catchment necessarily leading to graywater reuse or vice versa, both practices arose from a desire to reduce personal water use and be a good ecological citizen. However, as built, most rainwater systems probably save less water than graywater systems do. Most respondents had less than 600 gallons of rainwater storage, while only six had more than 2,000 gallons, and only one respondent had a rain garden. All reported using rainwater for irrigation, but only three people reported using rainwater for washing, toilets, or as an emergency backup supply. In our Mediterranean climate, realizing the full potential of residential rainwater harvesting will require larger tanks and a system for using rainwater for indoor uses.

Does Decentralized Infrastructure Increase Hydro-Ecological Citizenship?

All respondents named one or more tangible benefits of home water alternatives, including water savings, time savings, reduced water bills, the ability to grow more plants, and improved plant growth. More than half of the respondents also described intangible benefits: satisfaction from doing their part as ecological citizens, knowing more about water, or feeling more connected to local waterways. Several Oakland residents said that having a graywater system made them more aware of

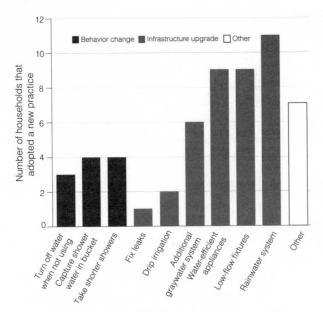

FIGURE 14.4. More than half of the households surveyed made further changes in water use after installing a graywater system.

their drinking-water source—the Mokolumne River, 90 miles distant in the foothills of the Sierra Nevada—for the first time. This awareness extended to the streams and estuaries that receive their treated sewage: people felt that keeping their graywater out of the sewer would make their local waters healthier.

Although our survey did not delve into the particularities of rainwater use, two people volunteered that they would boil and drink rainwater during an emergency, and thought that city codes should permit potable use with proper filtration. Several respondents said that they paid more attention to the timing and quantity of the rains, and noticed local weather patterns more, because they preferred to use rainwater for irrigation. This evidence corroborates responses in interviews I conducted at a more rural site in Sonoma County, where residents reported limiting irrigation toward the end of the dry season if rainwater supplies were low, even allowing some annual vegetables to die rather than irrigating them from the municipal supply (Woelfle-Erskine 2012). People who have rain tanks also described feeling more connected to local water bodies, and understanding that water left their property and flowed through the storm drain to the bay, river, or ocean.

Interpreting the Findings

Our study yielded several important findings for water policy, but more importantly for my task here, it also points to an emergent water culture. I can see these implications in our quantitative results because of observations I have made in other circumstances, and I bring these two lines of inquiry together to explore broader implications. In conversation, many rainwater and graywater users report a feeling of closer connection—even entanglement—with their local streams and sewer pipes and the distant rivers that supply their water. This feeling of connection would be unusual in a group of people who, for example, installed low-flow toilets through a utility rebate program. These simple graywater and rainwater systems share key characteristics that foster new attitudes toward water. Their proximity, draining directly to the home garden; their materiality, a network of pipes that is visible and tangible; and the interaction they require from their users, all keep the water itself in people's minds.

How people learn about and come to install these systems in their homes also matters socially and influences how cultural practices of water use change. Imagine yourself as a city-dwelling Californian who has heard that the state is in drought, but hasn't given it much thought. A friend tells you they are going to build a graywater system at their house through a workshop hosted by a local organization. They explain, more or less, what graywater is, and invite you to come to the workshop prepared to get dirty. You arrive at their house on a weekend morning along with twenty other people from various walks of life. Two instructors first ask if anyone knows the source of your local tap water, then discuss the municipal reservoir's impacts on river-based cultures and other creatures. You spend the next three hours learning how to drill holes through the walls of the house, fit pipe together, and install irrigation line (figure 14.5). Over a home-cooked lunch your host has prepared, you talk with new acquaintances about your garden and the lack of rain. You may plan to meet up with someone who lives in your neighborhood to build your own systems together.

Since 1999, have I co-led such workshops, which hundreds (if not thousands) of people have attended. Our curriculum, which I and other Greywater Action members developed and which has been taken up by other educators, encourages mutual aid among participants and explicitly connects home water use to environmental justice and interspecies solidarity. We didn't track how many participants have gone on to

FIGURE 14.5. A rainwater-harvesting workshop in Berkeley, California, in 2009. Photo by Tom Lux Photo.

install graywater or rainwater systems, but we did find that more than half of our sixty-six study households first learned about graywater from a friend or colleague or at one of our workshops. A third also said that someone they knew installed a graywater system after seeing or using theirs. Many of these workshop participants also engaged in California's turbulent water politics for the first time after hearing about a regulatory hearing through the Greywater Action mailing list. Dozens turned out for stakeholder meetings during the 2009 code-change process, prompting the Building Standards Commission to cite high stakeholder participation and the lived experience of many graywater users in their decision to legalize simple systems (Department of Housing and Community Development 2009a).

The people who came to the workshops were mostly middle-class professionals, homeowners or aspiring homeowners, and racially diverse; many probably self-identified as environmentalists. However, we made a deliberate attempt to separate graywater and rainwater systems from subcultures such as the permaculture and local-food scenes and eco-consumerist trends. We didn't want people to see their graywater system or rain tank as an eco-chic accessory that, once purchased, absolved them of the responsibility to critically engage with water politics. Our approach has had broad appeal for students in local community colleges and green-jobs programs, for recent immigrants in landscaper training programs, for youth from California tribal communities, for tradespeople, and for working-class do-it-yourselfers in California and beyond. Immigrants from Mexico and Central America and people who grew up rurally often comment that these simple systems echo the home water infrastructure of their places of origin and embody an ecological logic that resonates deeply.

These observations suggest that household graywater systems may spread differently in different communities. Where middle-class professionals might attend a hands-on workshop and then hire a professional landscaper to install their system, others with hands-on skills but less disposable cash might install their own system, or one for a friend—especially if subsidized plumbing kits are available. Such is the approach that several water districts (among them Santa Rosa, San Francisco, and Soquel) have recently adopted. They sponsor free workshops (usually led by someone who has been through a Greywater Action course) where they give out plumbing kits; residents can also request a free inspection once they install the system. This approach acknowledges the importance of social networks and hands-on learning in developing residents' capacities to install and maintain their system. It follows explicitly from the provisions of the new code, which create, via regulations, an unregulated space in which non-engineers can modify household plumbing. Ordinary people are granted a certain amount of responsibility for the public health, because regulators recognize—as public health department officials testified in the decisive 2009 hearing—that, in the balance, the risk of too much dependence on centralized water systems under a future drier, hotter climate is larger than the risk of many small graywater irrigation systems.

INFRASTRUCTURAL CONSTRAINTS AND CRACKS

As I have shown, small-scale water technologies are well developed and are starting to propagate into common use. Residential rainwater and graywater systems exemplify the kinds of small-scale infrastructure that might make up a hybrid water system of the twenty-first century. As Karen Bakker notes in *Privatizing Water* (2010), hybrid water systems involving a mix of centralized piped water and decentralized wells, rain cisterns, and tanker deliveries are the norm in many cities in the global South. Technologies such as point-of-use treatment, rainwater harvesting, and waterless sanitation that are part of sustainable development in "developing" countries need not be rejected out of hand in "developed" ones. If new, hybrid water infrastructures are to emerge in the United States, what better place than California's urban water systems, where the twentieth-century drive to tap distant rivers in search of ever-expanding water supply began?

California engineers launched the twentieth-century project of relent-lessly damming and diverting rivers to support urban (and agricultural) development, and exported this project—which I call Dam Nation—around the world. (This phenomenon is similar to the idea of the "hydrau-lic society," which Worster describes in his 1985 *Rivers of Empire* as the state's use of water works to consolidate power in the hands of elites.) This project has driven population growth and development but also fun-neled economic benefits to a wealthy few, at the expense of river-based cultures and rural communities that still lack access to clean and afford-able water. For the fishes, marsh plants, otters, and countless other spe-cies that depend on free-flowing rivers, Dam Nation's colonization has been devastating.

In a long-term view, the West's water works have yet to stand the test of prolonged drought. As paleoclimatologists Ingram and Malamud-Roam note in *The West without Water* (2013), the 150 years since European settlement in California have been unusually wet compared to the previous 10,000 years. Since the end of the last ice age, multi-decade droughts have occurred regularly. Zoe Sofoulis's observations about Big Water (i.e. large centralized water systems), made during Aus-tralia's recent 10-year drought, apply equally well to California's urban water supply systems.

> Although Big Water's infrastructure was created to supply drinking quality water to meet demands for cleanliness, flushing toilets, and green suburbs, in a "water crisis," domestic users are suddenly saddled with blame for this situation. They are castigated for being enthralled by the fantasy of endless supply embodied in the water faucet; criticised for lacking detailed knowl-edge of water used in different household processes (not that this informa-tion is easily available), and expected to make sacrifices in the amenity of their yards—quintessential icons of Australian suburbia—as their cherished gardens died. (Sofoulis 2005, 456)

As extended drought and anthropogenic climate change melt away the state's snowpack reservoir and drive accelerated pumping from stressed aquifers, strategies discarded during the height of the dam-building era in the mid-twentieth century are once again under consid-eration. Injecting treated sewage into depleted aquifers, breaching levees to reflood Delta floodplains, restoring mountain meadows, and even recreating the lost Tulare Lake in the southern San Joaquin Valley are all on the table (Hanak et al. 2011; Ingram and Malamud-Roam 2013). Such medium-scale projects cope with California's age-old cycles of flood and drought by protecting or creating ecosystems that capture water and store it in the soil.

The legacy of twentieth-century urban water infrastructure, in California and across the United States, is literally cemented in the supply lines, storm drains, and sewer pipes buried under city streets. Can rainwater harvesting and graywater reuse disrupt Dam Nation's insistent expansion, currently exemplified by the Bay Delta Conservation Plan's oversized Delta tunnels, which will enable much larger diversions south of the Delta in the future (*San Jose Mercury News* 2014)? If so, how? This question concerns the urban landscape and how it constricts certain patterns of water flow and water use. In California cities, pipes supply water, and streams flow through culverts or underground. Because city-dwellers rarely interact with local streams and untrammeled water cycles, we lose the chance to form complex, entangled relationships with watery ecosystems. My question also concerns people and their social networks. How might practices of procuring, using, and disposing of water evolve in the interstices of existing infrastructures and regulatory systems? If new cultural practices accompany decentralized infrastructure, then how much water people use, and how that use of water connects them with streams, fishes, and trees, will evolve as well.

These questions resonate at a time when California water managers acknowledge that few, if any, new large dams can be built, and existing dams will hold back less water in a future warmer, less snowy world (Cooley et al. 2010; Natural Resources Defense Council 2013; State Water Resources Control Board 2010). Recent urban water planning documents show that many cities are already diversifying water supplies by recharging aquifers, restoring forests in reservoir catchment areas to store more water in the soil, and reusing treated sewage. The recent ReNUWit collaboration between UC Berkeley, Stanford, and the Colorado School of Mines exemplifies an engineering approach to rethinking water infrastructure. Some of its researchers have proposed that water supply and wastewater treatment could be decentralized spatially, but managed by a central utility. One example of this approach is so-called green stormwater infrastructure—large vegetated basins that infiltrate stormwater, installed and maintained by municipalities in the Pacific Northwest in city parks and rights of way (figure 14.6). In another exam-

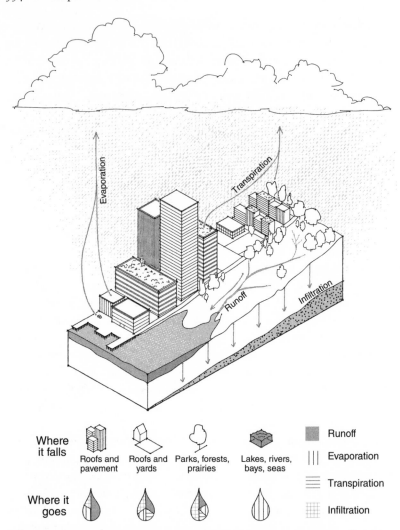

FIGURE 14.6. A typical urban water cycle. The urban raindrop can provide more water to cities of the future. Illustration by Arthur Mount, taken from *Creating Rain Gardens: Capturing the Rain for Your Own Water-Efficient Garden* by Cleo Woelfle-Erskine and Apryl Uncapher. Copyright 2012, published by Timber Press, Portland, OR. Used by permission of the publisher. All rights reserved.

ple, Albuquerque, New Mexico, is paying $2 million per year in city funds to a private company to maintain smart meters and software infrastructure, which let customers track their water use in near real time online (Strother 2013).

Green infrastructure approaches show significant water-saving potential, and they offer other benefits in the form of urban green space, reduced storm runoff, and lower discharges of treated sewage. (The smart meter approaches may also save water, but without any co-benefits, and with the side effect of shifting public dollars to private technology companies. One environmental-justice irony of Albuquerque's smart-meter bid is that Intel, which makes chips for smart-meter devices, diverts and then contaminates a large amount of the Rio Grande's flow—SouthWest Organizing Project 1995; Selcraig 1994.) However, compared to household systems managed by users, such projects are less likely to transform people's relationships to water sources—because they reinforce the "out of sight, out of mind" relationship to the rivers behind Big Water's taps. Approaches that focus on centralized green infrastructure projects, but dismiss household-scale graywater reuse, rainwater harvesting, and rain gardens, miss out on the savings that result from changes in social practices of water use.

In the 2009 graywater code change process, the California Department of Housing and Community Development (2009b) sought to reduce obstacles to social diffusion of graywater systems while promoting design standards that protect public and environmental health. "Californians have installed over a million graywater systems, most without permits," the Final Statement of Reasons read. "The homeowners do most of these installations. Creating simple guidelines without a permitting process for a clothes washer system installation allows the State of California to guide the installation of these low-tech systems much more effectively than the previous lack of control." The name of the Graywater Guerrillas was always a tongue-in-cheek challenge to governmental regulations, not a call to battle; nonetheless, the code-change process represented a "coming out of the jungle" moment, in which graywater rogues were able to effect change within the governmental arenas that regulate household water use. The new governance spaces that have emerged since then represent a new turn for municipal water governance. In conclusion, I look very briefly at what new water cultures might emerge in the century to come.

WATER IN FUTURE CITIES

As simple water systems proliferate in the interstices of municipal infrastructures, new policy debates are emerging. They represent a challenge to the orthodoxies of twentieth-century water thinking which drove the construction of 1,500 large dams and hundreds of miles of aqueducts in the state. In 2009, two water governance debates unfolded simultaneously. The first debate, over the state plumbing code rewrite, concerned how much control local regulators should exert over home graywater systems—that is, whether direct landscape graywater systems would be allowed, and if so, whether a permit would be required to build one. The rewrite took place as debate raged over the Delta Plan and a resurrected peripheral canal to transport water through giant tunnels beneath the Sacramento–San Joaquin Delta. Coalitions formed and re-formed over the course of several months of working-group meetings. The plumbers' union split on the new plumbing code initiative, while the state Department of Public Health's position evolved from rejection to support in response to scientific evidence and testimonies from graywater proponents. The central points of contention regarded health risks to humans who ingest graywater, via one of three routes: eating raw vegetables, like carrots, touched by graywater; (children) playing in graywater that had pooled in the yard; or through a "crossed connection"—the chance that an amateur plumber would inadvertently connect a graywater pipe into their home's supply pipes, thereby contaminating drinking water with graywater. (Amendments addressed these concerns to regulators' satisfaction, but the underlying conflict resurfaced in the 2012 code update, which included rainwater harvesting regulations.)

As reservoir levels dropped in August 2009, Governor Schwarzenegger approved the new graywater code as an emergency measure statewide—meaning that people could begin rerouting their washing machines to water their fruit trees immediately, with no permit required; and simple bath and lavatory-sink "branched drain" systems became legal with a simple permit. Testimony from Dr. Linda Rudolph, chief deputy director for policy and programs for the California Department of Public Health, proved decisive for several regulators. "We believe that [graywater] may provide substantial water conservation benefits and decrease demand on existing water supplies under threat from drought and climate change," she said. "We think this guidance will improve public health protection" (*Graywater Hearing* 2009).

The second debate concerned the East Bay Municipal Utility District's new 40-year water planning document. Later that fall, the district's board of directors met to hear public comments on the proposed Pardee Dam expansion in the Sierra foothills. Hundreds of people packed the courtroom to oppose the expansion (Gammon 2009). Many argued that cities should replace dam water with reused graywater and local rain water, and strive to live within the limits of their local water supplies rather than import from distant rivers (East Bay Municipal Utility District 2011). Several speakers mentioned that they had installed rainwater cisterns or graywater irrigation systems in their own homes. This experience had given them new appreciation for the value of water and changed their view of the East Bay's water supply system. In this new view, the utility should invest in home rainwater and graywater systems (as well as other conservation strategies) *instead of* large, distant water supply schemes. Resident Carole Shemmerling argued: "Water storage, graywater reuse, partnering with Contra Costa County, conservation and rationing, dry farming practices, and dry landscaping are far better ways to manage water resources. Revegetating streams, tree planting, and stopping gravel and gold mining in our streams and rivers, will go far to save and increase our water supplies. Creating dams loses water through massive evaporation and destroys watersheds which need to be protected" (EBMUD 2011).

Many speakers noted the need to consider water's multiple values—not just its price tag on the water bill. "To destroy a vast an important economy such as wild salmon runs, in order to have a dependable water and electrical supply for human communities which have no responsible policies limiting their growth, is neither sustainable or justifiable morally," said resident Clay Cockrill. Some, like E. Johnson, argued that other species and the river itself should have standing in water decisions, because their fates are intimately interconnected: "We are opposed to flooding the Mokelumne River. We favor using less destructive natural means to preserve California's natural rivers which benefit all residents—crucially the natural world and environment that allows humans to coexist with and support our native plant and animal species as well as our air, land, and most precious water."

These passionate speeches were addressed to East Bay Municipal Utility District staff members. No board members attended the public comment meeting—evidence, for many present, of the utility's twentieth-century, business-as-usual attitude that accepted without question the need for the new dam. In picturing this tableau—the row of empty seats behind the supervisor's desk, the stuffy boardroom, the interminable wait for the microphone—what prospects emerge for twenty-first-century urban water?

The arguments at the Pardee Dam hearing challenge three key orthodoxies embedded in twentieth-century water engineering projects. By orthodoxies I mean patterns of thinking that are so entrenched in a way of doing things that they are hardly ever questioned. After looking at these orthodoxies, I will mention several unorthodox approaches that may transform twenty-first-century water works.

The first orthodoxy is that large water works and centralized water management schemes can expand without limit. The 1974 movie *Chinatown* illustrates the nefarious underbelly of this orthodoxy. In this cinematic narrative a few Southern California real estate speculators get wildly rich on a real estate boom that enables Los Angeles to grow from a sleepy farm town into a megalopolis; urban customers receive cheap and clean water from the mountains; Paiute tribes and Owens valley farmers lose their ability to sustain traditional livelihoods; and all of the organisms inhabiting Owens Lake and River are obliterated (Reisner 1993).

The second orthodoxy is that centralized water works can best protect California residents from droughts, floods, disease, and economic stagnation. This orthodoxy drove the twentieth-century public health efforts that created municipal water and wastewater treatment systems to treat, monitor, and control the flows of water and wastewater through urban pipes. Harris-Lovett and Sedlak's piece in this volume (chapter 10) also subscribes to this orthodoxy in arguing that centralized utilities should expand municipal supply portfolios to include treated wastewater.

The third orthodoxy is that decentralized water supply and wastewater treatment have no place in modern urban water provision. This orthodoxy assumes a linear narrative of progress in which communal and artisanal sources of water (wells, springs, streams) are over-appropriated, contaminated by sewage, and ultimately abandoned. In their place, urban water utilities import water from distant rivers, and build centralized plants to disinfect it, and other plants to separate the sludge and rechlorinate (and then dechlorinate) the effluent before release into the nearest water body (Blake 1956).

The deepest challenges to these orthodoxies cannot take the form of alternative markets or new technologies, because Dam Nation has proved strikingly adaptive in integrating many technological fixes without dismantling inequitable and environmentally destructive structures. The deepest challenges will come from material changes in people's daily water-use practices and in the social structures of water provision and wastewater treatment. Decentralized governance creates space for adaptive ingenuity "from below" that is different from that imposed from above. Allowing unregulated space for self-building and design is important, as James Scott argues in *Seeing Like a State* (1998), because such spaces allow us humans to exercise our capacity for common-sense improvisation. For these new water cultures to emerge, twenty-first-century water governance must open spaces for us ordinary people to plumb our houses with our friends and neighbors, and to coevolve water-use practices and infrastructure into arrangements that are locally optimal (even if not necessarily "scalable" or generalizable). If these challenges do succeed, then future urban water managers will not be able to discount the lifeways of humans, fishes, birds, insects, and trees along the rivers that feed their pipes. And urban residents may develop a taste for the rain that falls on their own homes.

REFERENCES

Allen, Laura, Sherry Bryan, and Cleo Woelfle-Erskine. 2012. *Residential Greywater Irrigation Systems in California: An Evaluation of Soil and Water Quality, User Satisfaction, and Installation Costs.* Oakland, CA: Greywater Action. http://greywateraction.org/content/greywater-study-0.

Bakker, Karen J. 2010. *Privatizing Water: Governance Failure and the World's Urban Water Crisis.* Ithaca, NY: Cornell University Press.

Blake, Nelson M. 1956. *Water for the Cities: A History of the Urban Water Supply Problem in the United States.* Syracuse, NY: Syracuse University Press.

Burkhard, Roland, Ana Deletic, and Anthony Craig. 2000. "Techniques for Water and Wastewater Management: A Review of Techniques and Their Integration in Planning." *Urban Water* 2(3):197–221. doi:10.1016/S1462-0758(00)00056-X.

California Department of Water Resources. 2005. *California Water Plan Update 2005.* www.waterplan.water.ca.gov/previous/cwpu2005/.

City of Los Angeles. 1992. *Graywater Pilot Project: Final Report.* Office of Water Reclamation.

Cohen, Yoram. 2013. *Graywater: A Potential Source of Water.* UCLA Institute of the Environment and Sustainability. www.environment.ucla.edu/report-card/article4870.html.

Cooley, Heather, Juliet Christian-Smith, Peter H. Gleick, Michael J. Cohen, and Matthew Heberger. 2010. *California's Next Million Acre-Feet: Saving Water, Energy, and Money.* Oakland, CA: Pacific Institute. www.pacinst.org/publication/californias-next-million-acre-feet-saving-water-energy-and-money/.

Department of Housing and Community Development. 2009a (July 1). *Emergency Rulemaking: California Plumbing Code Graywater Systems (Title 24, Part 5, Chapter 16A, Part I): Initial Statement of Reasons.* www.hcd.ca.gov/codes/shl/graywater_emergency.html.

———. 2009b (December 21). *Emergency Rulemaking: California Plumbing Code Graywater Systems (Title 24, Part 5, Chapter 16A, Part I): Final Statement of Reasons.* www.hcd.ca.gov/codes/shl/graywater_emergency.html.

Domènech, Laia, and David Saurí. 2010. "A Comparative Appraisal of the Use of Rainwater Harvesting in Single and Multi-Family Buildings of the Metropolitan Area of Barcelona (Spain): Social Experience, Drinking Water Savings and Economic Costs." *Journal of Cleaner Production* 19(6–7):598–608. doi:10.1016/j.jclepro.2010.11.010.

East Bay Municipal Utility District. 2011. *Scoping Report for the Revised Program Environmental Impact Report SCH #2008052006.* www.ebmud.com/sites/default/files/pdfs/WSMP%202040%20-%20Revised%20PEIR%20Scoping%20Report_0.pdf.

Ferguson, Jennifer. 2009. *Substituting Residential Rainwater Harvesting and Greywater Reuse for Public Water Supply: Tools for Evaluating the Public Cost.* Master's thesis, Cal Poly San Luis Obispo. http://digitalcommons.calpoly.edu/theses/109.

Gammon, Robert. 2009. "EBMUD Has Yet Another Option Besides a New Mokelumne Dam." *East Bay Express,* June 3. www.eastbayexpress.com/oakland/ebmud-has-yet-another-option-besides-a-new-mokelumne-dam/Content?oid=1369961.

Ghisi, Enedir, and Daniel F. Ferreira. 2007. "Potential for Potable Water Savings by Using Rainwater and Greywater in a Multi-Storey Residential Building in Southern Brazil." *Building and Environment* 42(7):2512–22. doi:10.1016/j.buildenv.2006.07.019.

Graywater Hearing at California Building Standards Commission Hearing, 7/30/2009. 2009. www.documents.dgs.ca.gov/bsc/cal_evnt/2009/July-30-09-Comm-Mtg-Min.pdf.

Greywater Action. 2013. "Cost of Greywater Systems." http://greywateraction.org/content/cost-greywater-systems.

Hanak, Ellen, Jay R. Lund, Ariel Dinar, Richard Howitt, Jeffrey F. Mount, Peter B. Moyle, and Barton "Buzz" Thompson. 2011. *Managing California's Water: From Conflict to Reconciliation.* San Francisco: Public Policy Institute of California.

Ingram, B. Lynn, and Frances Malamud-Roam. 2013. *The West without Water: What Past Floods, Droughts, and Other Climatic Clues Tell Us about Tomorrow.* Berkeley: University of California Press.

Jarrard, Leigh. 2013. "Greywater Corps: FAQs." http://greywatercorps.com/faqs01.html.

Li, Zhe, Fergal Boyle, and Anthony Reynolds. 2010. "Rainwater Harvesting and Greywater Treatment Systems for Domestic Application in Ireland." *Desalination* 260(1–3):1–8. doi:10.1016/j.desal.2010.05.035.

Little, Val L. 2000. *Residential Graywater Reuse: The Good, the Bad, the Healthy in Pima County, Arizona. A Survey of Current Residential Graywater Reuse.* Tucson, AZ: Water Resources Research Center.

Mitchell, V. Grace. 2006. "Applying Integrated Urban Water Management Concepts: A Review of Australian Experience." *Environmental Management* 37(5):589–605. doi:10.1007/s00267-004-0252-1.

Natural Resources Defense Council. 2013. "Tackling Water Scarcity: Five Southern California Water Agencies Lead the Way to a More Sustainable Tomorrow." www.nrdc.org/water/california-tackling-water-scarcity.asp.

Praeger, Dave. 2007. "Scatologically Correct." *New York Times,* April 15. www.nytimes.com/2007/04/15/opinion/nyregionopinions/15CIpraeger.html.

Rainwater Capture Act of 2012. AB 1750. Water Code. http://leginfo.legislature.ca.gov/faces/billNavClient.xhtml?bill_id=201120120AB1750.

Reisner, Marc. 1993. *Cadillac Desert: The American West and Its Disappearing Water.* New York: Penguin Books.

San Jose Mercury News. 2014, January 3. "Mercury News Editorial: Massive Delta Tunnels Could Destroy Fragile Estuary." www.mercurynews.com /opinion/ci_24840717/mercury-news-editorial-massive-delta-tunnels-could-destroy.

Scott, James C. 1998. *Seeing Like a State: How Certain Schemes to Improve the Human Condition Have Failed.* New Haven, CT: Yale University Press.

Selcraig, Bruce. 1994. "Albuquerque Learns It Really Is a Desert Town." *High Country News,* December 26. www.hcn.org/issues/26/728.

Sofoulis, Zoë. 2005. "Big Water, Everyday Water: A Sociotechnical Perspective." *Continuum* 19(4):445–63. doi:10.1080/10304310500322685.

SouthWest Organizing Project. 1995. *Intel Inside New Mexico: A Case Study of Environmental and Economic Injustice.* Albuquerque, NM: SouthWest Organizing Project.

State Water Resources Control Board. 2010. *20x2020 Water Conservation Plan.* www.swrcb.ca.gov/water_issues/hot_topics/20x2020/.

Strother, Neil. 2013. "Smart Water Emerges in New Mexico." www.smartgrid-news.com/artman/publish/Technologies_Smart_Water/Smart-water-emerges-in-New-Mexico-6002.html.

U.S. Census Bureau. 2013. "California QuickFacts from the US Census Bureau." http://quickfacts.census.gov/qfd/states/06000.html.

Whitney, Alison, Richard Bennett, and Marsha Prillwitz. 1999. *Monitoring Graywater Use: Three Case Studies in California.* Sacramento, CA: California Department of Water Resources.

Woelfle-Erskine, Cleo. 2012. *Do Salmon Want Humans to Catch More Rain? Finding Commons at the Human-Water Interface.* Paper presented at Tapping the Turn: Water's Social Dimensions, November 17, Canberra, Australia.

Woelfle-Erskine, Cleo, July O. Cole, and Laura Allen. 2007. *Dam Nation: Dispatches from the Water Underground.* Brooklyn, NY: Soft Skull Press.

Worster, Donald. 1985. *Rivers of Empire: Water, Aridity, and the Growth of the American West.* New York: Oxford University Press.

CHAPTER 15

California's Water Footprint Is Too Big for Its Pipes

JULIAN FULTON AND FRASER SHILLING

How much water do you use in a day? Let's start with this morning. If you took a five-minute shower you probably used about ten gallons, then three or four more gallons to run the faucet and flush the toilet. At breakfast, maybe you heated a few cups of water to brew your coffee or tea. This doesn't seem like much so far, but before we move on, let's look into our cups and steep our minds in how much water really went into this warm infusion. What about the water needed to grow the coffee beans—should we count that? It turns out that it takes more than a thousand times as much water to grow and prepare the beans as it does to brew the coffee. That's a week's worth of showers for an eight-ounce cup of coffee. Tea turns out to be about a quarter as water-intensive as coffee.

We start from the standpoint that, yes, all that water should be counted; not on our monthly water bills, but toward our individual and collective *water footprints*. Your water footprint is the total amount of water required to support your lifestyle—to grow all the food you eat, to make the clothes and other goods you use, and to produce the energy to power your home and means of transportation. And it's probably bigger than you think. If you live in California, your water footprint is (on average) about 1,500 gallons per day, more than ten times the state's per capita daily amount for direct use through piped delivery (that's for showering, watering the garden, etc.). That average is about the same for the rest of Americans but is nearly double the global average of about 800 gallons per capita daily (GPCD). Worldwide, people's water footprints vary widely depending on their habits and the societies they live in. The average water footprint in other developed countries is about 1,000 GPCD, while the average in China is about 600 GPCD.

Our individual and collective water footprints have important implications for sustainability, which we define on a global level as the capability of current and future generations—in other locations and of other species—to meet their social, economic, and ecological needs. It is crucial that sustainability be defined on a global level because of the increasingly interconnected nature of people and resources worldwide. Thus, when we think about the major challenges of water management in the twenty-first century—what has been called the *global water crisis* (see UNDP 2006)—we see a relationship between observations "over there" and actions "over here," in places like California. Water footprint assessments provide a framework for understanding these relationships. For California, because many of the goods consumed come from other places, improving water footprint sustainability entails considering impacts on social and ecological systems both within and beyond its borders.

As individuals, understanding the water implications of our daily habits and decisions can help us live less resource-intensive lifestyles. At larger scales, when we add up people's cumulative water footprints, there can also be implications for water resource management and planning. Multiply 1,500 GPCD by 38 million Californians and we see that over 20 trillion gallons of water are needed each year to support the state's population. That's more water than would flow unimpaired down all the state's rivers, meaning without diversions for human use (California Department of Water Resources 2014). In other words, if California tried to produce everything that it consumes within the state's borders, there wouldn't be enough surface water to do so. California has outgrown itself, and is becoming more and more dependent on water from elsewhere. Our collective water footprint is truly global. The sustainability of California's future social and natural systems depends on how sustainably water resources can be managed inside and outside its borders.

This chapter discusses California's water footprint from our particular perspectives on water sustainability. Though we each come from more focused disciplines—engineering and ecology—here we take an integrated and macroscopic approach to looking at sustainability in California, that is, from a statewide perspective while acknowledging global connections. We see relevance in our analysis for actors at all levels of decision-making, from individual residents, to businesses, to local and state planners. Water affects and is affected by everything we do as a society, and we are all connected by it. The methods we use provide a framework that helps in understanding the nature of these connections. It is a relatively new method that we try to present as transparently and reflexively as possible to help improve its relevance in sustainability science.

In the second section we work through the science of water footprint assessment: what it tries to measure, why, and how. In the third section we present a case study of California's water footprint, including what and where it relates to, and how it has changed over time. In the fourth section we reflect on the degree of uncertainty in the information that our assessment provides, as well as what is needed to improve certainty in the processes we are attempting to study. Lastly, we discuss what we see as the implications and possible applications of water footprint science for various actors in California, as well as possible responses.

WATER FOOTPRINT SCIENCE

So far we have introduced the water footprint as both a concept and a number: the quantity of water required to produce the things that we consume. But how does that number relate to actual water sustainability challenges, both within California and elsewhere? After all, a "footprint" is about more than just its shoe size; every foot makes a different print depending on its size, weight, and shape, as well as the place it is stepping on. With water, then, what would we want a footprint assessment to tell us about the impacts those gallons have on sustainability concerns? To begin answering this question, we need to review the hydrologic cycle, thinking about how people and ecosystems engage with it and experience it.

From previous classes, you may recall the standard description of the hydrologic cycle (or just the "water cycle"). Solar energy drives evaporation from oceans and lakes, as well as transpiration from plants on land, accumulating water molecules in the atmosphere until they condense and fall as rain, sleet, or snow—which, when it falls on land, eventually flows as ground or surface water back to the sea through river basins (figure 15.1). Globally, this cycle operates on a fixed budget of water: about 1.4 billion cubic kilometers. While this may seem like a lot, currently less than one one-hundredth of 1 percent is readily available freshwater; the rest is locked up in ice, seawater, and inaccessible groundwater (Shiklomanov 2000).

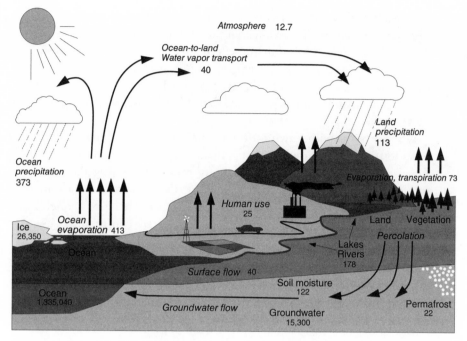

FIGURE 15.1. Simplified global water budget. Stocks in thousand cubic kilometers, flows in thousand cubic kilometers per year. Adapted from Trenberth et al. (2007).

While this may all sound familiar, notably missing from the standard description are the ways that humans and animals modify the stocks and flows of Earth's water. Human societies have always engaged with and modified the hydrologic cycle through their activities. Figure 15.1 depicts some of these activities—groundwater pumping for agriculture and industrial uses—though notably missing are domestic systems, as well as dams, canals, and other infrastructure that alters the hydrologic cycle in order to sustain human populations. People also modify the hydrologic cycle indirectly through land-use practices like urbanization, and most notably, by emitting massive amounts of greenhouse gases that cause global warming (see chapter 1, this volume).

Human activities that use water, depicted at the center of figure 15.1, make use of about a quarter of available freshwater annually, through rain-fed agriculture and pasturage, ground and surface water withdrawals, and in-stream uses such as transportation and waste assimilation (Postel, Daily, and Ehrlich 1996). These uses can be thought of as alterations or temporary interventions in the hydrologic cycle, since the water is ultimately put back into the cycle through evaporation, transpiration, or return flow. These interventions, while seemingly small in the global scheme, can have dramatic effects at the scale of a river basin when they are continuous or large relative to available flow patterns. This is particularly important because the river basin is the scale where people, plants, and animals grow and depend on water, a point we will return to later.

When water is withdrawn for human use, the term *consumptive use* refers to the portion of water that is made unavailable for reuse in the same basin, either through evaporation or transpiration (together, *evapotranspiration*) or through contamination (Gleick 2003). For example, in California, when a farmer draws water to irrigate his or her field, the portion that is evapotranspired is considered consumptive use because it is more likely to fall as rain somewhere much further east, like the Great Plains. The portion that returns to ground or surface water can also be considered consumptively used if it is contaminated with pollution to levels that exceed regulatory limits. Industrial processes like canning, making paper, manufacturing electronics, and producing electricity also use water for a variety of purposes that can result in consumptive use through evaporation and contamination.

In measuring the water footprint of any product (agricultural, industrial, etc.) we include only the consumptive portion of water use, since uncontaminated return flow can be used for other purposes in the basin. The water footprint accounting scheme divides consumptive use into three components, represented by three colors: blue, green and grey. *Blue water* is the managed surface and groundwater that flows through rivers, aqueducts, and pipes to where it is used, at homes, parks, factories, and farms. This is the water that we see; it has been described as more "charismatic" than less visible green water (Schneider 2013). *Green water* is the precipitation and soil moisture used directly by plants without being collected and applied by users. *Grey water* (not to be confused with graywater, discussed in chapter 14) is an indicator of contamination from a production process and is defined as the quantity of water needed to dilute pollution to levels that are not harmful to ecological and social needs (which may or may not correspond with regulatory standards).

The three water footprint components have different implications for sustainability in the location where water use occurs. Blue water, because it can be used for several alternative purposes, has an opportunity cost for each use. In other words, we should ask: What else could that water have been used for? In California, for example, blue water allocation often pits competing uses against one another, whether agricultural, urban, or environmental. Additionally, blue water typically has financial costs for treatment, pumping, or other infrastructure. Green water, on the other hand, is used directly on agricultural land, so it is important insofar as that water becomes unavailable for other land uses, alternative crops, or native vegetation. Changes in green water use can also affect blue water availability, and vice versa.

Grey water accounts for water quality impacts in a watershed, but it is an indicator estimate rather than a measurement, since it tells us how much additional water *would* be required to meet regulatory standards, regardless of whether that dilution actually occurs. In reality we would expect a farm, municipal waste discharger, or factory to figure out how to release less pollution rather than use more water to dilute it (though this does happen). While many water footprint practitioners add up blue, green, and grey water, we choose to report grey water separately from blue and green, so as not to double-count contaminated grey water as downstream blue water use. Still, grey water footprints provide an important indicator of the impact of production processes on water quality.

From a sustainability perspective, it is important to identify not just the different components of water use but how those types of water affect present and future conditions for social and ecological systems at the river-basin scale. Basins with fewer people and more water will likely be less impacted by consumptive water use than areas where larger populations already use much of the available water. Each river basin has its own particular issues related to blue, green, and grey water. The challenge for assessing the sustainability of water footprints is then to trace how much, what type, and where water was consumptively used, and to relate those uses to ongoing socio-ecological challenges at the river-basin scale. While this is a complex task, often deserving of a case-study approach, other methods exist that use indicators or combinations of indicators to help researchers understand the relative impacts and risks of water footprints at the basin scale. Examples of these indicators include the Watershed Sustainability Index (Chaves and Alipaz 2006) and the Water Poverty Index (Sullivan, Meigh, and Giacomello 2003).

Blue, green, and grey water footprints can also occur at multiple stages in making a product. Returning to the cup of coffee and the thousand cups of water required to make it, most of it was probably green water used by the coffee plant, since most coffee farming is mostly rainfed. Some of it, however, was probably blue water used (consumptively) to

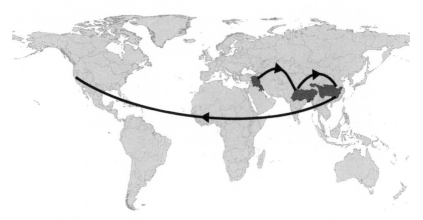

FIGURE 15.2. Water footprints can have impacts in multiple river basins. For example, here is the hypothetical supply chain of a cotton T-shirt, with impacts in the Tigris-Euphrates, Ganges-Brahmaputra, and Yangtze river basins.

remove the pulp from the bean and for other wet-processing steps. Other products, like cotton clothing, may require water at many more steps, from green water on the farm, to blue water at the factory for washing and dyeing fabrics, which may in turn also create grey water footprints. Industrial products like electronics require water for mining metals and manufacturing parts like semiconductors, and often have grey water impacts as well. Energy sources also entail mining fuels like oil and gas, but even more blue water is required for cooling at thermoelectric power plants.

These production stages can also take place in multiple locations, with differing impacts on river basins in far-off places. For example, cotton for a T-shirt purchased in a California shop might have been grown under rainfed conditions in India, processed into cloth and washed with municipal water in Bangladesh, then shipped to China, where further processing and dyeing may impact the water quality of a nearby river, before being finished and taken to market. Such a T-shirt would have a green, blue, and grey water footprint covering three different countries and many river basins (see figure 15.2). On average, a single T-shirt requires 290 gallons of green water, 320 gallons of blue water, and 100 gallons of grey water from river basins around the world (Chapagain et al. 2006).

Production in different locations can also be more or less efficient depending on production conditions, methods, and technologies; so the geographic routing of a product's supply chain can make a big difference in its overall water footprint. For example, the water footprint of a cotton

T-shirt made with U.S. cotton is about one-fourth the size of one made with Indian cotton (Chapagain et al. 2006). Globally averaged water footprints have been calculated for a range of products using United Nations agricultural and industrial water-use statistics, which are in turn derived from regional surveys and models. These results have been worked into several useful online tools for comparing the average water footprints of products and estimating one's individual water footprint based on personal diet, habits, and income (see www.waterfootprint.org).

National differences in production and consumption can also provide more detailed information about the water footprints of products and people. National production and trade statistics (also from the United Nations) indicate whether a product was likely produced domestically or imported, and from where. Using this information, the water footprint has been calculated for nearly every country, including the United States, allowing us to compare per capita water footprints as we did in the beginning of the chapter. However, within larger countries like the United States, we may expect regions and states to have different water footprint dynamics. Therefore, we use more locally tailored production and trade statistics to evaluate California's water footprint, which we turn to in the next section.

WATER FOOTPRINT AND CALIFORNIA

Data that may be used in calculating California's water footprint include state-level statistics on production of agricultural and industrial goods, surveyed and modeled information on how much water was consumptively used to make those goods, and trade statistics on whether those products were exported to other states or countries. If goods are not exported we assume that they are consumed and count toward California's water footprint. We also account for imported goods and the water footprint associated with their production in the country from where we import them. Goods that are imported and re-exported (which happens a lot in California ports) do not count toward California's water footprint. These readily available data sources allow us to calculate a first approximation of California's water footprint.

Before we present the results, we consider why a water footprint assessment might be important for a state like California. Historically, California has had a relationship with water that is unique among U.S. states. Given the temporal variability of its Mediterranean climate, combined with the geographic variability of its rainy north and desert

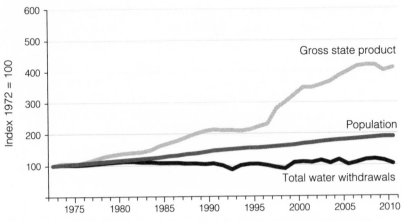

FIGURE 15.3. Trends in California's real gross domestic product, population, and freshwater withdrawals, 1972–2010. State-level real GDP is only available after 1986. GDP for 1972–1986 was estimated using a national-level inflation index. These data were collected by Department of Water Resources staff from older versions of Bulletin 160 (for 1972–1985), annual reports prepared by district staff (for 1989–1995), and the Water Portfolio from *California Water Plan Update 2013* (for 1998–2010). Sources: California Department of Finance (2011); California Department of Water Resources.

south, the management and manipulation of water flows have been integral to California's development and identity. Yet, as the state has grown to have the largest population in the nation and one of the highest-valued economies in the world, data show that total water withdrawals in California have remained relatively stable over the past 30 years (figure 15.3). This apparent "decoupling" of water use from population and state gross domestic product (GDP) growth raises two fundamental sustainability questions: How has growth in California been sustained? And can it be sustained in the future?

Regarding the first question, one answer is that Californians have come to use water in the state to do the things we want it to do much more efficiently. Indeed, to some extent, these trends reflect the adoption of more efficient technologies and practices by nearly all sectors of society, from households and businesses to farms, factories, and power plants (Gleick, Cooley, and Groves 2005; Rich 2009; Hanak et al. 2012). Many of these efforts have come about through technological innovation, strong policy, and behavioral change. All of these factors play roles and are interrelated in determining statewide water use and the management of California's water resources. Future water use within the state will continue to depend on these factors as well.

But another possible answer to the first question is that some of this growth has not in fact been decoupled but come to rely on water from elsewhere. The idea of relying on external sources of water should not be new to students of California water. After all, California has historically received up to 10 percent of its water supply from the Colorado River basin, as well as a few percent from Oregon and a tiny bit from Mexico. (Part of California is in fact within the Colorado River basin; however, this area is the smallest among the seven basin states.) These regional imports, however, have not increased in recent decades (in fact they are included in the water-use metric presented in figure 15.2), meaning that water to sustain California's growth might come from even farther afield.

The water footprint approach offers a possible explanation of this situation by acknowledging that water to sustain Californians and California's growth does not need to arrive in the state in liquid form. Rather than import water through pipes, products that use water can be imported. This has been referred to as *virtual water*, *water services*, and *embedded* or *embodied* water. All these mean roughly the same thing, and are an intrinsic part of California's water footprint. The water footprint method, as we have presented it here, helps identify where and how this water was brought into production for sustaining California. Thus, our assessment aims to offer deeper answers to fundamental sustainability questions about how water has figured in California's past and future development.

With these questions in mind, as well as the tools outlined in the previous section, we now consider what California's water footprint has looked like and what implications it has for future sustainability. The products accounted for in California's water footprint are shown in figure 15.4. Most of California's water footprint, like that of the average Californian, relates to food (85 percent). Meat and dairy products are especially water-intensive, making up half of the food portion. Energy products such as gasoline, ethanol, electricity, and natural gas make up the next-largest piece (8 percent) of California's water footprint, followed by industrial products (3 percent) and direct use of water for domestic, commercial, and institutional purposes (3 percent).

Next we look at how California's water footprint has changed over time. Lack of continuous data limited our results to snapshots at 5-year intervals starting in 1992, which makes for a kind of strobe-light view of the actual evolution of California's water footprint. Nevertheless, figure 15.5 reveals several interesting points. Total water footprint for

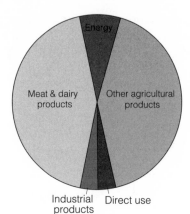

FIGURE 15.4. California's water footprint, by product class.

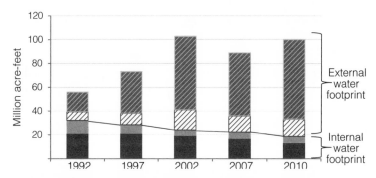

FIGURE 15.5. Evolution of California's blue and green water footprint since 1992. Blue water is represented in dark grey, green water in light grey.

each year is divided between external and internal, as well as blue and green components. In 1992, most of California's water footprint was internal, meaning that it came from the use of California's surface and groundwater. Just over half was also blue water. By 1997, at our next time step, we can see three trends developing that eventually proceed throughout the time series: an increase in the overall volume of the water footprint (overall bar height); increasing reliance on green water (relative size of the light-gray solid and hatched boxes to the total bar); and increasing reliance on external water resources (relative size of hatched boxes to the total bar).

Here we look at each of these trends individually and consider what might be driving it, whether it is policies, economics, environmental constraints, or simply the cumulative lifestyle choices of 38 million (and counting) Californians.

First, consider the trend in total water footprint quantity, as measured by the overall height of the bars. The average water footprint growth rate between 1992 and 2010 was 4.4 percent per year. This is slower than GDP growth (5.2 percent); however, it is more than twice the rate of population growth (1.4 percent). In fact, water footprint per capita has grown at 2.4 percent per year. Why is this? One explanation is that Californians are simply consuming more than in the past. Focusing on food consumption, this trend is somewhat supported by data from the U.S. Department of Agriculture's Food Availability Data System, which shows a 7-percent increase in daily calorie availability for all Americans between 1992 and 2010. These data do not include food losses, which in 2010 were 31 percent of available food (Buzby, Wells, and Bentley 2013). Another explanation could be a change in average per capita diet or consumer behavior toward more water-intensive goods. As shown above, meat consumption can make up a large proportion of an individual's water footprint. A final explanation could be that the amount of water used to make products consumed by Californians is increasing. As noted earlier, production in California has become more water-efficient; however, those efficiency gains will not affect California's water footprint if the water-efficient products are then exported and more water-intensive products imported. A final note on the overall height of the bars is that the spike that shows up in 2002 and then recedes by 2007 is related to the stockpiling of corn grain intended to be used to produce ethanol in subsequent years.

A second observation in California's changing water footprint is the increasing trend in the green water component. As California has very little rainfed agriculture, most of this green water comes from outside of its borders. The growing contribution of green water to California's water footprint raises concerns about the risk of relying on precipitation falling in other regions and the potential impacts of climate change. The 2012 droughts in the U.S. Midwest highlighted this concern when imported grain for livestock and ethanol were in short supply (U.S. Energy Information Administration 2012). This situation provided evidence of California's susceptibility to global climatic changes in regions outside its borders, which are only expected to become more dramatic in coming years. Incidentally, increased dependence on blue water could also expose California to potential impacts of climate change since, ultimately, sources of blue water such as surface water reservoirs, groundwater aquifers, rivers, canals, and streams are also directly dependent on the overall precipitation in an area. Nevertheless, management of blue water offers some flexibility to cope with year-to-year variations in precipitation.

Our third and perhaps most dramatic finding is that California's water footprint has become increasingly externalized, from 40 percent in 1992 to 80 percent in 2010. This means that to sustain itself California now relies far more on water resources from outside its borders than it did in the early 1990s. Most of this water is from other parts of the United States, but the percentage of virtual water from other countries has nearly doubled, from 21 to 41 percent, over this time period. The further externalization of California's water footprint raises concerns about our ability to manage water resource impacts and risks associated with our demand for goods and services. In the next section we look at how these impacts and risks are distributed in more geographic detail. At the same time, California's internal water footprint has decreased from 60 percent to 20 percent of the total, theoretically reducing its burden to water resources inside its borders. However, as shown in figure 15.3, water withdrawals have remained stable in California. Virtual water exports explain this apparent difference. The amount of water embedded in exported products has increased by an average of 6.2 percent per year since 1992.

Fundamental sustainability questions are raised by the relationship between water use and growth in population and GDP (figure 15.3). Summing up these findings, we see clear indications that sustaining California's growth has relied heavily on virtual water. On a per capita basis, virtual water has been used not just to sustain existing consumption habits but to support changing behavior toward higher levels of consumption as well as consumption of more water-intensive products. Most of the virtual water Californians have come to rely on in the past 20 years has been green water, which has quadrupled, although import of virtual blue water has at least doubled. Lastly, the sustained growth of California's water footprint over the past 20 years has been entirely permitted by the increased reliance on virtual water from outside its borders. While water resources actually managed within California have been increasingly applied to exports, a larger and larger share of California's growth has been sustained by water from elsewhere. In the next section we present a geographic analysis of California's water footprint.

To develop an understanding of the impacts and risks related to California's water footprint, the next step is to determine where water is used to sustain California's economy and population. As seen in the

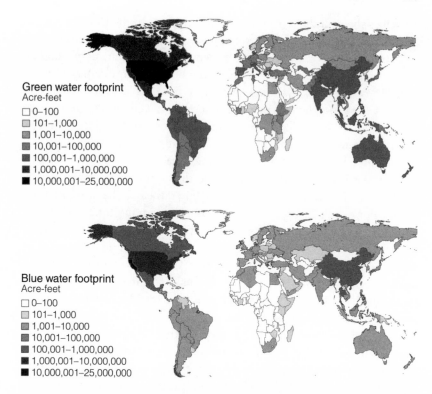

Green water footprint
Acre-feet
□ 0–100
▫ 101–1,000
▪ 1,001–10,000
▪ 10,001–100,000
■ 100,001–1,000,000
■ 1,000,001–10,000,000
■ 10,000,001–25,000,000

Blue water footprint
Acre-feet
□ 0–100
▫ 101–1,000
▪ 1,001–10,000
▪ 10,001–100,000
■ 100,001–1,000,000
■ 1,000,001–10,000,000
■ 10,000,001–25,000,000

FIGURE 15.6. California's green and blue water footprints, by country.

previous section, 80 percent of that water is outside its borders. Figures 15.6(a) and (b) depict where water is used to make products consumed in California. As can be seen in figure 15.6(a), most of the green water relates to production in other U.S. states, where we have already highlighted some of the risks associated with droughts. Similar risks appear in California's major trading partners, Mexico and Canada, where Californians depend on green water to the tune of several million acre-feet. (An acre-foot is the amount of water that would flood one acre—about the size of a football field—to a depth of one foot.) Further dependence on green water can be noted in drought-prone countries like Australia, parts of India and China, and East African and Mediterranean countries, as well as most of South America.

Figure 15.6(b) shows that California's largest dependence on blue water is within the state itself. Nevertheless, dependencies amounting to several hundred thousand acre-feet exist with neighboring Mexico and Canada, as well as the country providing the most imports to the United States of any trading partner: China. Despite massive investment in China's water infrastructure, there is increasing evidence of a "ruinous confrontation between water supplies and its increasing food and energy demands that is virtually certain to grow more dire over the next decade" (Schneider 2011). China also figures most prominently in California's grey water footprint (not shown) due to the quantity of goods imported from China's industrial sector, often cited as one of the most polluting worldwide.

Though figure 15.6 shows California's water footprint on a map of countries, it is important to understand that water impacts are usually experienced not at the national scale but rather at the scale of river basins. For example, China is a large country with multiple river basins and differing hydrologic conditions. Characterizing the sustainability of California's blue water footprint there would require first knowing in which basins goods were produced that California imports and then understanding the sustainability concerns associated with consuming that volume of blue water. Such techniques are possible but were beyond the scope of this initial assessment.

Concluding this section, we have shown that California's sustainability is intimately tied to water use and impacts in virtually every corner of the earth. The results presented here are a first-order analysis intended to raise awareness of the global nature of water sustainability for California and highlight the general structure and evolution of California's water footprint. We have attempted to take this step in a manner consistent with the intention of water footprint science, but much more analysis and effort are needed to characterize the nature of the connections we have drawn. Given the promise that we see in this scientific frontier, the next section is intended to highlight the major sources of uncertainty and variability in this and future water footprint assessments. We conclude by discussing how water footprint science can support sustainable water management going forward.

WATER FOOTPRINT VARIABILITY AND UNCERTAINTY

Because we advocate for water footprint being used as a scientific tool to assess water sustainability and develop actionable goals, an important consideration is how certain we can be in our calculations. There are several sources of variation in the calculated water footprint for a population (e.g. California). Some of these are related to variation in the behavior that leads to consumption of virtual water; some are related to geographic and other variation in water availability and use in goods production; and some are related to measurement error and accuracy of values used in calculations of virtual water use and impacts.

Agricultural/food production is the largest component of California's water footprint: 85 percent of it in 2007 (Fulton, Cooley, and Gleick 2012). Considering the importance of agricultural water demand in California, we estimated the impact of the variability in the water footprint of agriculture production on the total water footprint for the state. Blue water footprint and green water footprint of agricultural production describe the amount of water consumptively used in the growing of crops. Values for blue water footprints come from estimates of the total volume of evapotranspiration of applied water from agricultural crops, while green water footprints are derived from the total volume of effective precipitation. To assess variability in the water footprint estimates, a range of values for evapotranspiration of applied water and effective precipitation were used from the smallest-scale units used in the analysis. We found that variability in evapotranspiration of applied water among nine major crops resulted in the water footprint of agricultural production ranging by about 30 percent around the mean water footprint across 4 years of analysis. This is a result of a combination of differences in water use for the same crop in different places and at different times. If all other sources of variation are ignored, this variation results in the California water footprint varying by about 13 percent around the mean of 1,500 GPCD. This means that our estimate of the water footprint is pretty good.

Another source of variation in water footprint is in individual choices of consumable goods and services. One factor that seems to be a strong determinant of water footprint of consumption is income, with people who make more money tending to consume more goods and thus have a larger water footprint (Hoekstra and Chapagain 2006). To estimate the impact of variation in income within California on calculated water footprint, we assumed that the influence of national income levels on water footprint was approximately correct when used at finer scales, such as for a county within California. We used income data from the Census Bureau for select counties (Orange, Riverside, and San Bernadino) and the Water Footprint Network's online calculator (www .waterfootprint.org/?page=cal/waterfootprintcalculator_indv) to estimate individual water footprints. There was considerable variation around the mean income values, with 4.5–6.9 percent of households occupying the lowest income category (<$10,000/y) and 2.9–9.4 percent the highest ($200,000/y). This variation in income resulted in a range of calculated water footprints from ~640 GPCD for the poorest households to ~4,200 GPCD for the wealthiest. This indicates that if households in California act similarly to households around the world, then one large source of variation in water footprint will be rates and types of consumption, based on income.

A third factor causing variation in water footprint is diet, with vegetarian and vegan diets having lower water footprints than meat-containing diets (da Silva et al. 2013; Vanham 2013). This is because it takes more water to produce meat than the caloric or weight equivalent of vegetables and grains. Using the Water Footprint Network's online calculator, we found that for a moderate individual income of $30,000/y a vegetarian diet resulted in a 27-percent lower water footprint than a meat-containing diet. There is no similar calculator for a vegan diet, but it is likely that the water footprint for a person with a vegan diet will be considerably lower than for someone with a meat-containing or vegetarian diet.

Measuring uncertainty in water footprint calculations is useful because it helps to build confidence in the footprint as a tool to inform decisions. In our analysis we found uncertainty in much of the data. For example, first calculating the water footprint of products produced within California, we found uncertainty stemming from the production statistics and surveys/models of consumptive water use that were used to derive blue and green water footprint factors. And with respect to virtual water imports and exports, there is a great deal of uncertainty in trade data as far as how products are categorized and how the magnitudes of their trade flows are calculated.

Measuring uncertainty is also useful to find out how much individual and collective water footprint can vary due to environmental and consumption patterns, because these patterns often involve choices. This means that people can decide to change their water footprint by changing their consumption of water-intensive foods and goods. Because a lot of the variation within the water footprints of individual crops is related to where they are grown, this also means that decisions about crop production among subregions (e.g. within California) can include information about water intensity, which provides a role for water managers in improving sustainability of water use.

WATER FOOTPRINT AND MANAGING WATER

A key question is whether or not the water footprint can be used by individuals and water managers in making decisions about water sustainability. Global climatechange will affect regions in different ways and is likely to affect the reliability of receiving imported goods and services. This will in turn affect water management in geographic areas that are importers of these goods, such as California, as domestic sources either make up for shortfalls in imports through increased production, or reduce their water use due to international trade pressures. Calculating and using the water footprint in water planning and assessment is an acknowledgement that we participate both in global trade and in a single water cycle.

One interesting thing about the use of water footprints in water-related decision-making is that it can be done across spatial and organizational scales. Individuals may choose to reduce their consumption patterns or increase their support for broader water management efforts based on improved understanding of the relative sustainability of water used in particular goods and services. Companies can improve their understanding of how components of their supply chain may be at risk from variations in water availability and take actions to minimize those risks. Water managers can improve their understanding of how regional or state-scale water use for goods production may change in response to the swings in water use in globally traded products.

We find that 80 percent of California's water footprint is associated with products made outside the state's borders, including other U.S. states and other countries. This is dramatically different from 20 years ago, when California's water footprint related mostly to products made inside California. This means that California is becoming increasingly dependent upon goods from other states and countries and therefore dependent upon and vulnerable to water availability and management in those regions. Over the next century, virtually all of California's current trading partners will have from mild to severe water stress, suggesting risk to California's supply chain from global and U.S. sources.

Agricultural production is the largest component of an individual's or region's water footprint. Coupling virtual water with economic information describing the production value of a crop can strengthen agricultural water management. Spain was the first country in the European Union to include water footprint analysis in its river basin management plan (in 2009). The analysis included questions on when and where water footprints exceed water availability, how much of a catchment's total water footprint is used in producing exports, and the volume and value of crops produced per unit of water (Hoekstra and Mekonnen 2012). "Water economic productivity," expressed in terms of crop market value per cubic meter of water used, has been derived, for example, for the Mancha Occidental region of Spain (Aldaya, Martínez-Santos, and Llamas 2009). That study distinguished "low virtual water, high economic value" crops from "high virtual water, low economic value" alternatives, in a semi-arid region characterized by irrigated agriculture. The study found that "high virtual water, low economic value" crops, such as cereals, are widespread in the region, in part due to the legacy of subsidies. An expansion of low-water-consumption and high-economic-value crops such as grapevines was identified as a potentially important measure for more efficient allocation of water resources. The study concluded that to achieve significant water savings and environmental sustainability, potentially difficult decisions will have to be made regarding crop choice and water allocation. Pricing and regulation of allocation could be used as complementary mechanisms to allocate water to those crops that generate the highest economic value at low water demand.

One of the most promising advances in using water footprint assessments to measure sustainability is the increasingly fine scale at which calculations can be made, which increases the range of uses and users of this index. For example, people can choose food and other consumables based on the calculated and reported water footprints of these items. Just as importantly, producers and regional trade groups can use the total size and intentional improvements (reductions) in water footprints of their products to improve their competitive status with informed consumers. Because of the amount of fine-scale water and economic data available and the prevalence and familiarity of online tools, both users and producers can estimate the water footprint of products and act in concert to improve overall water sustainability.

CONCLUSION

Thinking back to the cup of coffee that started the chapter, what else do you understand about it after reading this chapter? Hopefully you think there is more than just a cup of coffee, possibly even revealing a new way to consider water sustainability for places like California. First we said that a thousand times as much water was used to bring that coffee to you, and that millions of times as much is needed to provide everything else you use and consume in a year. Second, we tried to convince you that you should care about that water because your sustenance, your family's and neighbors' sustenance, and California's sustainability are all wrapped up in the story of that water and how it touches other places and people. Third, we walked through how water footprint science tries to tell that story in a consistent framework that gets at the impacts and risks associated with interventions in the hydrologic cycle to produce the products we depend on. Fourth, we described how we used that science to understand, in macro terms, California's evolving story with water inside its borders as well as in other places, outside of its borders. Fifth, we qualified those understandings with a discussion of uncertainty in water footprint science and how it can be improved. Lastly, we highlighted possible applications of water footprint science and information to various levels of decision-making among individuals, companies, and government agencies. Water footprint assessments and their application to policy are finding firmer footing around the world and in various sectors. California is a place where sustainability science has had a profound impact on political action and social organization. Water footprint science can help build on these achievements and continue to make California and Californians leaders in water sustainability into the twenty-first century.

ACKNOWLEDGEMENTS

The research presented in this chapter was supported by the Pacific Institute's Water Program as well as the California Department of Water Resources and the U.S. Environmental Protection Agency (agreement #4600007984, task order no. SIWM-8 to UC Davis; agreement #201121440–01 to Pacific Institute). We acknowledge our collaborators, Heather Cooley of the Pacific Institute and Susana Cardenas of UC Davis, for their valuable contributions. We also thank the other authors of this edited volume, as well as members of Sabrina Soracco's 2014 Editing, Academic Writing, and Publishing course at UC Berkeley, for their comments and suggestions.

REFERENCES

Aldaya, Maite M., Pedro Martínez-Santos, and M. Ramón Llamas. 2009. "Incorporating the Water Footprint and Virtual Water into Policy: Reflections from the Mancha Occidental Region, Spain." *Water Resources Management* 24(5):941–58. doi:10.1007/s11269-009-9480-8.

Buzby, Jean C., Hodan Farah Wells, and Jeanine Bentley. 2013. "USDA Economic Research Service: ERS Food Loss Data Help Inform the Food Waste Discussion." www.ers.usda.gov/amber-waves/2013-june/ers-food-loss-data-help-inform-the-food-waste-discussion.aspx.

California Department of Finance. 2011. *California Statistical Abstract.* www
.dof.ca.gov/HTML/FS_DATA/STAT-ABS/Statistical_Abstract.php.

California Department of Water Resources. 2014. "Runoff Data for Water Year
2014." http://cdec.water.ca.gov/cgi-progs/stages/FLOWOUT.

Chapagain, A. K., A. Y. Hoekstra, H. H. G. Savenije, and R. Gautam. 2006.
"The Water Footprint of Cotton Consumption: An Assessment of the Impact
of Worldwide Consumption of Cotton Products on the Water Resources in
the Cotton Producing Countries." *Ecological Economics* 60(1):186–203.

Chaves, Henrique M.L., and Suzana Alipaz. 2006. "An Integrated Indicator
Based on Basin Hydrology, Environment, Life, and Policy: The Watershed
Sustainability Index." *Water Resources Management* 21(5):883–95.

da Silva, Vicente de Paulo Rodrigues, Kettrin Farias Bem Maracajá, Lincoln
Eloi de Araújo, José Dantas Neto, Danilo de Oliveira Aleixo, and João Hugo
Baracuy da Cunha Campos. 2013. "Water Footprint of Individuals with Dif-
ferent Diet Patterns." *Ambiente & Agua* 8(1).

Fulton, Julian, Heather Cooley, and Peter H. Gleick. 2012. *California's Water
Footprint.* Oakland, CA: Pacific Institute.

Gleick, Peter H. 2003. "Water Use." *Annual Review of Environment and
Resources* 28(1):275–314.

Gleick, Peter H., Heather Cooley, and David Groves. 2005. *California Water
2030: An Efficient Future.* Oakland, CA: Pacific Institute. www.pacinst.org
/wp-content/uploads/sites/21/2013/02/ca_water_20303.pdf.

Hanak, Ellen, Jay Lund, Barton Thompson, W. Bowman Cutter, Brian Gray,
David Houston, Richard Howitt, et al. 2012. *Water and the California
Economy.* San Francisco, CA: Public Policy Institute of California. www
.ppic.org/content/pubs/report/R_512EHR.pdf.

Hoekstra, Arjen Y., and Ashok K. Chapagain. 2006. "Water Footprints of
Nations: Water Use by People as a Function of Their Consumption Pattern."
Water Resources Management 21(1):35–48.

Hoekstra, Arjen Y., and Mesfin M. Mekonnen. 2012. "The Water Footprint
of Humanity." *Proceedings of the National Academy of Sciences* 109(9):
3232–37.

Postel, Sandra L., Gretchen C. Daily, and Paul R. Ehrlich. 1996. "Human
Appropriation of Renewable Fresh Water." *Science* 271(5250):785-88.

Rich, Jim. 2009. "Comparing Changes in Applied Water Use and the Real
Gross Value of Output for California Agriculture: 1967 to 2007." *California
Water Plan Update.* Vol. 4. Sacramento, CA: California Department of
Water Resources. www.waterplan.water.ca.gov/docs/cwpu2009/0310final
/v4c07a01_cwp2009.pdf.

Schneider, Caroline. 2013. "Three Shades of Water: Increasing Water Security
with Blue, Green, and Gray Water." *CSA News* 58(10):4.

Schneider, Keith. 2011. "Choke Point: China—Confronting Water Scarcity and
Energy Demand in the World's Largest Country." *Circle of Blue*, February 15.
www.circleofblue.org/waternews/2011/world/choke-point-chinaconfronting-
water-scarcity-and-energy-demand-in-the-worlds-largest-country/

Shiklomanov, Igor. 2000. "Appraisal and Assessment of World Water Resources."
Water International 25(1):11–32.

Sullivan, C. A., J. R. Meigh, and A. M. Giacomello. 2003. "The Water Poverty Index: Development and Application at the Community Scale." *Natural Resources Forum* 27:189–99.

Trenberth, Kevin E., Lesley Smith, Taotao Qian, Aiguo Dai, and John Fasullo. 2007. "Estimates of the Global Water Budget and Its Annual Cycle Using Observational and Model Data." *Journal of Hydrometeorology* 8 (August): 758–69.

UNDP. 2006. *Human Development Report 2006: Beyond Scarcity: Power, Poverty and the Global Water Crisis.* New York: United Nations Development Programme.

U.S. Energy Information Administration. 2012. "Drought Has Significant Effect on Corn Crop Condition, Projected Ethanol Production." *Today in Energy*, August 30. www.eia.gov/todayinenergy/detail.cfm?id=7770.

Vanham, Davy. 2013. "The Water Footprint of Austria for Different Diets." *Water Science and Technology* 67(4):824–30.

Afterword

A Soft Path for California Water

PETER H. GLEICK

California water issues have always been complex and contentious; they always will be. The state is a remarkably complicated place. It has a large and growing population, a diverse economy with extensive urban, agricultural, and ecological water demands, and a highly variable climate and water system. As the chapters in this book so clearly demonstrate, we are a long way from a reliable and sustainable water system—indeed, it is hard to imagine that we can actually solve our water problems. These problems are not rocket science—they are far, far more difficult.

Our water disputes take many forms. Cities, farms, and ecosystems vie for limited supplies. Groundwater overdraft and uncontrolled and unmonitored pumping pit neighbor against neighbor. Water-exporting counties and watersheds have different perspectives from water-importing areas. Senior water rights holders have far different worries from junior rights holders. And a mishmash of competing, overlapping, and confusing regulations, institutions, and organizations add to the mix.

It is relatively easy to find different ways to characterize the nature of California water problems, and the authors here do a good job, for their particular issues, priorities, and foci. But perhaps the most straightforward way to describe the problem is to observe that California has reached "peak water": the real limits of our renewable and sustainable supply (Gleick and Palaniappan 2010). Evidence of these limits is all around us, from the core controversy over water flows to and from the Sacramento–San Joaquin Delta, to the large and growing overdraft of groundwater, to collapsing ecosystems and fisheries, to inadequate access to safe water, especially in low-income Central Valley communities. But these physical, social, and hydrologic indicators of peak water are further intensified by clear and growing institutional problems, including the inability to properly measure, monitor, report, and manage groundwater, or to enforce water rights laws and punish violations, or to adequately price and finance water deliveries and infrastructure. Discussion of most of these challenges and limits is tackled in the previous chapters.

As of 2015, a severe drought has once again highlighted these limits and shown us the stark reality of a water system in need of new thinking, new strategies, and new answers. Fortunately, successful alternatives exist and are waiting for the opportunity to blossom. But they will require that policymakers and water managers move beyond the limited kinds of options pursued in the twentieth century.

Traditional water policy was focused around the simple assumption that the purpose of water policy was to continuously develop new water supplies to satisfy exponentially growing demand for populations, farming, and industry. As a result, California water policy meant building hundreds of dams and reservoirs in the Sierra Nevada and Coast Range regions to capture water for delivery through large aqueduct systems to the centers of demand along the coast and in Southern California. That resulted in the extensive spaghetti-like diagrams of water supply infrastructure (see for example the figure just before the introduction to this volume) and an exponential growth in reservoir storage in the twentieth century (figure 16.1).

That development brought enormous advantages to the state, supporting a large population, a major agricultural industry, and a booming economy in a semi-arid region. But it also brought unexpected problems that must now be addressed. Here is the good news. There are proven, cost-effective, environmentally sound water options for California (Cooley, Gleick, and Christian-Smith 2009; Cooley et al. 2010; Hanak et al. 2012). In the past several years, despite the inability of the state as a whole to develop a sound and sustainable water system, incremental steps have been taken to identify and implement alternative water policies, technologies, and strategies. While these have not resolved the state's water problems, they point the way to a different future—a future in which the state can continue to maintain a strong agricultural sector

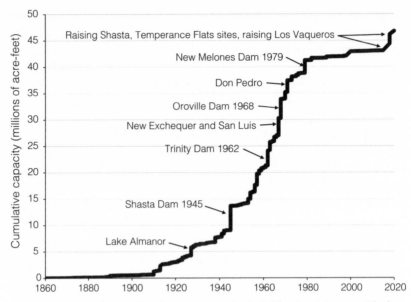

FIGURE 16.1. Cumulative reservoir storage capacity in California, 1860–2020. The last increments of new storage are hypothetical, including proposed (but not yet approved) expansion of Shasta Dam and the construction of reservoirs at Sites and Temperance Flats. Source: Pacific Institute, using data from the California Department of Water Resources (2014a).

and satisfy urban demands with high-quality reliable water deliveries, while also restoring and protecting some vestige of the natural ecological legacy that is also a core part of the California story.

As a capstone to the chapters in this book, here is a picture of the future—a positive vision for what a sustainable water future could be for California. This vision is not a prediction, but a possibility. The future is unclear, because it depends on the actions and decisions of individuals, local communities, public and private agencies, state policymakers, and conditions in the wider world of federal law and international trade.

CALIFORNIA WATER 2050

It is the year 2050. It is expected that global population will peak within the next ten to twenty years for the first time in human history, raising a variety of new economic, social, and political complications; but until that happens, the recent challenges of satisfying growing populations with limited resources continue to dominate all the strategies around the issue of sustainable development. For water resources, satisfying growing populations with a limited—albeit renewable—resource continues to vex most governments and communities.

But after two hundred years of effort, California has solved most of its severe water problems. As California approaches the bicentennial of its founding as a state, the water problems that vexed politicians, managers, and communities are finally under control. International scholars and visitors from around the world come to California to discover how the state finally achieved and implemented sustainable water management, planning, and use.

Urban Transformation

In the last few years, the population of the state has leveled off, peaking around 50 million (California Department of Finance 2014). Most people—more than 95 percent of California residents—live in the state's major coastal and Central Valley urban areas, where almost all of the economic activity occurs. Meeting urban needs has always been challenging, yet today California leads in providing reliable, high-quality urban water supply to all residents. Twenty percent of California's developed water supply goes to its cities—about the same percentage as at the turn of the twenty-first century (California Department of Water Resources 2014b). But that water supply meets the needs of many more people than half a century ago.

California's cities started rethinking water supply and use toward the end of the twentieth century, after a series of both severe droughts and floods threatened the reliability and quality of urban supplies. The first hint of trouble came in 1976 and 1977, when the most severe two-year drought on record opened the eyes of water managers to the possibility of long-term shortages. Subsequent droughts between 1987 and 1991, and then more persistently starting in the first decade of the 2000s, made it apparent that traditional patterns of water use and strategies of water supply were no longer adequate to ensure water for the state's cities and industries. By 2015, multi-year droughts were increasingly common, punctuated by increasingly severe rainfall and flooding events and disappearing Sierra Nevada snowpack—the first signs that human-caused climatic changes were now an unavoidable fact.

Combined with limited options for new supplies from the state's over-tapped rivers and aquifers, cities had to look to new ideas. A few abortive efforts to build large coastal desalination facilities demonstrated that expanding supply with costly centralized infrastructure could not completely satisfy urban needs. Fortunately, new ideas were being tested and implemented here in California and around the world, which offered alternatives to the old ways of thinking.

Today, California cities have among the lowest per capita water use of any cities in the United States—below 100 gallons per person per day. At the same time, nearly 100 percent of wastewater is captured in small, regional distribution systems, treated to the highest standard of purity, and reused for a wide range of urban needs, including direct potable reuse, following the lead of places like Singapore and Namibia, which pioneered this approach. In 2010, as much as five million acre-feet of wastewater was discharged into the ocean after treatment. Today most of that water is treated to a high standard and made available for groundwater recharge, ecosystem flows, local agriculture, and a wide range of urban demands. This water is the largest source of new supply since the major rivers of the state were channelized, piped, and diverted for the old-style federal and state water projects in the 1950s, '60s, and '70s.

New, sophisticated remote-sensing systems have cut California's urban water leakage rate—"unaccounted-for" water—to below 5 percent, and when leaks are detected in distribution systems, they are identified and repaired promptly. All water uses are monitored, with new sophisticated water meters on 100 percent of single- and multi-family homes and apartments, and all urban rate structures provide incentives for efficient use, full cost recovery, and protections for low-income communities.

New homes built in California are models of efficiency, often with dual-piping systems permitting collection and reuse of wastewater and graywater, highly efficient water-using appliances, and beautiful, low-water-using landscaping. Smart water meters give residents and urban water managers real-time information on water use and on strategies for further cutting waste and inefficiencies. Water rates are low for basic needs and higher for secondary, discretionary needs. At the same time, older homes are increasingly being retrofitted with more efficient water fixtures through a combination of economic incentives, new standards for appliances, and local ordinances.

Green grass lawns, once ubiquitous in California cities, are now anathema. Where a green monoculture lawn was once seen as a status symbol, today it is seen as a sign of waste. Just as no one today would drive drunk or smoke cigarettes in a restaurant or fail to use a seat belt, the social stigma of a green lawn has led to innovative and beautiful gardens of efficient, low-water-using plants and a significant drop in outdoor urban water use.

Beginning around 2015, efforts to collect, treat, and reuse stormwater runoff expanded statewide. A substantial amount of urban stormwater runoff had previously been diverted into the oceans during winter months, wasting a valuable resource and contributing to coastal pollution problems. Efforts to collect and treat this resource and use it for groundwater recharge and local uses led to an expansion of local water supply and an increase in supply reliability for many cities.

Agricultural Sustainability

California continues to be a national leader in the production of a wide range of agricultural products, and 80 percent of the state's water that is used by humans sustains this production. Yet the agricultural sector of 2050 looks very different from that of 1950, or even 2000. For the last 30 years, investment in improving agricultural water use has transformed the sector into the most advanced and productive in the world. Efficient irrigation technologies have been applied statewide. Soil-moisture monitors are standard, tied into computerized water-delivery systems that permit farmers to apply water when and where it is needed and to maximize crop yield and quality. Old, leaky distribution canals have been repaired and sometimes replaced with more flexible distribution systems capable of satisfying needs on demand. As a result of these advances, productivity measures of crop per drop and dollars per acre-foot have soared, even as total water use in the agricultural sector has dropped.

The trend away from water-intensive crops toward more water-efficient, higher-valued crops began in the 1980s and has continued. Production of some field crops, such as cotton, alfalfa, irrigated pasture, and rice, has shifted from California to other parts of the United States with more reliable rainfall, while acreage has further changed toward fruits, vegetables, and nuts. Some marginal lands have also been permanently removed from production because of the lack of reliable water or concerns about soil quality, but in an average year irrigation continues on between seven and eight million acres, with supplemental acreage available when surplus water supplies are available.

Perhaps the most remarkable change statewide has been the complete elimination of groundwater overdraft. In the twentieth century, this problem exemplified more than any other the extent of unsustainable water policy in California, with millions of acre-feet of overdraft in some years and a consistent decline in groundwater levels (Famiglietti and Rodell 2013). After the severe droughts of the early twenty-first century, objections to groundwater monitoring and management finally dried up, and the state, together with local communities and watersheds, implemented comprehensive groundwater policies that included metering all uses, calculating long-term sustainable groundwater yields, allocating those yields to users, and requiring the development of recharge basins and advanced groundwater storage practices. This effort has expanded storage capacity by millions of acre-feet, raised groundwater levels in depleted basins, restored surface flows fed by groundwater, and built an effective buffer against persistent droughts. Today, water users look back on the days of unregulated groundwater as an embarrassing anachronism.

Equally remarkable has been the reform of California's water rights system, which was a complicated mix that evolved from riparian and appropriation laws, diverse and often conflicting court opinions, and inadequate institutional oversight and management. By the second decade of the twenty-first century, it was apparent that water rights reform was desperately needed, and state agencies began the process of clearly defining "reasonable and beneficial use" under the state's Constitution, reviewing and verifying water rights claims, and intensively monitoring actual withdrawals and usage under those claims. It quickly became apparent that illegal or inappropriate diversions were rampant, that uncertainty about rights was hindering efforts to improve the efficient and productive use of water, and that the mechanisms for adjudicating conflicts were all contributing to poor water management. As each of these problems was tackled, the ability to satisfy human demands improved, as did the ability to more successfully restore in-stream and ecosystem flows. Now, water rights priorities and actual water use are carefully aligned, transparent, and understood by all users.

Ecosystem Restoration and Revival

After decades of deterioration, California's aquatic ecosystems are on the mend. As urban and agricultural water policies changed over the past half-century, efforts to integrate ecological concerns into water policy expanded. In recent years, thousands of acres of wetlands have been restored, new flow standards for fisheries have been established and expanded as some endangered and threatened species rebounded, and political support for protecting the environment has become mainstream. Most Californians no longer remember that the San Joaquin River was, for many years, completely dewatered; yet, now it is returning to health. New protected wildlife refuges support resurging waterfowl populations along the Pacific Flyway. While the Salton Sea continues to shrink, causing new environmental worries in Southern California, permanent flow commitments for the newly restored Colorado River Delta in Mexico have begun to reestablish a diverse community of plants, fish, birds, and animals.

Some unresolved environmental challenges associated with the growing impacts of climate change continue to complicate statewide environmental management. The growing loss of snowpack and the change in the timing of runoff in California rivers cannot be overcome through river or reservoir management options, and some of the aquatic species that evolved in one kind of runoff regime are having trouble adapting. Rising temperatures in rivers and lakes are also a concern, as are rising sea levels, which threaten coastal wetlands and aquifers. Nevertheless, the "fish versus farms" debates of the early part of the century are gone, replaced by a spirit of cooperation and enthusiasm over the restoration of part of California's rich natural endowment.

Institutional Reform

One reason California water resources are in better shape today than in the 2010s is substantial reform of the state's institutions and water management. Many of the practices from the twentieth century that led to a failing water system have been eliminated, and many of the laws and organizations that determine water policy have been modified and updated. Among the most important of these was the recognition that archaic water rights policies were failing to protect actual water rights holders, the interests of the state's citizens, and Public Trust resources. By 2020, the State Water Resources Control Board had implemented an improved water rights monitoring program that included comprehensive and transparent data collection on water withdrawals and use, correlated with legal water rights allocations. As this program ramped up, illegal or inappropriate water withdrawals began to decrease rapidly, and useable flows in California rivers and streams increased to a surprising degree. At the same time, rulings clarifying "reasonable and beneficial" water uses provided new incentives for improving efficiency of water use and cutting waste and inefficient uses. These two changes alone dramatically increased water availability and water-use productivity statewide.

*When comprehensive groundwater law was finally passed and imple-
mented, similar improvements in water use and restoration of groundwater levels materialized. Groundwater overdraft began to decline, and
by 2035, all groundwater basins were fully evaluated for their sustainable yield. Water allocations were adjudicated so that intentional
recharge and groundwater management could be implemented, and
local conflicts over access to water greatly declined.*

MOVING TOWARD A SOFT PATH FUTURE

This story—the positive vision for California water in the year 2050—is
not a guaranteed future, but an outline of what the state might strive to
achieve. Such a sustainable future is far easier to imagine and describe
than to reach. In many ways, we know what needs to be done; the challenge is doing it in the face of ignorance, lack of concern, powerful
interests vested in maintaining the status quo, financial challenges, and
institutional complacency. That's why "California Water 2050" is not a
prediction. But by defining where we want to be, it is possible to identify the key elements in moving forward. It is time for the state's water
warriors to put them all together to follow what I have previously call
the "soft path" for water (Gleick 2002, 2003). There are many components to the soft path, many of which are touched upon by the authors
in this collection. Here is the basic idea.

Integrated Water Management

We must move beyond the old idea of simply expanding supply to meet
some hypothetical projected future demand. In particular, we must look
at a much broader set of integrated solutions—described by some as a
portfolio approach—that integrates new thinking around both supply
and demand, addresses cross-sectoral relationships such as the nexus
between water and energy, and integrates future climate change into
long-term water planning.

Hanak and Lund (chapter 7) discuss the important concept of
expanding the range of solutions considered by water policymakers in
the form of a portfolio of solutions. The danger of seeking a single
"silver bullet" has led to a far-too-narrow conversation about solutions. How often do we hear ideas like, if only we built new dams—or
desalination plants—or stopped growing alfalfa—or eliminated environmental protections? The real solutions are far more complex (like

California's real water problems), and they require a complex set of approaches that include, rather than exclude, options. One of those options is to rethink "new supply" and to realize that while new dams may not provide any significant new benefits, there are other supply alternatives, in particular expanding the use of highly treated wastewater (Harris-Lovett and Sedlak, chapter 10) and improved rainwater and stormwater capture and use (Woelfle-Erskine, chapter 14).

At the same time, the soft-path approach asks us to rethink water "demand" as something that can be changed though a focus on improving efficiency and cutting waste. Indeed, some of the most significant changes in California water in the past few decades have resulted from demand management programs rather than supply expansion, and evidence suggests substantial additional potential for improving efficiency (defined here as the ability to meet a demand for goods and services with less water). Christian-Smith and Heberger (chapter 5) address some of these possibilities in the urban sector. In a similar positive vein, Cantú (chapter 9) shows that some water agencies and watersheds are already moving in the right direction with integrated water resource management, and there are positive lessons to be learned from that experience.

Other chapters also reflect this theme of integration. Wilkinson, who has long championed the need for new water solutions, discusses (chapter 2) the critical importance of looking at energy and water together. In addition to addressing the potential for urban efficiency improvements, Christian-Smith and Heberger (chapter 5) tackle the role of better pricing and economic approaches as a tool for more sustainable water management. Foster and Radke (chapter 6) remind us that water scarcity and droughts are not the only challenges facing California—we are also vulnerable to flooding. Any water management system must address all forms of extremes, especially in the Sacramento–San Joaquin Delta region. Chapter 1 (Andrew) reminds us that no integrated water approach can afford to ignore the risks of climatic changes and the role such changes will play in altering the availability, timing, and quality of water statewide; this is also addressed in the Southwest regional analysis of the newly released *National Climate Assessment Report* (2014).

Institutional Reform

Because our water systems are the result of the evolution over many decades of laws, tradition, court decisions, and water rights policies, society is somewhat constrained by how our existing institutions operate. These institutions reflect the past far better than they prepare for the future. It is time to reform local, state, federal, and international water institutions in the context of new problems and solutions. In the California context, this will require a wide range of changes, but especially in the areas of water rights, groundwater reform, and equity and environmental justice.

For water rights reform, as noted by Hanemann, Dyckman, and Park (chapter 3) and Gray (chapter 4), the priorities must be to clarify and better define "reasonable" and "beneficial" in the context of the California Constitution, to improve the transparency of water rights data, to monitor actual water withdrawals and use, and to enforce those rights. These reforms must encompass not just total withdrawal volumes but also water timing—an increasingly critical issue, as Deitch and Kondolf (chapter 12) note in their work on salmon and wine-grape growers.

For groundwater reform, the state must pass and implement some form of groundwater regulation that requires monitoring and management of groundwater, quantifying sustainable yields, and encouraging local, watershed, or statewide allocation and management schemes. Chapter 8 (Wendell and Hall) addresses this question directly, but other authors here also note the unsustainability of current groundwater use.

Finally, traditional water agencies and institutions in California and elsewhere have rarely been widely inclusive in the voices they listen to or the communities they address. Yet, experience increasingly shows that the most successful and sustainable water strategies are those where community participation in decision-making is high. As Balazs and Ray (chapter 11) describe, inequities in access to safe and affordable drinking water in California are the result of these institutional failures, and a sustainable soft path for water in California must address and correct these inequities as a top priority. California has taken small steps in this direction by acknowledging the human right to water (International Human Rights Law Clinic 2013), but improvements in access and affordability must move forward.

Technology and Policy Tools

Past water policy has relied on a relatively small set of solutions, including traditional technology in the form of dams, aqueducts, groundwater pumping, and old-style irrigation systems, and on an equally limited set of policy tools, such as simplistic water-pricing strategies and *laissez-faire* regulations. The soft path requires expanding these tools. A new buzzword in resource management science is *adaptive management*—a valuable concept that acknowledges rapidly changing economic, social, environmental, and political conditions and tries to develop more flexible policies. Podolak and Yarnell (chapter 13) describe the application of this concept in their discussion of relicensing hydroelectric systems on major California rivers to reflect new science around fisheries, ecosystem health, and water allocations. Other chapters also offer insights into newer technologies and policies for more sustainable water use.

New forms of analysis are now available, too, that make it possible to assess and understand more of the complexities behind California water. The concepts of *virtual water* and the *water footprint* have broadened the discussion about what we use our water for, how much is used and consumed in the state compared to how much is exported or imported in goods and services that flow across California's borders, and what the implications might be of such flows. Fulton and Shilling (chapter 15) explore these concepts and put real numbers on them, revealing that California is in fact a net importer, rather than exporter, of water. The discussion about whether this is a good or bad thing (or neither) is just beginning.

CONCLUSIONS

Much more can, has, and will be said about California water as we move into the future. The soft-path positive vision described here for 2050 will be difficult to achieve. But I think that it is inevitable that the management and use of water in California—and in other regions of the world—will continue to improve as water problems intensify, as populations and economies put more pressure on a limited resource, and as climate changes worsen. There is scarcely an alternative.

REFERENCES

California Department of Finance. 2014. *State and County Total Population Projections, 2010–2060 (5-Year Increments)*. www.dof.ca.gov/research /demographic/reports/projections/P-1/.

California Department of Water Resources. 2014a. *Dams within the Jurisdiction of the State of California*. www.water.ca.gov/damsafety/damlisting/.

———. 2014b. *California Water Plan 2013 Update*. www.waterplan.water .ca.gov/cwpu2013/.

Cooley, Heather, Juliet Christian-Smith, and Peter H. Gleick. 2009. *Sustaining California Agriculture in an Uncertain Future*. Oakland, CA: Pacific Institute. http://pacinst.org/publication/sustaining-california-agriculture-in-an-uncertain-future/.

Cooley, Heather, Juliet Christian-Smith, Peter Gleick, Michael Cohen, and Matthew Heberger. 2010. *California's Next Million Acre-Feet: Saving Water, Energy, and Money*. Oakland, CA: Pacific Institute. http://pacinst.org /publication/californias-next-million-acre-feet-saving-water-energy-and-money/.

Famiglietti, James S., and Matthew Rodell. 2013. "Water in the Balance." *Science* 340(6138):1300–01. doi:10.1126/science.1236460.

Gleick, Peter H. 2002. "Soft Water Paths." *Nature* 418:373.

———. 2003. "Global Freshwater Resources: Soft-Path Solutions for the 21st Century." *Science* 302 (28 November): 1524–28.

Gleick, Peter H., and Meena Palaniappan. 2010. "Peak Water Limits to Freshwater Withdrawal and Use." *Proceedings of the National Academy of Sciences* 107(25):11155–62. www.pnas.org/cgi/doi/10.1073/pnas.1004812107.

Hanak, Ellen, Jay Lund, Barton "Buzz" Thompson, W. Bowman Cutter, Brian Gray, David Houston, Richard Howitt, et al. 2012. *Water and the California Economy*. San Francisco: Public Policy Institute of California. http://www .ppic.org/main/publication.asp?i=1015.

International Human Rights Law Clinic. 2013. *The Human Right to Water Bill in California: An Implementation Framework for State Agencies*. UC Berkeley School of Law. www.law.berkeley.edu/files/Water_Report_2013_Interactive_FINAL.pdf.

National Climate Assessment Report: Southwest. 2014. Washington, DC: U.S. Global Change Research Program. http://nca2014.globalchange.gov/report /regions/southwest.

Index